MICROEMULSIONS – AN INTRODUCTION TO PROPERTIES AND APPLICATIONS

Edited by **Reza Najjar**

Microemulsions – An Introduction to Properties and Applications
Edited by Reza Najjar

Published by InTech
Janeza Trdine 9, 51000 Rijeka, Croatia

Publishing Process Manager Ivona Lovric

Technical Editor Teodora Smiljanic

Cover Designer InTech Design Team

Additional hard copies can be obtained from orders@intechopen.com

Microemulsions – An Introduction to Properties and Applications,
Edited by Reza Najjar
p. cm.
ISBN 978-953-51-0247-2

For EU product safety concerns:
IN TECH d.o.o., Prolaz Marije Krucifikse Kozulić 3, 51000 Rijeka, Croatia,
info@intechopen.com or visit our website at intechopen.com.

Meet the editor

Dr Reza Najjar (born in Marvdizaj/Soufian, Iran) studied chemistry at the University of Tabriz (BSc, 1994), Sharif University of Technology (MSc, 1997) and Aachen University of Technology (RWTH), Germany (PhD, 2006). After completing a year of post-doctoral work at the University College Dublin (UCD), Ireland, he joined the Faculty of Chemistry at the University of Tabriz as assistant professor. His main research interest is focused now on the synthesis of polymeric surfactants and their applications in microemulsion systems for different purposes.

Contents

Preface

Microemulsions are thermodynamically stable mixtures composed of at least three components, namely, oil, water and surfactant, and sometimes a fourth component called as co-surfactant essentially comes into play. This field is one of the still rapidly growing fields of science and technology that have attracted a great attention of the scientists with various professions, ranging from pure physical chemists to oil engineers, pharmacologists and biologists. Besides many researchers investigating the structural and pure scientific aspects of these systems by different tools and techniques, there are a lot of others benefiting them as a powerful and vital tool to achieve own aims. Regarding these facts, the investigation and work either on the scientific or application aspects of microemulsions still is a high priority research theme.

This book was intended to provide an insight into the fundamentals and some applications of microemulsion systems. In the first Chapter, a very short introduction about the background and general aspects of microemulsions is given. The author has tried to introduce different types of related fields and provide some reviews and original references to enter the reader into the play.

The second Part of the book was assigned for investigating the influence of some determining parameters on the microstructure and phase behavior of these systems. In Chapter 2 of the book, Dr Buckin one of the prominent experts in this special field has provided a new aspects in investigation of microemulsion characteristics, such as size, microstructure and state of water in nano-droplets and … with high-resolution ultrasonic measurements. The third Chapter of the book by Dr Cid and co-authors deals with the effect of changing the size and type of co-surfactants on the electric percolation and percolation temperature of the system. In Chapter 4, Dr Rout and co-authors have discussed on another aspect of the microemulsion science, and used one of the well known theoretical models already proposed for the description of the microemulsions, to predict phase behavior and microstructure of a system in 1-phase region.

In Part 3 and Part 4 of the book different type of applications of the microemulsions have been explained. Among the applications of microemulsions, their use in drug delivery systems and vaccines have gained a great interest during the last decade. One of the challenges in this field is to develop some common "ready to use" drug delivery systems where one only needs to add an active component to formulation. Dr Balogh in Chapter 5 has presented if it is really as simple as to add any active component and use them or further work will be required with these formulations. The next Chapter deals with the work on a thermal reversible system for delivering of poorly water-soluble drugs by oral administration. Dr Gao has discussed on the different parameters affecting the oral delivery of the lypophilic drugs by the mentioned system. In the Chapter 7, the improvement of the vaccination methods using microemulsions has been demonstrated. Dr França and co-author have explained the beneficial effect of using nanoparticles for vaccine delivery via microemulsions resulting in good immune responses.

The last Part of the book is devoted to their use in oil industry as an old and important application, as well as the other main type of microemulsion applications, namely, use of them in preparation of the nanostructured materials, which has become very popular during last decade. Dr Santanna and co-authors have described how several microemulsion systems which may be used in different areas of the oil industry, such as wells hydraulic fracturing, corrosion inhibitors in pipelines, in the breakdown of water/oil emulsions, in enhanced oil recovery, and as alternative fuel. In Chapter 9, Dr Kunitake and co-authors have introduced use of microemulsions with bicontinuous microstructures as a template for preparation of the continuous porous polymeric materials and bicontinuous hybrid materials by polymerization or gelation. In Chapter 10, another type of microemulsion applications, i.e., use of microemulsions with water in oil droplets as nanoreactors for the preparation of the nanostructured materials is investigated. Dr Dromta and co-authors have discussed on the preparation of the magnetic core-shell nanoparticles with a narrow particle size distribution. As described in Chapter 11 by Dr Mohd Zabidi, use of microemulsion technique was able to produce Co-based nanocatalysts with a good selectivity in Fischer-Tropsch reactions. The use of microemulsion systems in preparation of the photocatalysts is one of the other widely used applications. In the Chapter 12 Dr Zielińska-Jurek and co-authors have reviewed this field and discussed about the different factors affecting the properties of prepared photocatalysts.

Finally, I would like to express my gratitude to Mrs Ivona Lovric, the publishing process manager and Intech Open Access publisher for their efforts in the publishing process. Many thanks also to my family and families of all other collaborators for their patience and acceptance of the lost evenings. I wish that this book can give an insight into the field.

With the best wishes
Marvdizaj, 2012

Dr Reza Najjar
Assistant Professor of Chemistry,
Faculty of Chemistry,
University of Tabriz,
Tabriz,
Iran

Part 1

Introduction

1

Microemulsions – A Brief Introduction

Reza Najjar
Polymer Research Laboratory, Faculty of Chemistry, University of Tabriz, Tabriz, Iran

1. Introduction

Nevertheless, the existence and application history of the microemulsions goes back to the very older times, but the oldest available reports in this field have been published by Schulman (Schulman & Hoar, 1943) and Winsor (Winsor, 1954). Their works are the starting point of efforts for the systematic understanding of the microemulsions. Meanwhile, the widespread generalization and applications of these systems have been started in the late 1970s, with their use for the enhanced oil recovery during the energy crisis.

The term of microemulsion applies to a mixture with at least three components; an oily phase, an aqueous phase and a surface active species, so called surfactants. Sometimes the forth component i.e., co-surfactant can/must be present (Saito & Shinoda, 1967. Saito & Shinoda, 1970). Depending on the ratios between the components, in the two extremes the microstructure of the microemulsions vary from a very tiny water droplets dispersed in oil phase (w/o microemulsion) to a oil droplets dispersed in water phase (o/w microemulsion). The microstructure of the mixture changes continuously from one to another extreme, namely, from a spherical to cylindrical, tubular and interconnected continuous oil and water phases separated with a very thin layer of surfactant molecules, in the middle, which is defined as bicontinues microemulsion (Scriven, 1976). The microemulsions of each kind are thermodynamically stable and transparent solutions. There are main differences between emulsions and microemulsions in terms of structure and stability. In contrast to the microemulsions, the emulsions are unstable systems and without agitation, phase separation will occur in them. The other difference is that the size of droplets in emulsions are in the range of micrometers, while in microemulsions the size of micelles are in the range of 5-100 nm, depending on the some parameters such as surfactant type and concentration, the extent of dispersed phase (Prince 1977, Hou et al., 1988, Maitra, 1984). Hence, sometimes the microemulsion term is misleading, because it doesn't reflect the size of dispersed phase droplets in the system which, are in the nanometer range. Depending on the type of the surfactants employed in the preparation of the microemulsion, another important parameter that affects the main characteristics of a microemulsion is the presence of electrolytes in the aqueous phase.

2. Phase diagrams and types of microemulsions

The formation of the thermodynamically stable microemulsions require that an adequate amount of the corresponding components must be mixed. Determination of these proper compositions is an important issue in this field to obtain the microemulsions with required properties. For this purpose, one must prepare mixtures with different compositions of the components, and check them regarding the type and number of phases present in the system. The resulting diagrams, showing the number/or type of phases present in the system associated with each specific composition, are called phase diagrams. From the industrial and application point of view, this process is called formulation, which indicates the specific compositions of the components giving a stable mixture effective in the concerned property. A number of different methodologies have been used for determination of the phase diagrams. Almost the earliest studies about the phase diagrams of the microemulsions can be found in the 1960s (Ekwall et al., 1960). Using the phase diagrams, it has been confirmed that the Schulman's so-called micromulsion is not an emulsion but a solubilized solution (Shinoda & Kunieda, 1973). The mechanism of the microemulsion formation has been studied in connection with the phase diagrams and the relation between the amounts of components required to form a clear microemultion has been understood from the phase diagrams (Ahmad et al., 1974). They have studied the phase diagrams of different systems with anionic, cationic and non-ionic surfactants, and could obtain maximum solubillization with the optimum ratio of the surfactant and co-surfactant. By a detailed investigation on pseudoternary phase diagrams of two microemulsion systems it has been evidenced that a great variety of phases is present. They have concluded that the interaction between water and oil domains is an important parameter affecting the stability of microemulsions (Roux et al., 1983). The phase diagrams of the ternary system containing water-sodium alkylbenzene sulfonate (NaDBS)-hexanol and their quaternary system with xylene have been prepared at three different temperatures. The formation of different phases, such as microemulsion phase, reverse micelle phase was observed which have been qualitatively examined by optical (phase contrast and polarizing) microscopy or low angle X-ray diffraction. According to the results the amount of microemulsion phase was decreased by increasing of the temperature at surfactant concentrations of lower than 15% (Baker et al., 1984). The phase diagrams of the systems with alkyl polyether surfactants have been studied extensively in different aspects, (Zhao et al., 2011, Lang, 1999. Balogh, 2010, Selivanova et al., 2010, Magno et al., 2009, Boonme et al., 2006, Mitra & Paul, 2005, Lim et al., 2005).

The effect of addition of inorganic salts into the aqueous phase of the microemusions have been studied using phase diagrams. It has been observed that the added salts has a great influence on the solubilisation ability of the microemulsion system (Komesvarakul et al., 2006, Wei et al., 2005, Li et al., 2003, Van Nieuwkoop & Snoei, 1985, Yu et al., 2009, Chai et al., 2009, Qin et al 2008, Nedjhioui et al., 2007, Koyanagi et al., 2007, Mitra & Paul, 2005, Shinoda, 1967, Shinoda & Saito, 1968). As an example, it has been observed that the addition of salt shifts the fish diagram towards more hydrophobic oil systems and higher surfactant concentrations will be required (Komesvarakul et al., 2006). Determination of the phase diagrams has been used also as the bases for the applications of the microemulsion for the preparation of the nanoparticles (Najjar & Stubenrauch, 2009, Magno et al., 2009a). Here, the phase diagrams have been used to select the proper compositions of the microemulsions to get spherical well defined micelles, and consequently resulting nanoparticles.

3. Thermodynamics of microemulsions

The microemulsions are thermodynamically stable mixtures of oil, water and one /or more surface active agents (surfactants). For understanding of the thermodynamics of the microemulsions one must consider all kind of the interactions existing between the components present in the system, i.e. oil, water, surfactant (and co-surfactant) and the microstructures (micelles, globules, lamellar and ...) formed in the system with each other and the media. The theoretical aspects of the stability of microemulsions is a well known issue (Kumar & Mittal, 1999).

Nevertheless, the nature of the interactions between oil and water are repulsive forces, the presence of the surfactant molecules changes the balance between the forces towards the attractive forces. The stabilizing effect of the surfactants is exerted by the formation of the different types of microstructures to favour the stabilizing interactions. Hence, the understanding of the microstructure of the microemulsion systems is of prime importance.

Almost the first speculations about the microstructures of the microemulsion consisting of the surfactant, oil and water have been made in 1950s (McBain, 1950, Philipoff, 1951, Becher, 1968, Shinoda, 1970). In the meantime, the first accurate thermodynamic data about a microemulsion system have been reported in 1960s. Based on those data Shinoda has developed an acceptable model, which could reasonably explain this dissolution phenomenon by formation of the structures such as log-boom (Becher, 1968), lamellar (McBain, 1950), cylindrical, spherical, ellipsoidal, or rodlike micelles (Shinoda, 1970). Many reports can be found in the literature about the thermodynamic stability considerations of the different microemulsion systems (Ruckenstein, 1981, Bennett et al., 1981, Bellocq et al., 1982, Prouvost et al., 1985, Biais et al., 1987, Mukherjee et al., 1997, García-Sánchez et al., 2001, Fu et al., 2002, 2003).

For modeling of phase behavior, Bennett et al was presented a mathematical framework in a way consistent with the thermodynamically required critical tie lines and regarding critical endpoints. The modeling of surfactant-rich third phase evolution were extended to satisfy these requirements and also Hand's scheme for modeling of binodals and Pope and Nelson's approach was regarded (Bennett et al., 1981). It has been presented that the model-generated progressions of ternary phase diagrams gives a better understanding of the experimental data and reveals correlations of relative phase volumes (volume uptakes) with other phsae diagrams parameters.

In recent years, kartsev et al have used a two-phase model to approach to the thermodynamics of microemulsions (Kartsev et al., 2010). They proposed dispersion medium as one phase and the sum of disperse phase nanodrops as the second phase.

The performance of model was evaluated with experimental data and it was proved that the use of this model to solve microemulsion thermodynamics problems quantitatively gives satisfactory results with model inadequacy not more than 10%.

4. Techniques for investigation of microemulsions microstructure

In the course of development of the microemulsions, different techniques has played an important role in this process and helped scientists to understand the different aspects of microemulsion science.

Nuclear magnetic resonance (NMR) and infrared spectroscopy are among the oldest techniques used for the investigation of microemulsions. Using NMR measurements,

Gilberg and co-workers have indicated that in case of micelles with a larger water core the packing density of surfactant molecules is low, and consequently the stability of such micelles are lower than the micelles with higher packing density (Gillberg et al., 1970). Stilbs has demonstrated that by solubilization of the short-chain *n*-alcohols in microemulsions containing SDS micelles the 1H NMR line broadening occurs (Stilbs, 1982). He concluded that the results are indication of the highly disordered structures only, and the addition of the short-chain n-alcohols causes the breakdown of the micelles. Also, it was shown that by increasing of the surfactant concentration the growth in micellar size occurs progressively, and at higher concentrations long prolate-shaped aggregates form. The addition of water to the bicontinuous microemulsions, studied by 13C-NMR chemical shift trends of C8G2 and pentanol carbons, indicated a reduction in the mean surfactant film curvature towards water (Parker et at., 1993). The measurement of the rotational correlation time (τ) of a nitroxide labeled fatty acid probe, 5-doxyl stearic acid, versus cetyltrimethylammonium bromide (CTAB) (as surfactant) concentration in aqueous solution has been done via ESR spectroscopy (Li et al., 1997).

The CMC value obtained by this method has been in good agreement with the surface tenstion measurements. Using 2H NMR studies, it was indicated that five water molecule are tightly bound to each CTAB molecule. The chemical shifts and T1 relaxation time data obtained by 1H NMR measurements were used to investigate the microemulsion properties and structure (Waysbort et al., 1997, Bastogne et al., 1999, Kataoka et al., 2007, Causse et al., 2006).

Investigaton of the microemulsions containing didodecyldimethylammonium sulfate (as surfactant), water and dodecane /or hexadecane by NMR self-diffusion approach, has revealed that diffusion coefficients for the surfactant and oil are equal at high surfactant-to-oil ratios. This observation indicates that the structure is truly bicontinuous over distances on the order of μm in such a system (Söderman & Nydén, 1999). The existence of a worm-like structure in the intermediate water contents instead of the classical bicontinuous structure was proposed, which is confirmed by SAXS and SD-NMR analysis (Libster et al., 2006). There are many other reports in the literature about the studying of the microstructure and other properties of the microemulsion systems. Among them are the studyng of the competitive solubilization of cholesterol and phytosterols in nonionic microemulsions by pulse gradient spin-echo NMR (Rozner et al., 2008), study of the microstructure of four-component sucrose ester microemulsions using SAXS and NMR measurements (Fanun 2001), solution properties of C18:1E10/oil/water system by PGSE-NMR self-diffusion (Ko et al., 2003), reverse micelles of di-isobutylphenoxyethoxyethyl-dimethylbenzylammonium methacrylate in benzene (Emin et al., 2007).

Another type of techniques that have played an important role in understanding of the microemulsions, is the methods developed for visualization of microemusion micro-structures. These techniques based on the transmission electron microscopy (TEM) images prepared from a very thin film of the samples. This type of techniques is consisted of three different methods: a) freeze fracture electron microscopy (FFEM) (Jahn & Strey, 1988, Burauer et al., 2003), b) Cryo-Direct Imaging (Cryo-DI) (Talmon, 1999, Bernhein-Grosswasser et al., 1999) and c) freeze-fracture direct imaging (FFDI) (Belkoura et al., 2004). The first of these techniques was introduced by Jahn and Strey in 1988 (FFEM) (Jahn & Strey, 1988, Jian et al., 2001). Development of these techniques along with the other techniques helped the scientists in well understanding of microstructure of microemulsions. Later on, these techniques has been developed/used by other researchers to investigate the microemulsions (Agarwal et al., 2004, Ponsinet & Talmon, 1997, Hellweg et al., 2002, Yan et al., 2005, Mondain-Monval, 2005, Zhang et al., 2010, Klang et al., *2012*)

The use of cryo-field emission scanning electron microscopy (cryo-FESEM), in combination with the other techniques, has been reported by Boonme et al. for investigation and characterization of microemulsion structures in the pseudoternary phase diagram of isopropyl palmitate/water/Brij 97:1-butanol system (Bonne et al., 2006, Krauel et al., 2007). According to the photomicrographs made using cryo-FESEM technique, in microemulsions with higher than 15% wt/wt water contents the formation of globular structures have been observed (Sai et al., 2006, Lu et al., 2006, Anouti et al., 2012, Krauel et al., 2005, Kapoor et al., 2009, Lutter et al., 2007, Holland & Warrack, 1990).

The other type of methods which have played a significant role in the characterization of the microemulsions is the scattering techniques, such as dynamic light scattering and neutron spin-echo spectroscopy (Hellweg & Langevin, 1998, Nagao et al., 1997, Nagao et al., 2006, Geyer et al., 2004, Gradzielski & Langevin, 1996, Hellweg et al., 2001, Hellweg et al., 2001, Magid, 1986, Tabony et al., 1983, Atkinson et al., 1988, Magid et al., 1983, Chen, 1986, De Geyer & Tabony, 1986), light scattering (Attwood & Ktistis, 1989, Guest & Langevin 1986, Aoudia et al., 1991, Zhang & Michniak-Kohn, 2011, Li et al., 2010, Xie et al., 2007, Ben Azouz et al., 1992, Zemb, 2009, Magid, 1986, Kljajić et al., 2011, Tan et al., 2011, Wadle et al., 1993, Dave et al., 2007, Kataoka et al., 2007, Silas & Kaler, 2003, Wines & Somasundaran, 2002, Fanun, 2008, Hellweg et al., 2001, Fanun et al., 2001).

5. Surfactants

The surfactants are molecules with at least two parts, one part soluble in polar solvents (hydrophilic) and the other part insoluble in the polar solvent (hydrophobic). Because of this double character, the term amphiphile is also used as synonym with surfactant (Holmberg et al., 2002). The polar part of the surfactant molecule is referred as head, and the non-polar part of the molecule as tail. Having these two parts with opposing solubilisation abilities, gives the surfactant molecules unique capabilities, such as tendency to adsorb at the surfaces and interfaces, which results in the decrease of the surface tension, and also formation of the aggregates inside the solutions, resulting in the formation of the microemulsions. This double character of the surfactant molecule enables it to orient in desired way while in contact with the two phases with different hydro/lypophilic properties, or to make aggregates inside of the solution with hydro- or lypophilic parts directed towards the media. Such aggregates can solubilise an oil in aqueous phase (micelles) or water in the oily phase (reversed micelles).

The polar nature of the head group of surfactants vary from non-ionic to ionic character. Depending on the nature of this part, the surfactants are categorized into non-ionic, anionic, cationic and amphoteric (zwitterionic) surfactants (Tadros, 2005, Rosen et al., 2000, Os, 1998). Versatile types of functional groups have been utilized as the head group for the surfactants. Among them carboxylates, sulphates, phosphates, sulfonates, quaternary amines, polyethers have a great importance in many different applications. Commercially used surfactants can be obtained from synthetic or natural resources

(Hayes et al., 2009, Nace et al., 1996, Goodwin 2004, Holmberg 1998). Regarding the structure, surfactants can be simple molecules, like sodium or potassium salt of the carboxylic acids generally with 12-18 carbon atoms, or polymers with various molecular weights (Kwak 1998, Hill 1999, Malmsten, 2002,). For any application of the surfactants one should evaluate the issues concerned with that special case, as the toxicity, stability and performance of the surfactants is closely related to its structure (Esumi & Ueno, 2003, Dias & Lindman 2008).

6. Microemulsions in non-conventional systems

Ionic liquids and supercritical fluids are widely used non-conventional systems which have been used in many different fields of applications, as well as microemulsion research (Zhang et al., 2006, Tingey et al., 1991, Eastoe et al., 1991, Johnston et al., 1996). Among the supercritical fluids, carbon dioxide, because of its non-toxicity, cheapness and easy availability, has been used as solvent for different purposes, such as extraction and polymerizations. One of the main issues in using the supercritical carbon dioxides as solvent is that for the high molecular weight polymeric compounds it has lower solubilising power. Hence, for this type of applications design and synthesis of special surfactants is of prime importance (Najjar 2006, Beginn et al., 2006). The effect of supercritical conditions on the microemulsions and formation of the micelles in this type of solvents such as sc-ethane, sc-propane has been studied (Kumar & Mittal, 1999, McFann & Johnston, 1993, Beckman et al., 1991, Bartscherer et al., 1995, Schwan et al., 2010). Use of general low molecular weight surfactants such as sodium bis-2-ethylhexyl sulfosuccinate (AOT) (Kotlarchyk et al., 1985, Olesik & Miller, 1990, Zulauf & Eicke, 1979, Yazdi et al., 1990), poly(ethylene oxide) alkyl ethers (C_nE_m) (Johnston et al., 1989, Yee et al., 1992, Eastoe et al., 1990, Klein & Prausnitz, 1990), fluorinated analogues of AOT (Park et al., 2006) and flouropolymers based surfactants are the most widely used surfactants in supercritical hydrocarbon fluids (ethane and propane) (Eastoe et al., 2001, Eastoe et al., 2006, Hoefting et al., 1993).

Supercritical carbon dioxide is also one the widely used sc-fluids in microemulsion science. Because of the lack of favourable interactions between CO_2 and most of compounds, the commercially available general surfactants have not performed well in sc-CO_2. Meanwhile formation of the some microstructures similar to micelles has been evidenced in sc-CO_2 (Randolph et al., 1987, Ritter & Paulaitis, 1990, Iezzi et al., 1989, Oates 1989, Consani & Smith, 1990, Eastoe et al., 2001, Sagisaka et al., 2003, Klostermann et al., 2011). Many research activities have been done examining the performance of a lot of commercially available surfactants in sc-CO_2, mostly showing very low effect on the increasing of solubilisation of polar compounds (such as water) in the solvent. As mentioned previously, by the design of special surfactants, mainly based on the flouropolymers one can improve the micelle formation, and consequently solubilisation of the more polar compounds in sc-CO_2 phase (Eastoe 2006). Various techniques have been used for the investigation of the microstructure of microemulsions in supercritical fluids. Among them are the time-resolved fluorescence spectroscopy using coumarin 480 (Pramanik et al., 2010), pyridine N-oxide (Simón de Dios & Díaz-García, 2010, Yazdi et al., 1990, Zhang & Bright, 1992a, 1992b, López-Quintela et al., 2004), pyrene (Nazar et al., 2009) or Ti(IV) complexes (Chem et al., 2009) as fluorescence probe, FT-IR spectroscopy (Yee et al., 1991, 1992, Takebayashi et al., 2011), small-angle neutron scattering (SANS) measurements (Eastoe et al., 1996, Zielinski et al.,

1997, Cummings et al., 2012, Torino et al., 2010, Frielinghaus et al., 2006), electron paramagnetic resonance (EPR) spectroscopy (Johnston et al., 1996), small-angle X-ray scattering (SAXS) (Fulton et al., 1995, Kometani et al., 2008, Akutsu et al., 2007), near-infrared spectroscopy (Takebayashi et al., 2011), and high-pressure NMR (Thurecht et al., 2006), which have been extensively employed for investigation of these systems.

The earliest theoretical studies of microemulsions in supercritical fluids was reported in 1990s showing a good agreement between theory and experimental results, among them are the works of Johnston et al. (Peck & Johnston, 1991). Different models have been examined and the results compared to be in different levels of agreement with the experimental data (García-Sánchez et al., 2001, Taha et al., 2005, Ganguly & Choudhury, 2012).

A lots of reports can be found in the literature about the application of microemulsions prepared in supercritical fluids for different purposes, such as, investigation of sustained release of nucleic acids from polymeric nanoparticles prepared in sc-CO_2 microemulsions (Ge et al., 2010), biocatalysis using lipase encapsulated in microemulsion-based organogels in sc-CO_2 (Blattner et al., 2006), continuous tuning of size CdS and ZnS nanoparticles in a water-in-sc-CO_2 microemulsion (Fernandez & Wai, 2007), and synthesis of nanoporous clusters of zirconia (ZrO_2) (Lee et al., 2010).

7. Some applications of microemulsions

7.1 Industrial applications

Microemulsions play a great role in the everyday life of human body. There are many final products which, in principle based on the microemulsions and/or they are somehow in very close relation with the microemulsions. Sometimes the microemulsion formation is the important process that, occur at the final stage of the application. However, in every case the formation of the microemulsion results in the solubilisation of the chemicals which may be the active agent or the unwanted compound that its removal is the first task of the process. Or in some cases this solubilisation helps to deliver the active agents to the required sites. Any formulation which intended to be used in industrial scale should be economical. Various types of cleaning process are one of the main areas that relates to the application of the microemulsions in big scales. Some of the other areas include: agrochemicals formulations (Mulqueen 2003, Chen et al., 2000) (solubilizing organic agrochemicals in water), preparation of the vaccine adjuvants to improve the effectiveness of the active compounds (O'Hagan et al., 1997, Hariharan & Hanna, 1998), micro and -emulsion polymerization (Xu & Gan, 2005, Capek 2001), floatation process in the pulp and paper industry, concrete and asphalt, petroleum industry (for example in enhanced oil recovery, natural gas dehydration and etc.) (Santanna et al., 2009, Austad & Taugbøl 1995), firefighting foams, defogging agents, decontamination of the media from chemical and biological agents and many more other application (Solans & Kunieda, 1997). Also, formulation of the cosmetics (Valenta & Schultz, 2004), medicals and food additives are the other important areas that require very exact control and analysis. According to the statitistics, in 2003 about 2 milion m^3 of surfactants, made from fatty alcohols has been consumed (Brackmann & Hager, 2004). There are many more reports in the literature that indicate the importance and level of the applications of this type of compounds.

7.2 Applications in biological and health sciences

Because of many unique properties, such as stability, ability for solubilisation of the lypo- or hydrophilic compounds and etc, microemulsions has attracted a great attention for a several type of applications. Therefore, these mixtures have entered in many fields of research and applications, ranging from advanced oil recovery to delivering genes into the cells. The ability to formulate such mixtures with biocompatible ingredients has made possible to use them in biological and health related areas extensively. Hence, microemulsions have found wide application in delivering drugs with different physical and chemical characteristics and different ways of delivering such as, oral delivery of protein drugs (Sarciaux et al., 1995, Ke et al., 2005, Cheng et al., 2008, Kim et al., 2005,), ophthalmic (Lv et al., 2005., Gulsen & Chauhan, 2005), transdermal (Kantaria et al., 1999, Kreilgaard et al., 2000, Sintov & Botner 2006, Neubert, 2011, Zhu et al., 2008), amphiphilic drugs (Djordjevic et al., 2004, Oh et al., 2011), internasal (Cho et al., 2012) or other ways (Zhou et al., 2011, Lawrence & Rees, 2000).

Besides drug delivery, microemulsions have many other applications in these fields. Among them is the delivering of genes into the cells (Gupta et al., 2004), with the aim of treatment of disease or diagnosis (Pedersen et al., 2006, Peng et al., 2006, Xu et al., 2011, Rossi et al., 2006, Santra et al., 2005), targeting cancer cell (Tao et al., 2011, Lu et al., 2008, Reithofer et al., 2011), act as vaccine (Sun et al., 2009, Mumper & Cui, 2003), biocatalyss (Stamatis & Xenakis, 1999, Zoumpanioti et al., 2010, Blattner et al., 2006), cosmetics (Chiu & Yang, 1992, Förster et al., 1995, Valenta & Schultz, 2004, Teichmann et al., 2007) or changing the genetic to improve the performance of the targeted cells (Kitamoto et al., 2009, Courrier et al., 2004).

Use of food grade components, oil and surfactant allows to prepare microemulsions which, can be added to foods and beverages (Rao & McClements, 2011, Zhang et al., 2008, Zhang et al., 2009, De Campo et al., 2004, Feng et al., 2009, Rao et al., 2012, Ziani et al., 2012, Flanagan et al., 2006).

7.3 Preparation and processing of nanomaterials

One of the other important fileds of applications of microemulsions is their use as media for the synthesis of different materials. Regarding the microstructure of the microemulsions one can choose a special formulation to have a well defined microstructure in the system. The microstructures which, are widely used in this respect are the water droplets in oil phase or oil droplets in water phase. As the size of this droplets are in the range of nanometer (about 5 to 50 nm), they act as nanoreactors dispersed in the oil or water media. This concept has been used extensively for the preparation of the many different type of materials, such as organic, inorganic, oxides, polymers and etc. in nanometer dimentions. Since last two decades and even now, this field is one of the hot research topics, specially for preparation of the nanomaterials. By this methodology, many researchers have prepared metallic and intermetallic nanoparticles (Aubery et al., 2011, Hosseini et al., 2011a, 2011b, Shokri et al., 2011, García-Diéguez et al., 2010), metal oxides (Tian et al., 2012, Du et al., 2012, Lin et al., 2012), metal salts (Dromta et al., 2012, Guleria et al., 2012, Esmaeili et al., 2011), polymers (Ouadahi et al., 2012, Ma et al., 2012, Mishra & Chatterjee, 2011, Elbert, 2011), luminescent nanoparticles (Probst et al., 2012, Darbandi et al., 2012), magnetic nanoparticles (Jing et al., 2011, Xu et al., 2011) and lipid nanoparticles for drug delivery (Puri et al., 2010, Seyfoddin et al., 2010, Das & Chaudhury, 2011). One of the other applications of microemulsions is to polymerize bicontinous microemulsions to obtain porous materials. This has been done by use of polymerizable surfactants, which their polymerization reaction can be started photochemically (Schwering, 2008, Ye, 2007, Stubenrauch et al., 2008, Chow & Gan, 2005, Magno et al., 2009b).

8. Theoretical studies on microemulsions

The microemulsions have been also studied theoretically and some models have been introduced for the theortical discussion on the microemulsions. The models such as continuum and lattice models, and models based on phenomenological free energy densities as well as those based on microscopic Hamiltonians have been proposed. These models mainly discuss about the parameters such as interfacial tensions, progression of microemulsion phase equilibria, and the wetting or non-wetting of the interfaces (Widom 1996). Boyden et al. has used the Monte Carlo method to simulate the oil/water miscibility gap, the coexistence of various phases, kinetics of micelle formation (Boyden et al., 1994). Most of the observed properties have been described very well by the proposed model. The phase behaviour of the microemulsions has been simulated by a mathamatical framework (Bennett et al., 1981), excess Gibbs energy models (García-Sánchez et al., 2001) and lattice Monte Carlo simulation (Behjatmanesh-Ardakani et al., 2008). In another study, Acosta has used the net-average curvature (NAC) model, introduced by his own research group (Acosta et al., 2003) to prepare the equation of state to fit and predict the phase behavior of microemulsions formulated with ionic surfactants (Acosta et al., 2008). In a work reported by Kiran et al. the morphology and viscosity of microemulsions has been studied using the HLD-NAC model (Kiran et al., 2010). They have introduced a new shapebased NAC model, relating the net and average curvatures to the length and radius of microemulsion droplets, with a hypothesized cylindrical core with hemispherical end caps.

Many of the reactions and processes in the microemulsions have been simulated using different models. The investigation of the drug release from drug loaded microemulsions (Grassi et al., 2000, Sirotti et al., 2002), nanoparticle precipitation using a population balance model (Niemann & Sundmacher, 2010), the formation of nanoparticles in mixing of two microemulsion systems by a Monte Carlo model (Jain & Mehra, 2004), time-evolution of the polymer particle size distribution (Suzuki & Nomura, 2003) and solubilization of oil mixtures in anionic microemulsions (Szekeres et al., 2006) have been perfomed theoretically.

9. Conclusions

In the same way as other fields, the science and technology of the microemulsions is a rapidly growing area, which gained a very high importance during last two decades. According to the scopus database, the number of papers published in the area of microemulsions, 1960-70s (112 papers), 1980s (974 papers), 1990s (2762 papers), 2000s (6933 papers) and 2011 (843 papers) shows a very high increasing rate. Besides the other parameters, the finding of many novel applications is one of the reasons for this fast delevolment in the area. The use of microemulsions in drug delivery systems and also in nanoscience and nanotechnology are among the most important applications, which attracted a high attention of the researchers. The possibility and easiness of the tuning of microemulsion properties with different parameters has allowed the scientists to use them in many interdisciplinary fields of research and applications. The future need for the developing of systems and materials with sustainablilty and biodegradability requires that biodegradable surfactants and compounds must be developed. Doing this, will be another reason increasing the importance of the microemulsions which can be used in bio systems.

10. Acknowledgment

I acknowledge the supports of the University of Tabriz during the time of writing of this book chapter. Also, I would like to express my gratitude to Prof. Dr. Cosima Stubenrauch, University of Stuttgart, Germany, who taught me the first words of the microemulsions.

11. References

Acosta, E. J., (2008). The HLD-NAC equation of state for microemulsions formulated with nonionic alcohol ethoxylate and alkylphenol ethoxylate surfactants. *Colloids Surf. A Physicochem. Eng. Asp.*, Vol. 320, Nos. 1-3, pp. 193-204.

Acosta, E., Szekeres, E., Sabatini, D. A., and Harwell, J. H., (2003). Net-Average Curvature Model for Solubilization and Supersolubilization in Surfactant Microemulsions. *Langmuir*, Vol. 19, No. 1, pp. 186–195.

Agarwal, V., Singh, M., McPherson, G., John, V., and Bose, A. (2004). Freeze fracture direct imaging of a viscous surfactant mesophase. *Langmuir*, Vol. 20, No. 1, pp. 11–15.

Ahmad, S. I., Shinoda, K. and Friberg, S., (1974). Microemulsions and phase equilibria. Mechanism of the formation of so-called microemulsions studied in connection with phase diagram. *J. Colloid Interface Sci.*, Vol. 47, No. 1, pp. 32-37.

Akutsu, T., Yamaji, Y., Yamaguchi, H., Watanabe, M., Smith Jr., R., L. and Inomata, H., (2007). Interfacial tension between water and high pressure CO_2 in the presence of hydrocarbon surfactants. *Fluid Phase Equilibria*, Vol. 257, No. 2, pp. 163-168.

Anouti, M., Sizaret, P.-Y., Ghimbeu, C., Galiano, H., and Lemordant, D., (2012). Physicochemical characterization of vesicles systems formed in mixtures of protic ionic liquids and water, *Colloids Surf. A: Physicochem. Eng. Asp.*, Vol. 395, pp. 190-198.

Aoudia, M., Rodgers, M.A.J. and Wade, W.H. (1991). Light scattering and fluorescence studies of o/w microemulsion: The sodium 4-dodecylbenzene-sulfonate –butanol-water–NaCl–Octane system. *J. Coll. Int. Sci.*, vol. 144, pp. 353–362.

Atkinson, P. J., Grimson, M. J., Heenan, R. K., Howe, A. M., Mackie, A. R., Robinson, B. H., (1988). Microemulsion-based gels: A small-angle neutron scattering study, *Chem. Phys. Lett.*, Vol. 151, No. 6, pp. 494-498.

Attwood, D. and Ktistis, G. (1989). A light scattering study on oil-in-watermicroemulsions. *Int. J. Pharma.*, Vol. 52, pp. 165–171.

Aubery, C., Solans, C., Sanchez-Dominguez, M., (2011). Tuning high aqueous phase uptake in nonionic water-in-oil microemulsions for the synthesis of Mn-Zn ferrite nanoparticles: Phase behavior, characterization, and nanoparticle synthesis. *Langmuir*, Vol. 27, No. 23, pp. 14005-14013.

Austad, T., and Taugbøl, K., (1995). Chemical flooding of oil reservoirs 2. Dissociative surfactant-polymer interaction with a negative effect on oil recovery. *Colloids Surf. A: Physicochem. Eng. Asp.*, Vol. 103, No. 1–2, pp. 73-81.

Azouz, I. B., Ober, R., Nakache, E., and Williams, C.E., (1992). A small angle X-ray scattering investigation of the structure of a ternary water-in-oil microemulsion, *Colloids Surf.*, Vol. 69, No. 2-3, pp. 87-97.

Baker, R.C., Florence, A.T., Tadros, Th.F., Wood, R.M. (1984). Investigations into the formation and characterization of microemulsions. I. Phase diagrams of the ternary system water—sodium alkyl benzene sulfonate—hexanol and the quaternary system water—xylene—sodium alkyl benzene sulfonate—hexanol. *J. Colloid Interface Sci.*, Vol. 100, No. 2, pp. 311-331.

Balogh, J., (2010). Determining scaling in known phase diagrams of nonionic microemulsions to aid constructing unknown. *Adv. Colloid Interface Sci.*, Vol.159, No.1, pp. 22-31.

Bartscherer, K. A., Minier, M., and Renon, H., (1995). Microemulsions in compressible fluids — A review. *Fluid Phase Equilib.* Vol. 107, pp. 93.

Bastogne, F., Nagy, B. J., and David, C., (1999). Quaternary 'N-alkylaldonamide-brine-decane-alcohol' systems. Part II: microstructure of the one-phase microemulsion by NMR spectroscopy, *Colloids Surf. A: Physicochem. and Eng. Asp.*, Vol. 148, No. 3, pp. 245-257.

Becher, P., and Arai, H., (1968). Nonionic surface-active compounds. XI. Micellar size, shape, and hydration from light-scattering and hydrodynamic measurements. *J. Colloid Interface Sci.* Vol. 27, pp. 634.

Beckman, E. J., Fulton, J. L., and Smith, R. D., 1991, in *Supercritical Fluid Technology* (Bruno T. J., and Ely, J. E., Eds.), CRC, Boca Raton, FL, pp. 405-449.

Beginn, U., Najjar, R., Ellmann, J., Vinokur, R., Martin R., and Möller, M., (2006). Copolymer -ization of vinylidene difluoride with hexafluoropropene in supercritical CO_2, *J. Polymer Sci. A, Polymer Chem.*, Vol. 44, No. 3, pp. 1299-1316.Behjatmanesh-Ardakani, R., Karimi, M.A., Nikfetrat, M., (2008). Monte Carlo simulation of microemulsion phase transitions by solvent accessible surface area. *J. Chin. Chem. Soc.*, Vol. 55, No. 4, pp. 716-723.

Belkoura, L., Stubenrauch, C. and Strey, R. (2004). Freeze fracture direct imaging: A hybrid method in preparing specimen for Cryo-TEM. *Langmuir*, Vol. 20, pp. 4391–4399.

Bellocq, A.M., Bourbon, D., Lemanceau, B., Fourche, G. (1982). Thermodynamic, interfacial, and structural properties of polyphasic microemulsion systems. *J. Colloid Interface Sci.*, Vol. 89, No. 2, pp. 427-440.

Bennett, K.E., Phelps, C.H.K., Davis, H.T., and Scriven, L.E. (1981). Microemulsion phase behavior - observations, thermodynamic essentials, mathematical simulation. *Soc. Petroleum Eng. J.*, Vol. 21, No. 6, pp. 747-762.

Bernhein-Grosswasser, A., Tlusty, T., Safran, S.A. and Talmon, Y. (1999). Direct observation of phase separation in microemulsion networks. *Langmuir*, Vol. 15, pp. 5448–5453.

Biais, J., Trouilly, J. L., Clin, B., and Lalanne, P. (1987). New model for microemulsion stability. *Prog. Colloid Polym. Sci.*, Vol. 73, pp. 193.

Blattner, C., Zoumpanioti, M., Kröner, J., Schmeer, G., Xenakis, A., and Kunz, W., (2006). Biocatalysis using lipase encapsulated in microemulsion-based organogels in supercritical carbon dioxide, *J. Supercritical Fluids*, Vol. 36, No. 3, pp. 182-193.

Boonme, P., Krauel, K., Graf, A., Rades, T., Junyaprasert, V.B. (2006). Characterization of microemulsion structures in the pseudoternary phase diagram of isopropyl palmitate/ water/Brij 97:1-butanol. *AAPS Pharm Sci Tech*, Vol. 7, No. 2, pp. E99-E104.

Boyden, S., Jan, N., Ray, T., (1994). Monte carlo simulations of microemulsions. *Il Nuovo Cimento D*, Vol. 16, No. 9, pp. 1439-1445.

Brackmann, B., and Hager, C.-D., (2004). The statistical world of raw materials, fatty alcohols and surfactants, Proceedings 6th World Surfactant Congress CESIO, Berlin Germany.

Burauer, S., Belkoura, L., Stubenrauch, C. and Strey, R. (2003). Bicontinuous microemulsions revisited: A new approach to freeze fracture electron microscopy (FFEM). *Colloids Surf. A*, Vol. 228, pp. 159–170.

Capek I., (2001). Microemulsion polymerization of styrene in the presence of a cationic emulsifier. *Adv. Colloid Interface Sci.*, Vol. 92, No. 1–3, pp. 195-233.

Causse, J., Lagerge, S., de Menorval, L.C., and Faure, S. (2006). Micellar solubilization of tributylphosphate in aqueous solutions of Pluronic block copolymers: Part II. Structural characterization inferred by 1H NMR, *J. Colloid Interface Sci.*, Vol. 300, No. 2, pp. 724-734.

Chai, J.-L., Liu, J., Li, H.-L. (2009). Phase diagrams and chemical physical properties of dodecyl sulfobetain/alcohol/oil/water microemulsion system. *Colloid J.*, Vol. 71, No. 2, pp. 257-262.

Chen, F., Wang, Y., Zheng, F., Wu, Y., and Liang W., (2000). Studies on cloud point of agrochemical microemulsions. *Colloids Surf. A: Physicochem. Eng. Asp.*, Vol. 175, No. 1–2, pp. 257-262.

Chen, S.-H. (1986). Interactions and phase transitions in micellar and microemulsion systems studied by small angle neutron scattering, *Physica B+C*, Vol. 137, No. 1–3, pp. 183-193.

Chen, X., Wei, Q., Cai, Y., Han, Y., Zhao, Y., Du, B. (2009). Determination of ultra trace amounts of protein by 4-chlorosulfo-(2′-hyaroxylphenylazo)-rhodanine-Ti(IV) complex [ClSARP-Ti(IV)] as the fluorescence spectral probe in AOT micro-emulsion. *Spectrochimica Acta - Part A: Molecular and Biomolecular Spectroscopy*, Vol. 72, No. 5, pp. 1047-1053.

Cheng, M.-B., Wang, J.-Ch., Li, Y.-H., Liu, X.-Y., Zhang, X., Chen, D.-W., Zhou, Sh.-F., and Zhang, Q., (2008). Characterization of water-in-oil microemulsion for oral delivery of earthworm fibrinolytic enzyme. *J. Controlled Release*, Vol. 129, No. 1, 2, pp. 41-48.

Chiu, Y. C., and Yang, W. L., (1992). Preparation of vitamin E microemulsion possessing high resistance to oxidation in air. *Colloids Surf.*, Vol. 63, No. 3–4, pp. 311-322.

Cho, H.-J., Ku, W.-S., Termsarasab, U., Yoon, I., Chung, Ch.-W., Moon, H. T., and Kim, D.-D., (2012). Development of udenafil-loaded microemulsions for intranasal delivery: *In vitro* and *in vivo* evaluations. *Inter. J. Pharm.*, Vol. 423, No. 2,. Pp. 153-160.

Chow, P.Y., Gan, L.M., (2005). Microemulsion polymerizations and reactions. *Adv. Polym. Sci.*, Vol. 175, pp. 257-298.

Consani, K. A., and Smith, R. D., (1990). Observations on the solubility of surfactants and related molecules in carbon dioxide at 50°C. *J. Supercrit. Fluid*, Vol. 3, pp. 51.

Courrier, H. M., Vandamme, Th. F., and Krafft, M. P., (2004). Reverse water-in-fluorocarbon emulsions and microemulsions obtained with a fluorinated surfactant. *Colloids Surf. A: Physicochem. Eng. Asp.*, Vol. 244, No. 1–3, pp. 141-148.

Cummings, S., Enick, R., Rogers, S., Heenan, R., and Eastoe, J., (2012). Amphiphiles for supercritical CO_2. *Biochimie*, Vol. 94, No. 1, Pages 94-100.

Darbandi, M., Urban, G., Krüger, M., (2012). Bright luminescent, colloidal stable silica coated CdSe/ZnS nanocomposite by an in situ, one-pot surface functionalization. *J. Colloid Interface Sci.*, Vol. 365, No. 1, pp. 41-45.

Das, S., Chaudhury, A., (2011). Recent advances in lipid nanoparticle formulations with solid matrix for oral drug delivery. *AAPS PharmSciTech*, Vol. 12, No. 1, pp. 62-76.

Dave, H., Gao, F., Schultz, M., and Co, C. C., (2007). Phase behavior and SANS investigations of edible sugar–limonene microemulsions, *Colloids Surf. A: Physicochem. Eng. Asp.*, Vol. 296, No. 1–3, pp. 45-50.

De Campo, L. Yaghmur, A., Garti, N., Leser, M. E., Folmer, B., and Glatter, O., (2004). Five-component food-grade microemulsions: structural characterization by SANS. *J. Colloid & Interface Sci.*, Vol. 274, No. 1, pp. 251-267.

De Geyer, A., and Tabony, J. (1986). Small-angle neutron scattering evidence for a bicontinuous structure in a microemulsion containing equal volumes of oil and water, *Chem. Phys. Lett.*, Vol. 124, No. 4, pp. 357-360.

de Geyer, A., Molle, B., Lartigue, C., Guillermo, A., and Farago, B., (2004), Dynamics of caged microemulsion droplets: a neutron spin echo and dynamic light scattering study, *Physica B: Condensed Matter*, Vol. 350, No. 1–3, pp. 200-203.

Dhar, N., Akhlaghi, S.P., Tam, K.C., (2012). Biodegradable and biocompatible polyampholyte microgels derived from chitosan, carboxymethyl cellulose and modified methyl cellulose. *Carbohydr. Polym.*, Vol. 87, No. 1, pp. 101-109.

Dias R., and Lindman, B., (2008). *DNA Interactions with Polymers and Surfactants*, John Wiley & Sons, Inc., Hoboken, New Jersey.

Djordjevic, L., Primorac, M., Stupar, M., and Krajisnik, D., (2004). Characterization of caprylocaproyl macrogolglycerides based microemulsion drug delivery vehicles for an amphiphilic drug. *Inter. J. Pharm.*, Vol. 271, No. 1–2, pp. 11-19.

Drmota, A., Drofenik, M., Žnidaršič, A., (2012). Synthesis and characterization of nano-crystalline strontium hexaferrite using the co-precipitation and microemulsion methods with nitrate precursors. *Ceramics International*, Vol. 38, No. 2, pp. 973-979.

Du, Y., Wang, W., Li, X., Zhao, J., Ma, J., Liu, Y., Lu, G., (2012). Preparation of NiO nanoparticles in microemulsion and its gas sensing performance. *Mater. Lett.*, Vol. 68, pp. 168-170.

Eastoe, J., Bayazit, Z., Martel, S., Steytler, D. C., and Heenan, R. K., (1996). Droplet Structure in a Water-in-CO_2 Microemulsion. *Langmuir*, Vol. 12, pp. 1423.

Eastoe, J., Gold, S. and Steytler, D. C. (2006). Surfactants for CO_2. *Langmuir*, Vol. 22, pp. 9832–9842.

Eastoe, J., Gold, S., Rogers, S., Wyatt, P., Steytler, D. C., Gurgel, A., Heenan, R. K., Fan, X., Beckman, E. J. and Enick, R. M., (2006). Designed CO_2-philes stabilize water-in-carbon dioxide microemulsions. *Angewandte Chemie – Internat. Ed*, Vol. 45, pp. 3675–3677.

Eastoe, J., Paul, A., Nave, S., Steytler, D. C., Robinson, B. H., Rumsey, E., Thorpe, M. and Heenan, R. K., (2001). Micellization of hydrocarbon surfactants in supercritical carbon dioxide. *J. Am.Chem. Soc.*, Vol. 123, pp. 988–989.

Eastoe, J., Robinson, B. H., Visser, A. J. W. G., and Steytler, D. C. (1991). Rotational dynamics of AOT reversed micelles in near-critical and supercritical alkanes. *J. Chem. Soc. Faraday Trans.*, Vol. 87, pp. 1899–1903.

Eastoe, J., Young, W. K., Robinson, B. H., and Steytler, D. C., (1990). Scattering studies of microemulsions in low-density alkanes. *J. Chem. Soc. Faraday Trans.* 1, Vol. 86, pp. 2883.

Ekwall, P., Danielsson, I. and Mandell, L., (1960). Assoziations- und Phasengleichgewichte bei der Einwirkung von Paraffinkettenalkoholen an wässrigen Lösungen von Assoziations-kolloiden. *Kolloid-Z.* Vol. 169, pp. 113 .

Elbert, D.L., (2011). Liquid-liquid two-phase systems for the production of porous hydrogels and hydrogel microspheres for biomedical applications: A tutorial review. *Acta Biomaterialia*, Vol. 7, No. 1, pp. 31-56.

Emin, S. M., Denkova, P. S., Papazova, K. I., Dushkin, C. D., and Adachi, E., (2007). Study of reverse micelles of di-isobutyl phenoxyethoxyethyl dimethylbenzyl ammonium methacrylate in benzene by nuclear magnetic resonance spectroscopy, *J. Colloid Interface Sci.*, Vol. 305, No. 1, pp. 133-141.

Esmaeili, N., Kazemian, H., Bastani, D., (2011). Synthesis of nano particles of LTA zeolite by means of microemulsion technique. *Iran. J. Chem. Chem. Eng.*, Vol. 30, No. 2, pp. 1-8.

Esumi K., and Ueno, M., (2003). *Structure-Performance Relationships in Surfactants*, Marcel Dekker, Inc., New York.

Fanun, M., (2008). A study of the properties of mixed nonionic surfactants microemulsions by NMR, SAXS, viscosity and conductivity, *J. Molecular Liquids*, Vol. 142, No. 1–3, pp. 103-110.

Fanun, M., Wachtel, E., Antalek, B., Aserin, A., and Garti, N., (2001). A study of the microstructure of four-component sucrose ester microemulsions by SAXS and NMR, *Colloids Surf. A: Physicochem. and Eng. Asp.*, Vol 180, Iss 1-2, pp. 173-186.

Feng, J.-L., Wang, Zh.-W., Zhang, J., Wang, Zh.-N., and Liu, F., (2009). Study on food-grade vitamin E microemulsions based on nonionic emulsifiers. *Colloids Surf. A: Physicochem. Eng. Asp.*, Vol. 339, No. 1–3, pp. 1-6.

Fernandez, C.A., Wai, C.M. (2007). Continuous tuning of cadmium sulfide and zinc sulfide nanoparticle size in a water-in-supercritical carbon dioxide microemulsion. *Chem. - A Eur. J.*, Vol. 13, No. 20, pp. 5838-5844.

Flanagan, J., Kortegaard, K., Pinder, D. N., Rades, Th., and Singh, H., (2006). Solubilisation of soybean oil in microemulsions using various surfactants. *Food Hydrocolloids*, Vol. 20, No. 2–3, pp. 253-260.

Förster, T., Von Rybinski, W., and Wadle, A., (1995). Influence of microemulsion phases on the preparation of fine-disperse emulsions. *Adv. Colloid and Interface Sci.*, Vol. 58, No. 2–3, pp. 119-149.

Frielinghaus, H., Maccarrone, S., Byelov, D., Allgaier, J., Richter, D., Auth, T., and Gompper, G., (2006). SANS studies of confined diblock copolymers in microemulsions. *Physica B: Condensed Matter*, Vol. 385–386, Part 1, pp. 738-741.

Fu, D., Lu, J., Liu, J., Li, Y. New equation of state for microemulsion system (2003). *J. Chem. Indus. Eng. (China)*, Vol. 54, No. 6, pp. 725-730.

Fu, X., Xiong, Y., Qingli, W., Shuyun, X., Shaona, Z., Hu, Z. Study on the thiophosphinic extractants. II. Thermodynamic functions and structural parameters of the w/o microemulsion of the saponified acid systems (2002). *Colloids Surf. A: Physicochem. Eng. Asp.*, Vol. 211, No. 2-3, pp. 249-258.

Fulton, J. L., Pfund, D. M., McClain, J. B., Romack, T. J., Maury, E. E., Combes, J. R., Samulski, E. T., Desimone, J. M. and Capel, M. (1995). Aggregation of amphiphilic molecules in supercritical carbon-dioxide - a small-angle X-ray-scattering study. *Langmuir*, Vol. 11, pp. 4241–4249.

Ganguly, R., and Choudhury, N., (2012). Investigating the Evolution of the Phase Behavior of AOT-Based w/o Microemulsions in Dodecane as a Function of Droplet Volume Fraction. *J. Colloid Interface Sci.*, in press, http://dx.doi.org/10.1016/j.jcis.2012.01.037

García-Diéguez, M., Pieta, I.S., Herrera, M.C., Larrubia, M.A., Alemany, L.J., (2010). Improved Pt-Ni nanocatalysts for dry reforming of methane. *Applied Catalysis A: General*, Vol. 377, Nos. 1-2, pp. 191-199.

García-Sánchez, F., Eliosa-Jiménez, G., Salas-Padrón, A., Hernández-Garduza, O., and Ápam-Martínez, D., (2001). Modeling of microemulsion phase diagrams from excess Gibbs energy models. *Chem. Eng. J.*, Vol. 84, No. 3, pp. 257-274.

Ge, J., Jacobson, G. B., Lobovkina, T., Holmberg, K., and Zare, R.N. (2010). Sustained release of nucleic acids from polymeric nanoparticles using microemulsion precipitation in supercritical carbon dioxide. *Chem. Commun.*, Vol. 46 No. 47, pp. 9034-9036.

Gillberg, G. Lehtinen, H. and Friberg, S. (1970). NMR and IR Investigation of the Conditions Determining the Stability" of Microemulsions, *J. Colloid Interface Sci.*, Vol. 33, No. 1, pp.40-53.

Goodwin, J. W., (2004). *Colloids and interfaces with surfactants and polymers: an introduction*, John Wiley & Sons Ltd, West Sussex, England.

Goodwin, J. W., (2009). *Colloids and interfaces with surfactants and polymers*, John Wiley & Sons Ltd, West Sussex, England.

Gradzielski, M., and Langevin, D., (1996). Small-angle neutron scattering experiments on microemulsion droplets: relation to the bending elasticity of the amphiphilic film, *J. Molecular Structure*, Vol. 383, No. 1–3, pp. 145-156.

Grassi, M., Coceani, N., Magarotto, L., (2000). Mathematical modeling of drug release from microemulsions: Theory in comparison with experiments. *J. Colloid Interface Sci.*, Vol. 228, No. 1, pp. 141-150.

Guest, D. and Langevin, D. (1986). Light scattering study of amultiphase microemulsion system. *J. Coll. Inter. Sci.*, Vol. 112, pp. 208–220.

Guleria, A., Singh, S., Rath, M.C., Singh, A.K., Adhikari, S., Sarkar, S.K. (2012). Tuning of photoluminescence in cadmium selenide nanoparticles grown in CTAB based quaternary water-in-oil microemulsions. *J. Luminescence*, Vol. 132, No. 3, pp. 652-658.

Gulsen, D., and Chauhan, A., (2005). Dispersion of microemulsion drops in HEMA hydrogel: a potential ophthalmic drug delivery vehicle. *Inter. J. Pharm.*, Vol. 292, No. 1–2, pp. 95-117.

Gupta, A. K., Gupta, M., Yarwood, S. J., and Curtis, A. S. G., (2004). Effect of cellular uptake of gelatin nanoparticles on adhesion, morphology and cytoskeleton organisation of human fibroblasts. *J. Controlled Release*, Vol. 95, No. 2, pp. 197-207.

Hariharan, K., and Hanna, N., (1998). Development and application of PROVAX™ adjuvant formulation for subunit cancer vaccines. *Adv. Drug Delivery Rev.*, Vol. 32, No. 3, pp. 187-197.

Hayes, D. G., Kitamoto, D., Solaiman, D. K. Y., and Ashby, R. D., (Eds.), (2009). *Biobased Surfactants and Detergents Synthesis, Properties, and Applications*, AOCS Press, Urbana, IL.

Hellweg, T., Gradzielski, M., Farago, B., Langevin, D., and Safran, S., (2001) Reply to the comment on "Shape fluctuations of microemulsion droplets. A neutron spin-echo study" by V. Lisy, *Colloids Surf. A: Physicochem. Eng. Asp.*, Vol. 221, No. 1–3, pp. 257-262.

Hellweg, Th. and Langevin, D. (1998). Bending elasticity of the surfactant film in droplet microemulsions: Determination by a combination of dynamic light scattering and neutron spin-echo spectroscopy. *Phys. Rev. E*, Vol. 57, pp. 6825–6834.

Hellweg, Th., (2002). Phase structures of microemulsions, *Current Opinion in Colloid & Interface Sci.*, Vol. 7, No. 1–2, pp. 50-56.

Hellweg, Th., Gradzielski, M., Farago, B., and Langevin, D., (2001). Shape fluctuations of microemulsion droplets: a neutron spin–echo study, *Colloids Surf. A: Physicochem. Eng. Asp.*, Vol. 183–185, pp. 159-169.

Hill, R. M., (1999). *Silicone Surfactants*, Marcel Dekker, Inc., NY.

Hoefting, T.A., Beitle, R. R., Enick, R. M., and Beckman, E. J., (1993). Design and synthesis of highly CO_2-soluble surfactants and chelating agents. *Fluid Phase Equilib*. Vol. 83, pp. 203.

Holland, S. J., and Warrack, J. K., (1990). Low-temperature scanning electron microscopy of the phase inversion process in a cream formulation, *Inter. J. Pharmaceutics*, Vol. 65, No. 3, pp. 225-234.

Holmberg, K., (1998). *Novel Surfactants: Preparation, Applications, and Biodegradability*, CRC Press, Marcel Dekker, Inc., New York.

Holmberg, K., Jönsson, B., Kronberg B., and Lindman, B., (2002). *Surfactants and Polymers in Aqueous Solution*. John Wiley & Sons, Ltd, West Sussex, England.

Hosseini, M. G., Shokri, M., Khosravi, M., Najjar, R., and Sheikhy, Sh., (2011). Fabrication of Highly Stable Silver, Platinum and Gold Nanoparticles via Microemulsions: Influence of Operational Parameters. *J. Mater. Sci. Eng. A*, Vol. 1, pp. 268-278.

Hosseini, M.G., Shokri, M., Khosravi, M., Najjar, R., Darbandi, M., (2011). Photodegradation of an azo dye by silver-doped nano-particulate titanium dioxide. *Toxicological Environm. Chem.*, Vol. 93, No. 8, pp. 1591-1601.

Hou, M., and Shah, D.O. (1988). A light scattering study on the droplet size and interdroplet interaction in microemulsions of AOT—oil—water system. *J. Colloid Interf. Sci.*, Vol. 123, pp. 398–412.

Iezzi, A., Enick, R., and Brady, J., (1989). *in Supercritical Fluid Science and Technology* (Johnston K. P., and Penninger, J. M. L., Eds.), (ACS Syrup Ser. No, 406), American Chemical Society, Washington, DC, pp. 122-139.

Jahn, W. and Strey, R. (1988), Microstructure of microemulsions by freeze fracture electron microscopy. *J. Phys. Chem.*, Vol. 92, pp. 2294–2301.

Jain, R., Mehra, A., (2004). Monte Carlo models for nanoparticle formation in two microemulsion systems. *Langmuir*, Vol. 20, No. 15, pp. 6507-6513.

Jian, X., Ganzuo, L., Zhiqiang, Zh., Guowei, Z., and Kejian, J., (2001). A study of the microstyructure of CTAB/1-butanol/octane/water system by PGSE-NMR, conductivity and cryo-TEM, *Colloids Surf. A: Physicochem. Eng. Asp.*, Vol. 191, No. 3, pp. 269-278.

Jing, L., Li, Y., Ding, K., Qiao, R., Rogach, A.L., Gao, M., (2011). Surface-biofunctionalized multicore/shell CdTe@SiO2 composite particles for immunofluorescence assay. *Nanotech.*, Vol. 22, No. 50, art. no. 505104.

Johnston, K. P., Harrison, K. L., Clarke, M. J., Howdle, S. M., Heitz, M. P., Bright, F. V., Carlier, C. and Randolph, T. W. (1996). Water in carbon dioxide microemulsions: An environment for hydrophiles including proteins. *Science*, Vol. 271, pp. 624–626.

Johnston, K.P., McFann, G. J., and Lemert, R. M., (1989). *Supercritical Fluid Science and Technology* (Johnston, K. P., and Penninger, J. M. L., Eds.), Am. Chem. Soc., Washington, DC, pp. 140-164.

K Shinoda, H Saito. (1968). The effect of temperature on the phase equilibria and the types of dispersions of the ternary system composed of water, cyclohexane, and nonionic surfactant. *J Colloid Inter Sci.*, Vol. 26 pp. 70.

Kantaria, Sh., Rees, G. D., and Lawrence, M. J., (1999). Gelatin-stabilised microemulsion-based organogels: rheology and application in iontophoretic transdermal drug delivery. *J. Controlled Release*, Vol. 60, No. 2–3, pp. 355-365.

Kapoor, Y,. Thomas, J. C., Tan, G., John, V. T., and Chauhan, A., (2009). Surfactant-laden soft contact lenses for extended delivery of ophthalmic drugs, *Biomaterials*, Vol. 30, No. 5, pp. 867-878.

Kartsev, V. N., Polikhronidi, N. G., Batov, D. V., Shtykov, S. N., Stepanov, G. V. (2010). A model approach to the thermodynamics of microemulsion systems: Estimation of adequacy of the two-phase model of microemulsions. *Russian J. Phys. Chem. A*, Vol. 84, No. 2, pp. 169-175.

Kataoka, H., Ueda, T., Ichimei, D., Miyakubo, K., Eguchi, T., Takeichi, N., and Kageyama, H., (2007). Evaluation of nanometer-scale droplets in a ternary o/w microemulsion using SAXS and 129Xe NMR, *Chem. Phys. Lett.*, Vol. 441, No. 1-3, pp. 109-114.

Ke, W.-T., Lin, S.-Y., Ho, H.-O., and Sheu, M.-Th., (2005). Physical characterizations of micro -emulsion systems using tocopheryl polyethylene glycol 1000 succinate (TPGS) as a surfactant for the oral delivery of protein drugs. *J. Controlled Release*, Vol. 102, No. 2, pp. 489-507.

Kim, S. K., Lee, E. H., Vaishali, B., Lee, S., Lee, Y.-k., Kim, Ch.-Y., Moon, H. T., and Byun, Y., (2005). Tricaprylin microemulsion for oral delivery of low molecular weight heparin conjugates. *J. Controlled Release*, Vol. 105, No. 1–2, pp. 32-42.

Kiran, S.K., Acosta, E.J., (2010). Predicting the morphology and viscosity of microemulsions using the HLD-NAC model. *Ind. Eng. Chem. Res.*, Vol. 49, No. 7, pp. 3424-3432.

Kitamoto, D., Morita, T., Fukuoka, T., Konishi, M,-a., and Imura, T.,Self-assembling properties of glycolipid biosurfactants and their potential applications. (2009). *Current Opinion in Colloid & Interface Sci.*, Vol. 14, No. 5, pp. 315-328.

Klang, V., Matsko, N., B. Valenta, C., and Hofer, F., (2012). Electron microscopy of nanoemulsions: An essential tool for characterisation and stability assessment, *Micron*, Vol. 43, No. 2–3, pp. 85-103.

Klein T., and Prausnitz, J. M., (1990). Phase behavior of reverse micelles in compressed propane at 35.degree.C and pressures to 30 MPa: solubilization of poly(ethylene glycol). *J. Phys. Chem.*, Vol. 94, pp. 8811.

Kljajić, A., Bešter-Rogač, M., Trošt, S., Zupet, R., and Pejovnik, S., (2011). Characterization of water/sodium bis(2-ethylhexyl) sulfosuccinate/sodium bis(amyl) sulfosuccinate/*n*-heptane mixed reverse micelles and w/o microemulsion systems: The influence of water and sodium bis(amyl) sulfosuccinate content, *Colloids Surf. A: Physicochem. Eng. Asp.*, Vol. 385, No. 1–3, pp. 249-255.

Klostermann, M., Foster, T., Schweins, R., Lindner, P., Glatter, O., Strey, R., Sottmann, T., (2011). Microstructure of supercritical CO_2-in-water microemulsions: A systematic contrast variation study. *Phys. Chem. Chem. Phys.* Vol. 13, No. 45, pp. 20289-20301.

Ko, C.J., Ko, Y.J., Kim, D. M., and Park, H. J., (2003). Solution properties and PGSE-NMR self-diffusion study of C18:1E10/oil/water system, *Colloids Surf. A: Physicochem. and Eng. Asp.*, Vol 216, No.s 1-3, pp. 55-63.

Komesvarakul, N., Sanders, M. D., Szekeres, E., Acosta, E. J., Faller, J. F., Mentlik, T., Fisher, L. B., Nicoll, G., Sabatini, D. A., Scamehorn, J. F. (2006). Microemulsions of triglyceride-based oils: The effect of co-oil and salinity on phase diagrams. *J. Cosmetic Sci.*, Vol. 57, No. 4, pp. 309-325.

Kometani, N., Kaneko, M., Morita, T., and Yonezawa, Y., (2008). The formation of photolytic silver clusters in water/supercritical CO_2 microemulsions. *Colloids Surf. A: Physicochem. and Eng. Asp.*, Vol. 321, No. 1–3, pp. 301-307.

Kotlarchyk, M., Huang, J. S., and Cheng, S. H., (1985). Structure of AOT reversed micelles determined by small-angle neutron scattering. *J. Phys. Chem.* Vol. 89, pp. 4382.

Krauel, K., Davies, N. M., Hook, S., and Rades, T., (2005). Using different structure types of microemulsions for the preparation of poly(alkylcyanoacrylate) nanoparticles by interfacial polymerization, *J. Controlled Release*, Vol. 106, No. 1–2, pp. 76-87.

Krauel, K., Girvan, L., Hook, S., and Rades, T. (2007). Characterisation of colloidal drug delivery systems from the naked eye to Cryo-FESEM. *Micron*, Vol. 38, No. 8, pp. 796–803.

Kreilgaard, M., Pedersen, E. J., and Jaroszewski, J. W., (2000). NMR characterisation and transdermal drug delivery potential of microemulsion systems. *J. Controlled Release*, Vol. 69, No. 3, pp. 421-433.

Kumar, P. and Mittal, K. L. (Ed(s).). (1999). *Handbook of Microemulsion Science and Technology*, Marcel Dekker, Inc. ISBN: 0-8247-1979-4, New York. (and references sited therin).

Kwak, J. C. T., (1998). *Polymer-surfactant Systems*, CRC Press, Marcel Dekker, Inc., Halifax.

Lang, P. (1999). The Surface Phase Diagram of the Hexagonal Phase of the C12E5/Water System. *J. Phys. Chem. B*, Vol. 103, No. 24, pp. 5100-5105.

Lawrence, M. J., and Rees, G. D., (2000). Microemulsion-based media as novel drug delivery systems. *Adv. Drug Delivery Rev.*, Vol. 45, No. 1, pp. 89-121.

Lee, M.-H., Lin, H.-Y., Thomas, J. L. (2010). Synthesis of zirconia with nanoporous structure by a supercritical carbon dioxide microemulsion route, *Inter. J. Appl. Ceramic Tech.*, Vol. 7, No. 6, pp. 874-880.

Li, F., Li, G.-Z., Wang, H.-Q., and Xue, Q.-Ji, (1997). Studies on cetyltrimethyl-ammonium bromide (CTAB) micellar solution and CTAB reversed microemulsion by ESR and ^{2}H NMR, *Colloids Surf. A: Physicochem. and Eng. Asp.*, Vol 127, pp. 89-96.

Li, J., Zhang, J., Han, B., Gao, Y., Shen, D., and Wu, Z. (2006). Effect of ionic liquid on the polarity and size of the reverse micelles in supercritical CO_2. *Colloids Surf. A* Vol. 279, pp. 208–212.

Li, X., He, G., Zheng, W., and Xiao, G., (2010). Study on conductivity property and microstructure of TritonX-100/alkanol/*n*-heptane/water microemulsion, *Colloids Surf. A: Physicochem. Eng. Asp.*, Vol. 360, No. 1–3, pp. 150-158.

Li, Y.-K., Zhao, F.-L., Yang, P. (2003). *Oilfield Chemistry*, Vol. 20, No. 1, pp. 50-53.

Libster, D., Aserin, A., and Garti, N. (2006). A novel dispersion method comprising a nucleating agent solubilized in a microemulsion, in polymeric matrix: II. Microemulsion characterization, *J. Colloid Interface Sci.*,Vol 302, No. 1, pp. 322-329.

Lim, K.-H., Zhang, W., Smith, G. A., Smith, D. H. (2005). Temperature dependence of emulsion morphologies and the dispersion morphology diagram: Two-phase emulsions of the system $C_6H_{13}(OC_2H_4)_2OH$/*n*-tetradecane/"water". *Colloids Surf. A: Physicochem. Eng. Asp.*, Vol. 264, No. 1-3, pp. 43-48.

Lin, J.-C., Lee, C.-P., Ho, K.-C., (2012). Zinc oxide synthesis via a microemulsion technique: Morphology control with application to dye-sensitized solar cells. *J. Mater. Chem.*, Vol. 22, No. 4, pp. 1270-1273.

López-Quintela, M. A., Tojo, C., Blanco, M. C., García Rio, L., and Leis. J. R., (2004). Microemulsion dynamics and reactions in microemulsions, *Current Opinion in Colloid and Interface Sci.*, Vol. 9, No. 3–4, pp. 264-278.

Lu, J., Chen, D., and Jiao, X., (2006). Fabrication, characterization, and formation mechanism of hollow spindle-like hematite via a solvothermal process, *J. Colloid Interface Sci.*, Vol. 303, No. 2, pp. 437-443.

Lu, J.-L., Wang, J.-Ch., Zhao, S.-X., Liu, X.-Y., Zhao, H., Zhang, X., Zhou, S.-F., and Zhang, Q., (2008). Self-microemulsifying drug delivery system (SMEDDS) improves anticancer effect of oral 9-nitrocamptothecin on human cancer xenografts in nude mice. *Eur. J. Pharm. Biopharm.*, Vol. 69, No. 3, pp. 899-907.

Lutter, S., Tiersch, B., Koetz, J., Boschetti-de-Fierro, A., and Abetz, V., (2007). Covalently closed microemulsions in presence of triblock terpolymers, *J. Colloid Interface Sci.*, Vol. 311, No. 2, pp. 447-455.

Lv, F.-F., Zheng, L.-Q., and Tung, Ch.-H., (2005). Phase behavior of the microemulsions and the stability of the chloramphenicol in the microemulsion-based ocular drug delivery system. *Inter. J. Pharm.*, Vol. 301, No. 1–2, pp. 237-246.

Ma, Y., Liang, J., Sun, H., Wu, L., Dang, Y., Wu, Y., (2012). Honeycomb micropatterning of proteins on polymer films through the inverse microemulsion approach. *Chem. Eur. J.*, Vol. 18, No. 2, pp. 526-531.

Magid, L. J., (1986). The elucidation of micellar and microemulsion architecture using small-angle neutron scattering, *Colloids Surf.*, Vol. 19, No. 2–3, pp. 129-158.

Magid, L.J., Triolo, R., Jones, R.M., and Johnson Jr., J.S., (1983). Small-angle neutron scattering from an oil-in-water microemulsion as a function of temperature, *Chem. Phys. Lett.*, Vol. 96, No. 6, pp. 669-673.

Magno, M., Angelescu, D. G., Stubenrauch, C. (2009). Phase diagrams of non-ionic microemulsions containing reducing agents and metal salts as bases for the synthesis of bimetallic nanoparticles. *Colloids Surf. A: Physicochem. and Eng. Asp.*, Vol. 348, No. 1-3, pp. 116-123.

Magno, M., Tessendorf, R., Medronho, B., Miguel, M.G., Stubenrauch, C., (2009). Gelled polymerizable microemulsions. Part 3 Rheology. *Soft Matter* Vol. 5, No. 23, pp. 4763-4772.

Maitra, A.N., (1984). Determination of size parameters of water-Aerosol OT-oil reverse micelles from their nuclear magnetic resonance data. *J. Phys. Chem.* Vol. 88, pp. 5122–5125.

Malmsten, M., (2002). *Surfactants and Polymers in Drug Delivery*, Marcel Dekker, Inc., NY.

McBain, J. W., 1950. "*Colloid Science*" Heath, Boston.

McFann, G. J., and Johnston, K. P., (1993). Phase behavior of nonionic surfactant/oil/water systems containing light alkanes. *Langmuir*, Vol. 9, pp. 2942.

Mishra, S., Chatterjee, A., (2011). Effect of nano-polystyrene (nPS) on thermal, rheological, and mechanical properties of polypropylene (PP). *Polym. Adv. Technol.*, Vol. 22, No. 12, pp. 1547-1554.

Mitra, R. K., Paul, B. K. (2005). Effect of temperature and salt on the phase behavior of nonionic and mixed nonionic–ionic microemulsions with fish-tail diagrams. *J. Colloid Interface Sci.*, Vol. 291, No. 2, pp. 550-559.

Mondain-Monval, O., (2005). Freeze fracture TEM investigations in liquid crystals, *Current Opinion in Colloid Interface Sci.*, Vol. 10, No. 5–6, pp. 250-255.

Mukherjee, K., Mukherjee, D.C., Moulik, S.P. (1997). Thermodynamics of microemulsion formation III. Enthalpies of solution of water in chloroform as well as chloroform in water aided by cationic, anionic, and nonionic surfactants. *J. Colloid Interface Sci.*, Vol. 187, No. 2, pp. 327-333.

Mulqueen, P., (2003). Recent advances in agrochemical formulation. *Adv. Colloid Interface Sci.*, Vol. 106, No. 1–3, pp. 83-107.

Mumper, R. J., and Cui, Zh., (2003). Genetic immunization by jet injection of targeted pDNA-coated nanoparticles. *Methods*, Vol. 31, No. 3, pp. 255-262.

Nace, V. M., (Ed.), (1996). Nonionic Surfactants Polyoxyalkylene Block Copolymers, Marcel Dekker, Inc., New York.

Nagao, M., Seto, H., Okuhara, D., Okabayashi, H., Takeda, T., and Hikosaka, M., (1997). A small-angle neutron-scattering study of the effect of pressure on structures in a ternary microemulsion system, *Physica B: Condensed Matter*, Vol. 241–243, pp. 970-972.

Nagao, M., Seto, H., Shibayama, M., and Takeda, T., (2006). Pressure effect on semi-microscopic structures in a nonionic microemulsion, *Physica B: Condensed Matter*, Vol. 385–386, part 1, pp. 783-786.

Najjar, R., (2006). *Polymerization studies of vinylidene difluoride in supercritical carbon dioxide*, Verlag Mainz, Aachen, Germany.

Najjar, R., Stubenrauch, C., (2009). Phase diagrams of microemulsions containing reducing agents and metal salts as bases for the synthesis of metallic nanoparticles. *J. Colloid Interface Sci.*, Vol. 331, No. 1, pp. 214-220.

Nazar, M. F., Khan, A. M., Shah, S. S. (2009). Microemulsion system with improved loading of piroxicam: A study of microstructure. *AAPS Pharm Sci Tech*, Vol. 10, No. 4, pp. 1286-1294.

Nedjhioui, M., Canselier, J. P., Moulai-Mostefa, N., Bensmaili, A., and Skender, A. (2007). Determination of micellar system behavior in the presence of salt and water-soluble polymers using the phase diagram technique. *Desalination*, Vol. 206, No. 1-3, pp. 589-593.

Neubert, R. H. H., (2011). Potentials of new nanocarriers for dermal and transdermal drug delivery.*Eur. J. Pharm. Biopharm.*, Vol. 77, No. 1, pp. 1-2.

Oates, J., (1989). Thermodynamics of solubilization in aqueous surfactant systems.*Ph.D. dissertation*, University of Texas, Austin, TX.

Oh, D. H., Kang, J. H., Kim, D. W., Lee, B.-J., Kim, J. O., and Yong, C. S., (2011). Comparison of solid self-microemulsifying drug delivery system (solid SMEDDS) prepared with hydrophilic and hydrophobic solid carrier. *Inter. J. Pharm.*, Vol. 420, No. 2, pp. 412-418.

O'Hagan, D. T. , Ott, G. S., and Van Nest G., (1997). Recent advances in vaccine adjuvants: the development of MF59 emulsion and polymeric microparticles. *Molecular Medicine Today*, Vol, 3, No. 2, pp. 69-75.

Olesik, S. V., and Miller, C. J., (1990). Critical micelle concentration of AOT in supercritical alkanes. *Langmuir*, Vol. 6, pp. 183.

Os, N. M. van., (1998). *Nonionic Surfactants: Organic Chemistry*, CRC Press, Marcel Dekker, INC., New York.

Ouadahi, K., Allard, E., Oberleitner, B., Larpent, C., (2012). Synthesis of azide-functionalized nanoparticles by microemulsion polymerization and surface modification by click chemistry in aqueous medium. *J. Polym. Sci., Part A: Polym. Chem.*, Vol. 50, No. 2, pp. 314-328.

Park, J.-Y., Lim, J. S., Lee, Y. W., Yoo, K. -P. (2006). Phase behavior of water-in-supercritical carbon dioxide microemulsion with sodium salt of bis(2,2,3,3,4,4,5,5-octafluoro-1-pentanol) sulfosuccinate. *Fluid Phase Equilibria*, Vol. 240, No. 1, pp. 101-108.

Parker, W. O., Genova, Jr., C., and Carignano, G. (1993). Study of micellar solutions and microemulsions of an alkyl oligoglucoside via NMR spectroscopy, *Colloids Surf. A: Physicochem. and Eng. Asp.*, Vol. 72, pp. 275-284.

Peck D.G., and Johnston, K.P., (1991). Theory of the pressure effect on the curvature and phase behavior of AOT/propane/brine water-in-oil microemulsions, *J. Phys. Chem.* Vol. 95. pp. 9549.

Pedersen, N., Hansen, S., Heydenreich, A. V., Kristensen, H. G., and Poulsen, H. S., (2006). Solid lipid nanoparticles can effectively bind DNA, streptavidin and biotinylated ligands. *Eur. J. Pharm. Biopharm.*, Vol. 62, No. 2, pp. 155-162.

Peng, J., He, X., Wang, K., Tan, W., Li, H., Xing, X., and Wang, Y., (2006). An antisense oligonucleotide carrier based on amino silica nanoparticles for antisense inhibition of cancer cells. *Nanomedicine: Nanotechnology, Biology and Medicine*, Vol. 2, No. 2, pp. 113-120.

Philipoff, W., (1951). Colloidal and polyelectrolytes. The micelle and swollen micelle on soap micelles. *Discussion Faraday Soc.*, Vol. 11, pp. 96.

Ponsinet, V. and Talmon, Y. (1997). Direct imaging of lamellar phases by cryo-transmission electron microscopy. *Langmuir*, Vol. 13, pp. 7287–7292.

Prakash, S. S., Francis, L. F., and Scriven, L. E., (2006). Microstructure evolution in dry cast cellulose acetate membranes by cryo-SEM, *J. Membrane Sci.*, Vol. 283, No. 1–2, pp. 328-338.

Pramanik, R., Sarkar, S., Ghatak, C., Rao, V. G., Setua, P., Sarkar, N. (2010). Microemulsions with surfactant TX100, cyclohexane, and an ionic liquid investigated by conductance, DLS, FTIR measurements, and study of solvent and rotational relaxation within this microemulsion. *J. Phys. Chem., B*, Vol. 114, No. 22, pp. 7579-7586.

Prince, L.M., (1977). *Microemulsions: Theory and Practice*, New York, Academic Press.

Probst, J., Dembski, S., Milde, M., Rupp, S., (2012). Luminescent nanoparticles and their use for in vitro and in vivo diagnostics. *Expert Rev. Mol. Diagn.*, Vol. 12, No. 1, pp. 49-64.

Prouvost, L., Pope, G. A., and Rouse, B. (1985). Microemulsion phase behavior: A thermo-dynamic modeling of the phase partitioning of amphiphilic species. *Soc. Pet. Engineers J.*, Vol. 25, No. 5, pp. 693-703.

Puri, D., Bhandari, A., Sharma, P., Choudhary, D., (2010). Lipid nanoparticles (SLN, NLC): A novel approach for cosmetic and dermal pharmaceutical. *J. Global Pharma Technol.*, Vol. 2, No. 9, pp. 1-15.

Qin, C., Chai, J., Chen, J., Xia, Y., Yu, X., Liu, J. (2008). Studies on the phase behavior and solubilization of the microemulsion formed by surfactant-like ionic liquids with ε–β -fish-like phase diagram. *Colloid and Polymer Sci.*, Vol. 286, No. 5, pp. 579-586.

Randolph, T. W., Clark, D. S., Blanch, H. W., and Prausnitz, J. M., (1988). Enzymatic Oxidation of Cholesterol Aggregates in Supercritical Carbon Dioxide. *Science*, Vol. 239, pp. 387.

Rao, J., and McClements, D. J., (2011). Food-grade microemulsions, nanoemulsions and emulsions: Fabrication from sucrose monopalmitate & lemon oil. *Food Hydrocolloids*, Vol. 25, No. 6, pp. 1413-1423.

Rao, J., and McClements, D. J., (2012). Lemon oil solubilization in mixed surfactant solutions: Rationalizing microemulsion & nanoemulsion formation. *Food Hydrocolloids*, Vol. 26, No. 1, pp. 268-276.

Reithofer, M. R., Bytzek, A. K., Valiahdi, S. M., Kowol, Ch. R., Groessl, M., Hartinger, Ch. G., Jakupec, M. A., Galanski, M., and Keppler, B. K., (2011). Tuning of lipophilicity and cytotoxic potency by structural variation of anticancer platinum(IV) complexes. *J. Inorg. Biochem.*, Vol. 105, No. 1, pp. 46-51.

Ritter, J. M., and Paulaitis, M. E., (1990). Multiphase behavior in ternary mixtures of carbon dioxide, water, and nonionic amphiphiles at elevated pressures. *Langmuir* Vol. 6, pp. 934.

Rosen, M., J., and Dahanayake, M. (2000). *Industrial utilization of surfactants: principles and practice*, mcs Press, Urbana, Illinois.

Rossi, L. M. , Shi, L., Rosenzweig, N., and Rosenzweig, Z., (2006). Fluorescent silica nanospheres for digital counting bioassay of the breast cancer marker HER2/nue. *Biosensors and Bioelectronics*, Vol. 21, No. 10, pp. 1900-1906.

Roux, D., Bellocq, A. M., Leblanc, M. S. (1983). An interpretation of the phase diagrams of microemulsions. *Chem. Phys. Lett.*, Vol. 94, No. 2, pp. 156-161.

Rozner, Sh., Aserin, A., and Garti, N. (2008). Competitive solubilization of cholesterol and phytosterols in nonionic microemulsions studied by pulse gradient spin-echo NMR, *J. Colloid Interface Sci.*, Vol 321, No. 2, pp. 418-425.

Ruckenstein, E. (1981). Evaluation of the interfacial tension between a micro-emulsion and the excess dispersed phase. *Soc. Petroleum Eng. J.*, Vol. 21, No. 5, pp. 593-602.

Sagisaka, M., Yoda, S., Takebayashi, Y., Otake, K., Kitiyanan, B., Kondo, Y., Yoshino, N., Takebayashi, K., Sakai, H., and Abe, M. (2003). Preparation of a W/scCO_2 microemulsion using fluorinated surfactants. *Langmuir*, Vol. 19, No. 2, pp. 220-225.

Saito, H. and Shinoda, K., (1967). The solubilization of hydrocarbons in aqueous solutions of nonionic surfactants. *J. Colloid Interface Sci.* Vol. 24, No. 1, pp. 10.

Saito, H. and Shinoda, K., (1970). The stability of W/O type emulsions as a function of temperature and of the hydrophilic chain length of the emulsifier. *J. Colloid Interface Sci.* Vol. 32, No. 4, pp. 647.

Santanna, V. C., Curbelo, F. D. S., Castro Dantas, T. N., Dantas Neto, A. A., Albuquerque, H. S., and Garnica, A. I. C., (2009). Microemulsion flooding for enhanced oil recovery. *J. Petroleum Sci. Eng.*, Vol. 66, No. 3–4, pp. 117-120.

Santra, S., Dutta, D., and Moudgil, B. M., (2005).Functional Dye-Doped Silica Nanoparticles for Bioimaging, Diagnostics and Therapeutics. *Food and Bioproducts Processing*, Vol. 83, No. 2, pp. 136-140.

Sarciaux, J. M., Acar, L., and P. A., Sado, (1995). Using microemulsion formulations for oral drug delivery of therapeutic peptides. *Inter. J. Pharm.*, Vol. 120, No. 2, pp. 127-136.

Schulman, J. H. & Hoar, T. P. (1943). Transparent water-in-oil dispersions: The oleopathic hydromicelle. *Nature*, Vol. 152, pp. 102.

Schwan, M., Kramer, L. G. A., Sottmann, T., Strey, R. (2010). Phase behaviour of propane- and scCO_2-microemulsions and their prominent role for the recently proposed foaming procedure POSME (Principle of Supercritical Microemulsion Expansion) *Phys. Chem. Chem. Phys.*, Vol. 12, No. 23, pp. 6247-6252.

Schwering, R., (2008). Polymerization of highly viscous bicontinuous and droplet Microemulsions. *PhD Thesis*, University of Cologne, Cologne, Germany.

Scriven, L. E. (1976). Equilibrium bicontinuous structure. *Nature*, Vol. 263, pp. 123.

Selivanova, N. M., Galeeva, A.I., Konov, A. B., Gnezdilov, O. I., Salikhov, K.M., Galyametdinov, Yu. G. (2010). Phase Diagram of the Liquid Crystal System of Water–Decanol–Lanthanum Nitrate–Decaethylene Glycol Monododecyl Ether. *Russian J. Phys. Chem. A*, Vol. 84, No. 5, pp. 802-807.

Seyfoddin, A., Shaw, J., Al-Kassas, R., (2010). Solid lipid nanoparticles for ocular drug delivery. *Drug Delivery*, Vol. 17, No. 7, pp. 467-489.

Shinoda, K. (1967.) The correlation between the dissolution state of nonionic surfactant and the type of dispersion stabilized with the surfactant. *J. Colloid Interface Sci.*, Vol. 24, pp. 4.

Shinoda, K. (1970). Thermodynamic aspects of non-ionic surfactant–water systems. *J. Colloid Interface Sci.*, Vol. 34, pp. 278.

Shinoda, K. and Kunieda, H., (1973). Conditions to produce so-called microemulsions: Factors to increase the mutual solubility of oil and water by solubilizer. *J. Colloid Interface Sci.*, Vol. 42, No. 2, pp. 381-387.

Shokri, M., Hosseini, M.G., Khosravi, M., Najjar, R., Sheikhy, S., (2011). The preparation of Pt-modified TiO_2 nanoparticles via microemulsions, and their application in photocatalytic removal of an azo dye (C.I. Acid Red 27). *Fresenius Environmental Bulletin*, Vol. 20, No. 4 A, pp. 1063-1068.

Silas, J. A., and Kaler, E. W., (2003) Effect of multiple scattering on SANS spectra from bicontinuous microemulsions, *J. Colloid Interface Sci.*, Vol. 257, No. 2, pp. 291-298.

Simón de Dios, A., and Díaz-García M. E., (2010). Multifunctional nanoparticles: Analytical prospects. *Analytica Chimica Acta*, Vol. 666, No. 1–2, pp. 1-22.

Sintov, A. C., and Botner, Sh., (2006). Transdermal drug delivery using micro-emulsion and aqueous systems: Influence of skin storage conditions on the in vitro permeability of diclofenac from aqueous vehicle systems. *Inter. J. Pharm.*, Vol. 311, No. 1–2, pp. 55-62.

Sirotti, C., Coceani, N., Colombo, I., Lapasin, R., Grassi, M., (2002). Modeling of drug release from microemulsions: A peculiar case. *J. Membr. Sci.*, Vol. 204, Nos. 1-2, pp. 401-412.

Smith, D. H., Sampath, R., Dadyburjor, D. B. (1996). Temperature Dependence of Emulsion Morphologies and the Dispersion Morphology Diagram. 3. Inversion Hysteresis Lines for Emulsions of Middle and Bottom Phases of the System $C_6H_{13}(OC_2H_4)_2OH$/*n*-Tetradecane/"Water". *J. Phys. Chem.*, Vol. 100, No. 44, pp. 17558-17562.

Söderman, O., and Nydén, M. (1999). NMR in microemulsions. NMR translational diffusion studies of a model microemulsion, *Colloids Surf. A: Physicochem. Eng. Asp.*, Vol. 158, No. 1-2, pp. 273-280.

Solans, C., and Kunieda, H., (1997), *Industrial Applications of Microemulsions*. Marcel Dekker Inc., New York, and references cited therein.

Stamatis, H., and Xenakis, A., (1999). Biocatalysis using microemulsion-based polymer gels containing lipase. *J. Molecular Cat. B: Enzymatic*, Vol. 6, No. 4, pp. 399-406.

Stilbs, P., (1982). Micellar breakdown by short-chain alcohols. A multicomponent FT-PGSE-NMR self-diffusion study, *J. Colloid Interface Sci.*, Vol 89, No 2, pp. 547-554.

Stubenrauch, C., (2009). *Microemulsions : background, new concepts, applications, perspectives*, 1st ed. Wiley-Blackwell Ltd.

Stubenrauch, C., Tessendorf, R., Salvati, A., Topgaard, D., Sottmann, Th., Strey, R., and Lynch, I., (2008). Gelled Polymerizable Microemulsions. 2. Microstructure. *Langmuir*, Vol. 24, pp. 8473-8482.

Sun, L., Zhou, Sh., Wang, W., Li, X., Wang, J., and Weng, J., (2009). Preparation and characterization of porous biodegradable microspheres used for controlled protein delivery. *Colloids Surf. A: Physicochem. Eng. Asp.*, Vol. 345, No. 1–3, pp. 173-181.

Suzuki, K., Nomura, M., (2003). A simulation method to predict time-evolution of particle size distribution in microemulsion polymerization of styrene. *J. Chem. Eng. Jpn.*, Vol. 36, No. 10, pp. 1242-1247.

Tabony, J., Drifford, M., and De Geyer, A., (1983). Structure of a microemulsion in the critical region: Neutron small-angle scattering results, *Che.l Phys. Lett.*, Vol. 96, No. 1, pp. 119 -125.

Tadros, Th., F. (2005). *Applied Surfactants Principles and Applications*, Wiley-VCH Verlag GmbH & Co. KGaA, Weinheim, Germany.

Taha, M. O., Abdel-Halim, H., Al-Ghazawi, M., and Khalil, E., (2005). QSPR modeling of pseudoternary microemulsions formulated employing lecithin surfactants: Application of data mining, molecular and statistical modeling. *Inter. J. Pharmaceutics*, Vol. 295, No. 1–2, pp. 135-155.

Takebayashi, Y., Mashimo, Y., Koike, D., Yoda, S., Furuya, T., Sagisaka, M., Otake, K., Sakai, H., Abe, M. (2008). Fourier transform infrared spectroscopic study of water-in-supercritical CO_2 microemulsion as a function of water content. *J. Phys. Chem., B*, Vol. 112, No. 30, pp. 8943-8949.

Takebayashi, Y., Sagisaka, M., Sue, K., Yoda, S., Hakuta, Y., Furuya, T. (2011). Near-infrared spectroscopic study of a water-in-supercritical CO_2 microemulsion as a function of the water content. *J. Phys. Chem., B*, Vol. 115, No. 19, pp. 6111-6118.

Talmon, Y. (1999). Cryogenic temperature transmission electron microscopy in the study of surfactant systems. In B.P. Binks (ed), *Modern Characterization Methods of Surfactant Systems*. Marcel Dekker, New York, pp. 147–178.

Tan, T. T. Y., Liu, S., Zhang, Y., Han, M.-Y., and Selvan, S. T., (2011). Microemulsion Preparative Methods (Overview), *Comprehensive Nanosci. Tech.*, Vol. 5, pp. 399-441.

Tao, G.-P., Chen, Q.-Y., Yang, X. Zhao, K.-D., and Gao, J., (2011). Targeting cancer cells through iron(III) complexes of di(picolyl)amine modified silica core–shell nanospheres. *Colloids Surf. B: Biointerfaces*, Vol. 86, No. 1, pp. 106-110.

Teichmann, A., Heuschkel, S., Jacobi, U., Presse, G., Neubert, R. H. H., Sterry, W., and Lademann, J., (2007). Comparison of stratum corneum penetration and localization of a lipophilic model drug applied in an o/w microemulsion and an amphiphilic cream. *Eur. J. Pharm. Biopharm.*, Vol. 67, No. 3, pp. 699-706.

Thurecht, K. J., Hill, D. J.T., and Whittaker. A. K. (2006). Investigation of spontaneous microemulsion formation in supercritical carbon dioxide using high-pressure NMR. *The J. Supercritical Fluids*, Vol. 38, No. 1, pp. 111-118.

Tian, H., He, J., Liu, L., Wang, D., Hao, Z., and Ma, C. (2012). Highly active manganese oxide catalysts for low-temperature oxidation of formaldehyde. *Microporous Mesoporous Mater.*, Vol. 151, pp. 397-402.

Tingey, J. M., Fulton, J. L., Matson, D. W., and Smith. R. D., (1991). Micellar and bicontinuous microemulsions formed in both in near critical and supercritical propane with didocyldimethylammonium bromide and water, *J. Phys. Chem.* Vol. 95, pp. 1445–1448.

Torino, E., Reverchon, E., and Johnston, K. P., (2010). Carbon dioxide/water, water/carbon dioxide emulsions and double emulsions stabilized with a nonionic biocompatible surfactant. *J. Colloid Interface Sci.*, Vol. 348, No. 2, pp. 469-478.

Valenta, C., and Schultz, K., (2004). (Influence of carrageenan on the rheology and skin permeation of microemulsion formulations. *J. Controlled Release*, Vol. 95, No. 2, pp. 257-265.

Van Nieuwkoop, J., and Snoei, G. (1985). Phase diagrams and composition analyses in the system sodium dodecyl sulfate/butanol/water/sodium chloride/heptane. *J. Colloid Interface Sci.*, Vol. 103, No. 2, pp. 400-416.

Wadle, A., Förster, Th., von and Rybinski, W., (1993). Influence of the microemulsion phase structure on the phase inversion temperature emulsification of polar oils, *Colloids Surf. A: Physicochem. Eng. Asp.*, Vol. 76, pp. 51-57.

Waysbort, D., Ezrahi, S., Aserin, A., Givati, R., and Garti, N. (1997). 1H NMR Study of a U-Type Nonionic Microemulsion, *J. Colloid Interface Sci.*, Vol. 188, No. 2, pp. 282-295.

Wei, Y.-B., Wu, J., Wu, S.-S., Zheng, C.-R. (2005). *Polymeric Materials Sci. Eng.*, Vol. 21, No. 1, pp. 141-144.

Widom, B., (1996). II. Theoretical modeling theoretical modeling: An introduction. *Ber Bunsenges Phys Chem Chem Phys.*, Vol. 100, No. 3, pp. 242-251.

Wines, T. H., and Somasundaran, P., (2002). Effects of Adsorbed Block Copolymer and Comb-like Amphiphilic Polymers in Solution on the Electrical Percolation and Light Scattering Behavior of Reverse Microemulsions of Heptane/Water/AOT, *J. Colloid Interface Sci.*, Vol. 256, No. 1, pp. 183-189.

Winsor, P.A. (1954). *Solvent Properties of Amphiphilic Compounds*. Butherworth &Co., London.

Wu, H., Zhou, A. Lu, C., and Wang, L., (2011). Examination of lymphatic transport of puerarin in unconscious lymph duct-cannulated rats after administration in microemulsion drug delivery systems. *Eur. J. Pharm. Sci.*, Vol. 42, No. 4, pp. 348-353.

Xie, Y., Ye, R., and Liu, H., (2007). Microstructure studies on biosurfactant-rhamnolipid/*n*-butanol/water/*n*-heptane microemulsion system, *Colloids Surf. A: Physicochem. Eng. Asp.*, Vol. 292, No. 2–3, pp. 189-195.

Xu, H., Cheng, L., Wang, Ch., Ma, X., Li, Y., and Liu, Zh., (2011). Polymer encapsulated upconversion nanoparticle /iron oxide nanocomposites for multimodal imaging and magnetic targeted drug delivery. *Biomaterials*, Vol. 32, No. 35, pp. 9364-9373.

Xu, X.-J., and Gan L. M., (2005). Recent advances in the synthesis of nanoparticles of polymer latexes with high polymer-to-surfactant ratios by microemulsion polymerization. *Current Opinion in Colloid & Interface Science*, Vol. 10, No. 5–6, pp. 239-244.

Yamada, T., Li, J., Koyanagi, C., Iyoda, T., Yoshida, H. (2007). Effect of lithium trifluoro-methanesulfonate on the phase diagram of a liquid-crystalline amphiphilic diblock copolymer. *J. Appl. Crystallography*, Vol. 40 (Suppl. 1), pp. s585-s589.

Yan, Y.-l., Zhang, N.-Sh., Qu, Ch.-T. And Liu, L., (2005). Microstructure of colloidal liquid aphrons (CLAs) by freeze fracture transmission electron microscopy (FF-TEM), *Colloids Surf. A: Physicochem. Eng. Asp.*, Vol. 264, No. 1–3, pp. 139-146.

Yazdi, P., McFann, G. J., Fox, M. A., and Johnston, K. P., (1990). Reverse micelles in supercritical fluids. 2. Fluorescence and absorption spectral probes of adjustable aggregation in the two-phase region. *J. Phys. Chem.*, Vol. 94, pp. 7224.

Ye, F., (2007). Porous polymeric materials derived from bicontinous microemulsions for drug delivery. MSc *Thesis*, University of Akron.

Yee, G. G., Fulton, J. L., and Smith, R. D., (1992). Aggregation of polyethylene glycol dodecyl ethers in supercritical carbon dioxide and ethane. *Langmuir*, Vol. 8, pp. 377.

Yee, G. G., Fulton, J. L., Blitz, J. P., and Smith, R. D., (1991). FT-IR investigation of the partitioning of sodium bis(2-ethylhexyl) sulfosuccinate between an aqueous and a propane phase. *J. Phys. Chem.*, Vol. 95, pp. 1403.

Yu, X. Y., Chai, J. L., Li, H. L., Xia, Y., Liu, J., Chen, J. F., Qin, C. K. (2009). Phase Diagrams and Solubilization of Chlorocarbons in Chlorocarbon/Water/Anionic Surfactant/Alcohol Microemulsion Systems. *J. Disper. Sci. Tech.*, Vol. *30,* No. 10, pp. 1506-1510.

Zemb, Th., (2009). Flexibility, persistence length and bicontinuous microstructures in microemulsions, *Comptes Rendus Chimie*, Vol. 12, No. 1–2, 218-224.

Zhang J., and Bright, E. V., (1992). Steady-state and time-resolved fluorescence studies of bis(2-ethylhexyl) sodium succinate (AOT) reserve micelles in supercritical ethane. *J. Phys. Chem.*, Vol. 96, pp. 5633.

Zhang, H., Cui, Y., Zhu, S., Feng, F., and Zheng, X., (2010). Characterization and antimicrobial activity of a pharmaceutical microemulsion. *Inter, J. Pharmaceutics*, Vol. 395, No. 1–2, pp. 154-160.

Zhang, H., Shen, Y., Bao, Y., He, Y., Feng, F., and Zheng, X., (2008). Characterization and synergistic antimicrobial activities of food-grade dilution-stable microemulsions against *Bacillus subtilis. Food Res. Inter.*, Vol. 41, No. 5, pp. 495-499.

Zhang, H., Shen, Y., Weng, P., Zhao, G., Feng, F., and Zheng, X., (2009). Antimicrobial activity of a food-grade fully dilutable microemulsion against *Escherichia coli* and *Staphylococcus aureus. Inter. J. Food Microbiology*, Vol. 135, No. 3, pp. 211-215.

Zhang, J., and Bright, E V., (1992). Probing the internal dynamics of reverse micelles formed in highly compressible solvents: aerosol-OT in near-critical propane. *J. Phys. Chem.*, Vol. 96, pp. 9068.

Zhang, J., and Michniak-Kohn, B., (2011). Investigation of microemulsion microstructures and their relationship to transdermal permeation of model drugs: Ketoprofen, lidocaine, and caffeine, *Inter. J. Pharmaceutics*, Vol. 421, No. 1, pp. 34-44.

Zhang, Sh., Gao, Y., Dong, B., and Zheng, L., (2010). Interaction between the added long-chain ionic liquid 1-dodecyl-3-methylimidazolium tetrafluoroborate and Triton X-100 in aqueous solutions, *Colloids Surf. A: Physicochem. Eng. Asp.*, Vol. 372, No. 1–3, pp. 182-189.

Zhao, Y.-G., Ding, W., Wei, J. (2011). Preparation of a bis-demethoxy curcumin microemulsion based on pseudo-ternary phase diagrams and an orthogonal test analysis. *J. Pesticide Sci*, Vol. 36, No. 2, pp. 248-251.

Zhu, W., Yu, A., Wang, W., Dong, R., Wu, J., and Zhai, G., (2008). Formulation design of microemulsion for dermal delivery of penciclovir. *Inter. J. Pharm.*, Vol. 360, No. 1–2, pp. 184-190.

Ziani, K., Fang, Y., and McClements, D. J., (2012). Fabrication and stability of colloidal delivery systems for flavor oils: Effect of composition and storage conditions. *Food Res. Inter.*, Vol. 46, No. 1, pp. 209-216.

Zielinski, R. G., Kline, S. R., Kaler, E. W. and Rosov, N. (1997). A small-angle neutron scattering study of water in carbon dioxide microemulsions. *Langmuir*, Vol. 13, pp. 3934–3937.

Zoumpanioti, M., Stamatis, H., and Xenakis, A., (2010). Microemulsion-based organogels as matrices for lipase immobilization. *Biotechnology Advances*, Vol. 28, No. 3, pp. 395-406.

Zulauf M., and Eicke, H. E., (1979). Inverted micelles and microemulsions in the ternary system water/aerosol-OT/isooctane as studied by photon correlation spectroscopy.*J. Phys. Chem.*, Vol. 83, pp. 480.

Part 2

Microstructures: Experimental and Modeling Studies

2

Ultrasonic Characterisation of W/O Microemulsions - Structure, Phase Diagrams, State of Water in Nano-Droplets, Encapsulated Proteins, Enzymes

Vitaly Buckin and Shailesh Kumar Hallone
School of Chemistry & Chemical Biology, University College Dublin, Belfield, Dublin 4 Ireland

1. Introduction

The unique properties of microemulsions, such as stability, ease of preparation and their ability to form spontaneously have attracted considerable interest in numerous applications including, synthesis of polymers, delivery of active compounds, solubilisation of enzymes in organic solvents and biocatalysis. Developments of microemulsion based formulations require routine analytical techniques that allow users to analyse their microstructure and particle size, transitions between different phases and phase diagrams, the state of solutes encapsulated in microemulsion droplets and their activity (e.g. enzymes), and to monitor chemical reactions curried in microemulsions.

Ultrasonic spectroscopy has a considerable potential in this area as demonstrated previously (Lang et al., 1980; Ballaro et al., 1980; Zana et al., 1983; Cao et al., 1997; Mehta & Kawaljit, 1998; Wines et al., 1999; Letamendia et al., 2001). This technique employs an ultrasonic wave (MHz frequency range), which probes the elastic characteristics of materials. As it transverses a sample, compressions and decompressions in the ultrasonic wave change the distance between molecules within the sample, which in turn respond by intermolecular repulsions and attractions. This ability of ultrasonic waves to probe intermolecular forces allows access to molecular levels of organisation of materials. Measurements of scattering effects of ultrasonic waves provide particle size analysis. Application of ultrasonic spectroscopy in microemulsion systems in the past was often limited by the resolution of the measurements, requirements of large sample volumes and limited range of measuring regimes. Introduction of high-resolution ultrasonic spectroscopy (HR-US) (Buckin & O'Driscoll, 2002; Buckin et al., 2003; Kudryashov et al., 2000; Buckin et al., 2002) overcame the above limitations and allowed applications of this technique for analysis of microemulsion structure and phase diagrams (Hickey et al., 2006) in a simple automatic or manual titration regimes. The technique is based on precision measurements of ultrasonic velocity and attenuation in the frequency range 1 to 20MHz. In w/o microemulsions it provided characterization of transitions between different phases, such as swollen micelles, microemulsion, coarse emulsion, liquid crystals, and evaluation of microstructural attributes of the phases, as well as characterisation of the state of water in emulsion droplets and the size of the droplets (Hickey et al., 2006, 2010). Previous applications of ultrasonic spectroscopy for analysis of microemulsions demonstrated an exceptional sensitivity of ultrasonic parameters to the structural characteristics of microemulsions and to the state of its components. However, utilisation of this information requires further developments of algorithms of processing of ultrasonic data and their relationship with microscopic and molecular characteristics of dispersions. One of the major focuses of this paper is interpretation of compressibility of microemulsions obtained from ultrasonic measurements. As a second derivative of thermodynamic potentials, compressibility is extremely sensitive to the state of components of complex mixtures. Compressibility represents the rigidity of the media, which is determined by intermolecular forces. In this sense, measurements of compressibility provide direct access to intermolecular interactions. As compression of medium is associated with energy transformations, including release of heat, compressibility is affected by the dynamics of this process, which in some cases results in its dependence of on the frequency or the time scale of the compression. This provides an additional tool for ultrasonic analysis of microstructural characteristics of mixtures and kinetics of molecular transformations. However, it also requires the correct theoretical algorithms for molecular interpretation of compressibility. The current paper combines the discussion of fundamental relationships between compressibility and structural and physical characteristics of microemulsions, and applications of these relationships for analysis of ultrasonic data on microemulsions, their microstructure and state of their components. Special attention was given to the assessment of the state of water in w/o microemulsion droplets, which plays a key role in regulation of the activity of water soluble ingredients encapsulated in the droplets (enzymes, drugs, etc.) and also to the algorithms for assessment of state of globular proteins encapsulated in w/o microemulsions. The results demonstrate that the ultrasonic technique allows efficient mapping of the areas of microemulsion phase diagrams with different states of water, including water in nano-droplets. The composition of microemulsion systems at the transition lines, surrounding the areas of presence of nano-droplets with well-defined water pool was correlated with the droplet size, obtained ultrasonically, and by other techniques. The 'upgraded' phase diagrams were used for successful (retaining of bioactivity) encapsulation of hydrolytic enzymes (e.g. cellobiase) in aqueous nano-droplets and for ultrasonic characterisation of enzyme activity in microemulsions.

2. Ultrasonic parameters and viscoelastic properties of liquids and gels

2.1 Ultrasonic wave and elasticity moduli

Ultrasonic wave is a wave of longitudinal deformations where compressions occur in direction of wave propagation. In isotropic materials, which are homogeneous within the scale of ultrasonic wave, the differential equation for a plain ultrasonic wave propagating over the axis x is given by (Litovitz & Davis, 1964; Thurston, 1964; Morse & Ingard, 1986; Powey, 1997):

$$\rho_I \frac{\partial^2 X}{\partial t^2} = M \frac{\partial^2 X}{\partial x^2} \tag{1}$$

where M is the longitudinal modulus of the medium, which represents the ratio of stress to strain in longitudinal deformation, X is the displacement of the medium at the point x, t is time, ρ_I is the effective inertial density of the medium, which represents the ratio between the force applied to a small part of the liquid and the acceleration caused by the force. For homogeneous on microscopic level (length scale of the shear wave) liquids ρ_I is equal to the normal (gravimetric) density, for non- homogeneous liquids ρ_I deviates from the gravimetric density and depends on the frequency. The Equation (1) is valid for ultrasonic waves of small amplitude, e.g. relative longitudinal deformation in the wave, strain, is << 1. At this condition the speed of the wave is much higher than the speed of particles of the medium (Thurston, 1964).

The strain (deformation) may occur with some time delay relative to the stress applied. Incorporation of these effects for the case of harmonically oscillating stress is often done through representation of strain, stress and other associated variables (e.g. change of pressure, δP , temperature, δT , and volume, δV) as complex values given by $\mathbf{Z}e^{i\omega t}$,where $\mathbf{Z}$ ($\equiv |\mathbf{Z}|e^{i\varphi_\mathbf{Z}} \equiv Z'+iZ"$, $|\mathbf{Z}|$ is the absolute amplitude and $\varphi_\mathbf{Z}$ the phase of $\mathbf{Z}$; Z' is the real and $Z"$ is the imaginary parts of $\mathbf{Z}$) is the complex amplitude of the corresponding parameter, ω ($\equiv 2\pi f$, f is frequency of ultrasonic wave) is the angular frequency of oscillations, t is time, and i ($\equiv \sqrt{-1}$) is the imaginary unit. In this case the longitudinal modulus can be represented by a complex constant, $\mathbf{M}$ ($\equiv M'+iM"$). The solution of Equation (1) in complex notations can be presented as $\mathbf{X} = \mathbf{X}_0 e^{-\alpha x + i\omega(t - \frac{x}{u})}$ where $\mathbf{X}_0$ is the amplitude of the wave at $x = 0$ and $t = 0$, u is the ultrasonic velocity (phase speed of the wave) and α is the ultrasonic attenuation. If the elastic response dominates, $M' >> M"$ (often expressed as a small value of attenuation per wavelength, $\alpha\lambda << 1$, $\lambda = \frac{u}{f}$), which is correct for most of liquids and gels, the velocity and attenuation of ultrasonic wave are given as:

$$u = \sqrt{\frac{M'}{\rho_I}}; \qquad \alpha = \frac{\omega M"}{2\rho_I u^3} \tag{2}$$

In the case of a medium, which consist of a liquid within a gel network, with the size of the networks comparable or exceeding the ultrasonic wavelength, the longitudinal module can be presented as a sum of contribution of the liquid, $\mathbf{M_{liquid}}$, and the gel, $\mathbf{M_{gel}}$, parts: $\mathbf{M} = \mathbf{M_{liquid}} + \mathbf{M_{gel}}$. The modulus of longitudinal deformation is a combination of the modulus of volume deformation, $\mathbf{K_V}(\equiv K_V'+iK_V")$, and the modulus of shear deformation, $\mathbf{G}(\equiv G'+iG")$ (see for example Litovitz & Davis, 1964): $\mathbf{M} = \mathbf{K_V} + \frac{4}{3}\mathbf{G}$. For most liquids the shear storage modulus, G'_{liquid} , is much smaller than the volume storage modulus, $K_V'_{liquid}$, ($G'_{liquid} << K_V'_{liquid}$) and its contribution to M' in Equation (2) can be neglected.

It is often convenient to substitute the volume modulus of the liquid, $\mathbf{K_{V\,liquid}}$, by the coefficient of adiabatic compressibility of the liquid $\boldsymbol{\beta} \equiv \beta'+i\beta" \equiv \frac{1}{\mathbf{K}_{\mathbf{V}\,liquid}}$, which represents a relative change of the volume per unit of pressure applied: $\boldsymbol{\beta} \equiv -\frac{1}{V}\frac{\boldsymbol{\delta}\mathbf{V}}{\boldsymbol{\delta}\mathbf{P}}$. For $\frac{K_V'_{liquid}}{K_V"_{liquid}} >> 1$

$$u = \sqrt{\frac{\frac{1}{\beta'} + M'_{gel}}{\rho_I}}; \qquad \alpha = \frac{\omega}{2\rho_I u^3}\left(-\frac{\beta''}{\beta'^2} + \frac{4}{3}G_{liquid}'' + M_{gel}''\right) \tag{3}$$

For liquids M'_{gel} and M''_{gel} shall be excluded from the above equations, which provides the well-known relationships for ultrasonic velocity: $u = \frac{1}{\sqrt{\beta' \rho_I}}$.

2.2 Compressibility and effects of heat exchange

Compression of a volume V caused by an applied pressure, δP, results in production of heat, which changes the temperature of the compressed volume by δT. The volume change, δV, caused by the applied pressure is a sum of two terms, compression of volume V at constant temperature, $\delta V_T = -K_T \delta P$, and the change of volume caused by its heat expansion at constant pressure, $\delta V_P = E\delta T$: $\delta V = \delta V_T + \delta V_P$, where K_T ($\equiv -\left(\frac{\delta V}{\delta P}\right)_T$) is the isothermal compressibility of volume V at constant temperature and E ($\equiv \left(\frac{\delta V}{\delta T}\right)_P$) is its heat expansibility at constant pressure. Therefore, the compressibility, K ($\equiv -\frac{\delta V}{\delta P}$), is given as $K = K_T - E\frac{\delta T}{\delta P}$.

The change of temperature δT is determined by the balance of the heat production in the compression of the volume V and the heat exchange between this volume and its environment. Two (major) processes of the heat exchange exist in the ultrasonic waves propagating through liquids. The first one is the heat flow between different parts of the ultrasonic wave, which have at any given time different compression and thus different temperature. The length scale of the volumes involved in heat exchange process is determined by the characteristics of the thermal wave, its wavelength, λ_T ($= \sqrt{\frac{4\pi\kappa}{f\rho c_P}}$, κ is the heat conductance of the medium, ρ is the gravimetric density) and is attenuation with distance $\sim e^{-\frac{x}{\delta_T}}$, where $\delta_T = \frac{\lambda_T}{2\pi}$ is the exponential decay length of the thermal wave. Therefore, the extent of the heat exchange within the ultrasonic wave is defined by the ratio of the wavelength of ultrasonic wave, λ, which controls the gradient of temperature within the wave and λ_T, which controls the length scale of the heat exchange during the time interval of compression. The functional dependence of the real and imaginary parts of compressibility (and ultrasonic velocity and attenuation) in ultrasonic waves in liquids on the ratio of λ_T to λ was analyzed previously (Thurston, 1964). Its application for such liquids as water or oils at room temperature shows that for ultrasonic frequencies, up to hundreds of MHz, the deviations of ultrasonic velocity and the real part of compressibility coefficient, β', from their values at pure adiabatic conditions of compression do not exceed 10^{-7} of their absolute values. This is negligible when compared with the existing limits of resolution in measurements of ultrasonic velocity. Thus, the compressibility of the media measured within normal ultrasonic frequency range, represents the adiabatic compressibility, K_S and the value of β' shall be taken as coefficient of adiabatic compressibility of the medium, β_S ($\equiv \frac{K_S}{V}$; $K_S \equiv -\left(\frac{\partial V}{\partial P}\right)_S$, S is entropy). The imaginary part of compressibility at this limit provides the well-known equation for contribution to ultrasonic attenuation caused by the heat losses in homogeneous liquids, α_{TH}, (Pierce, 1991): $\alpha_{TH} = \omega^2 T \frac{\kappa e^2 \rho}{2uc_P^{\ 2}}$, where e ($\equiv \frac{E}{m}$, m is the mass of the volume V) is the specific thermal expansibility of the medium.

The second process of the heat exchange in the ultrasonic wave is realised in dispersions. It is associated with the heat flow between microscopic parts of dispersions, which thermophysical properties are different. This process and its effect on compressibility of liquids are discussed in the Section 3.

2.3 Inertial density of microemulsions

Deviation of the effective inertial density, ρ_I, of a dispersion from its gravimetric density (mass of unit of volume), ρ, is determined by the different accelerations of the dispersed particles and of the continuous medium, in a gradient of pressure in ultrasonic wave. For an oscillating gradient of pressure this difference is controlled by the ratio of the size if the particles to the wavelength of the shear wave, λ_η ($= \sqrt{\frac{4\pi\eta}{f\rho}}$, η is viscosity of the medium). The parameter $\frac{\lambda_\eta}{2\pi}$ (exponential decay length of the shear wave) determines the effective thickness of the transition layer around the border between the particle and the continuous medium, within which the speed of the medium changes from the speed at the border to the speed of the bulk medium. Dependence of λ_η on frequency (Figure 1 (a)) results in a change of the value of ρ_I from the value equal to the gravimetric density at low frequencies, to a different value at high frequencies (Povey, 1997).

The effect of frequency on ρ_I can be illustrated qualitatively by considering a rigid container filled with a liquid dispersion, and a periodic force applied to the walls of the container. The effective inertial mass of the container is the ratio of the applied force to the acceleration of the container. At low frequencies, when the transition layer around the borders between the particles and the continuous medium exceeds the size of the container, the speed and acceleration of the particles are the same for all parts of the container. In this case the effective inertial mass of the container is the sum of the masses of its parts (gravimetric mass): the particles, the continuous medium and the walls. At high frequencies, when the transition layer is smaller than the size of particles, the acceleration of the particles and the continuous medium are different. This results in a deviation of the effective inertial mass of the container from its gravimetric mass.

Microemulsions droplets normally are much smaller than the wavelength of the shear wave at ultrasonic frequencies (Figure 1 (a)). Also, the 'density contrast' determined by the difference between the density of the continuous medium and the droplets is relatively small. Therefore, for microemulsions $\rho_I = \rho$, which will be assumed below. If required, the outlined equations can be upgraded by including an appropriate relationship between ρ and ρ_I. In further discussions we will use the specific volume, v ($\equiv \frac{1}{\rho}$) and specific adiabatic compressibility,

k_S ($\equiv v\beta_S$) instead of the density and the coefficient of adiabatic compressibility, β_S, of medium. In this case the equation for ultrasonic velocity in liquids is transformed to:

$$u = \frac{v}{\sqrt{k_S}} \tag{4}$$

3. Compressibility of solutes and solute particles in liquid mixtures

3.1 Apparent compressibility of solutes

Contribution of solute to the compressibility, K_S, and the volume, V, of a mixture composed of the mass of solute m_{solute} and the mass of solvent m_0 can be characterised by the specific apparent adiabatic compressibility, φK_S, and the specific apparent volume, φV, of the solute:

$$\varphi K_S \equiv \frac{K_S - K_{S_0}}{m_{solute}}; \qquad \varphi V \equiv \frac{V - V_0}{m_{solute}} \tag{5}$$

Here and below the subscript indexes *solute* and 0 are referred to the properties of the solute and pure solvent. The above equations provide the relationships linking the specific volume and compressibility of the mixture and of the pure solvent with the apparent characteristics of solute:

$$\begin{aligned} v &= (1-w)v_0 + w\varphi V; & \varphi V &= \frac{v - v_0}{w} + v_0 \\ k_S &= (1-w)k_{S_0} + w\varphi K_S; & \varphi K_S &= \frac{k_S - k_{S_0}}{w} + k_{S_0} \end{aligned} \tag{6}$$

where $w\left(\equiv \frac{m_{solute}}{m_0 + m_{solute}}\right)$ is the concentration (weight fraction) of the solute in the mixture.

The Equations (6) are often used for calculations of φK_S and φV from the measured compressibility and volume (or density) of the mixture and of pure solvent.

The apparent properties represent the change of compressibility and volume of liquid caused by addition of solute to solvent. They can be measured experimentally and do not depend on the initial state of the solute (prior to its addition to solvent). However, in many cases they do not represent the state of the solute in the mixture due to solvation effects, which affect the state of the solvent. These effects include a transfer of components of solvent from the free (bulk) to the solvation state, which contributes to the measured apparent characteristics. In w/o microemulsions the solvation effects are represented by the transfer of surfactant (and cosurfactant) from the oil phase to the surface of microemulsion droplet, when water is added as a solute. Some amount of oil can also be incorporated into the solute particle (water pool surrounded by surfactant), especially between the hydrophobic tails of surfactant. Below we will apply the term 'solute particle' for particulates of solute, which are dispersed in the bulk solvent, and which may include some omponents of solvent. In case of solutions this term shall be applied to solvated solute molecules. To characterise the compressibility and the volume of the solute particle in its physical state in the mixture the specific apparent volume and adiabatic compressibility of solute particle, v_{sp} and k_{sp}, can be introduced in the same way as the apparent characteristics of solute:

$$v_{sp} \equiv \frac{V - V_{0sp}}{m_{sp}}; \quad v = (1 - w_{sp})v_{0sp} + w_{sp}v_{sp}; \quad v_{sp} = \frac{v - v_{0sp}}{w_{sp}} + v_{0sp}$$
$$k_{S\,sp} \equiv \frac{K_S - K_{S_0\,sp}}{m_{sp}}; \quad k_S = (1 - w_{sp})k_{S_0\,sp} + w_{sp}k_{S\,sp}; \quad k_{S\,sp} = \frac{k_S - k_{S_0\,sp}}{w_{sp}} + k_{S_0\,sp} \tag{7}$$

where V and K_S are the volume and the adiabatic compressibility of the mixture containing the solvent and solute particles, m_{sp} ($\equiv m_{solute} + m_{solvent\,sp}$) is the mass of solute particles in the mixture, which includes the mass of the solvent incorporated into the solute particle, $m_{solvent\,sp}$, V_{0sp} and $K_{S_0\,sp}$ are the volume and the adiabatic compressibility of the bulk solvent in the mixture. In some cases the specific apparent volume and adiabatic compressibility of solute particles can be measured directly by adding all the components of solute particle to the solvent (measuring of v, v_{0sp}, k_S and $k_{S_0\,sp}$). They can also be calculated from the measured specific apparent volume and adiabatic compressibility of the solute as:

$$v_{sp} = \frac{\varphi V + w_s v_0 + \delta v_0 (1 + w_s - \frac{1}{w})}{(1 + w_s)}; \quad k_{S\,sp} = \frac{\varphi K_S + w_s k_{S_0} + \delta k_{S_0} (1 + w_s - \frac{1}{w})}{1 + w_s} \tag{8}$$

where w_s ($\equiv \frac{m_{solvent\,sp}}{m_{solute}}$, $m_{solvent\,sp}$ is the mass of solvent in the mixture incorporated into the solute particle) is the solvation level of the solute, δv_0 ($\equiv v_{0\,sp} - v_0$, $v_{0\,sp}$ is the specific volume of the bulk solvent in the mixture) is the difference between the specific volumes of the pure solvent and of the bulk solvent in the mixture and δk_{S_0} ($\equiv k_{S_0\,sp} - k_{S_0}$, $k_{S_0\,sp}$ is the specific adiabatic compressibility of the bulk solvent in the mixture) is the difference in between the specific adiabatic compressibilities of the pure solvent and of the bulk solvent in the mixture. Equation (8) is a consequence of Equations (6) and (7). The concentration of solute, w, in this equation reflects the dependence of δv_0 and δk_{S_0} on w at constant solvation level, w_s, when these values are not equal to 0. If solvation does not affect the composition of solvent the values of δv_0 and δk_{S_0} are equal to zero. In this case w vanishes from the above equation.

Similar to the above other apparent thermodynamic characteristics of solute and thermodynamic characteristics of solute particles can be introduced. This includes thermal expansibility, E ($\equiv \left(\frac{\partial V}{\partial T}\right)_P$), heat capacity, C_P ($\equiv \left(\frac{\partial H}{\partial T}\right)_P$, H is the enthalpy of the mixture), and isothermal compressibility, K_T ($\equiv -\left(\frac{\partial V}{\partial P}\right)_T$), thus producing the specific apparent thermal expansibility, φE, specific apparent heat capacity, φC_P, specific apparent isothermal compressibility, φK_T, specific isothermal compressibility of solute particle, $k_{T\,sp}$, specific thermal expansibility of solute particle, e_{sp}, and specific heat capacity of solute particle, $c_{P\,sp}$, of the solute particles defined as:

$$\varphi E = \frac{E - E_0}{m_{solute}};\quad \varphi C_P = \frac{C_P - C_{P_0}}{m_{solute}};\quad \varphi K_T = \frac{K_T - K_{T_0}}{m_{solute}};\quad e_{sp} = \frac{E - E_{0sp}}{m_{sp}};\quad c_{P sp} = \frac{C_P - C_{P_0 sp}}{m_{sp}};\quad k_{T sp} = \frac{K_T - K_{T_0 sp}}{m_{sp}}$$

where the abbreviations 0 and sp has the same meaning as above. Thus, the specific thermal expansibility, $e\ (\equiv \frac{E}{m})$, specific heat capacity, c_P and specific isothermal compressibility, $k_T\ (\equiv \frac{K_T}{m})$, of a mixture are presented as:

$$\begin{aligned} e &= (1-w)e_0 + w\varphi E;\quad c_P = (1-w)c_{P_0} + w\varphi C_P;\quad k_T = (1-w)k_{T_0} + w\varphi K_T \\ e &= (1-w_{sp})e_{0\,sp} + w_{sp}e_{sp};\quad c_P = (1-w_{sp})c_{P_0\,sp} + w_{sp}c_{P\,sp};\quad k_T = (1-w_{sp})k_{T_0\,sp} + w_{sp}k_{T\,sp} \end{aligned} \tag{9}$$

Parameters φK_S, φV, v_{sp} and k_{sp} characterise the state of the whole solute or the whole pool of solute particles in the mixture. If the state of the solute depends on its concentration it is often useful to interpret the data in terms of specific partial characteristics of solutes, $\bar{v}$ and $\bar{k}_S$, defined as a change of volume, V, or compressibility, K_S, of the mixture caused by addition of infinitely small mass of solute δm_{solute}:

$$\begin{aligned} \bar{v} &\equiv \left(\frac{\partial V}{\partial m_{solute}}\right)_{m_0} = \varphi V + w(1-w)\left(\frac{\partial(\varphi V)}{\partial w}\right)_{m_0}; \\ \bar{k}_S &\equiv \left(\frac{\partial K_S}{\partial m_{solute}}\right)_{m_0} = \varphi K_S + w(1-w)\left(\frac{\partial(\varphi K_S)}{\partial w}\right)_{m_0} \end{aligned} \tag{10}$$

where the symbol m_0 represents constant amount of other (than solute) components of the mixture. The partial characteristics can be calculated from the measured apparent characteristics, φK_S and φV, and the slope of their concentration dependence, $\left(\frac{\partial(\varphi V)}{\partial w}\right)_{m_0}$ and $\left(\frac{\partial(\varphi K_S)}{\partial w}\right)_{m_0}$. In the same way the specific partial volume and compressibility of solute particles, $\bar{v}_{sp}$ and $\bar{k}_{S\,sp}$, can be introduced:

$$\begin{aligned} \bar{v}_{sp} &\equiv \left(\frac{\partial V}{\partial m_{sp}}\right)_{m_0\,sp} = v_{sp} + w_{sp}(1-w_{sp})\left(\frac{\partial v_{sp}}{\partial w}\right)_{m_0\,sp}; \\ \bar{k}_{S\,sp} &\equiv \left(\frac{\partial K_S}{\partial m_{sp}}\right)_{m_0\,sp} = k_{S\,sp} + w_{sp}(1-w_{sp})\left(\frac{\partial k_{S\,sp}}{\partial w}\right)_{m_0\,sp} \end{aligned} \tag{11}$$

3.2 Thermal low and thermal high frequency limits for compressibility

Below we will consider mixtures of solute particles (molecules or molecular aggregates) surrounded by solvent. If the physical properties of the solute particles and the solvent are different, a change of pressure in the mixture results in a different change of the temperature within the solute particles and within the surrounding solvent. This causes a heat flow between the particles and the solvent, which affects the volume change during compression. Thus, the effects of solute particle - solvent heat exchange shall be included in interpretations of measured compressibilities.

Two major structural characteristics control the change of the volume of the mixture when an excessive pressure is applied. The first is the length scale of homogeneity of pressure within the mixture, which is determined by the wavelength of the ultrasonic wave, λ. For the practical frequency range of precision measurements of compressibility, 1 to 100 MHz, λ is between 10^6nm and 10^4nm (Figure 1 (a)). Therefore for the nano-scale particles of the size below 1000 nm the effects of non-homogeneity of pressure around them can be neglected. The second structural characteristic, is the wavelength of the thermal wave, λ_T, (or the decay length, $\delta_T = \frac{\lambda_T}{2\pi}$), which determines the portions of the solute particle and the

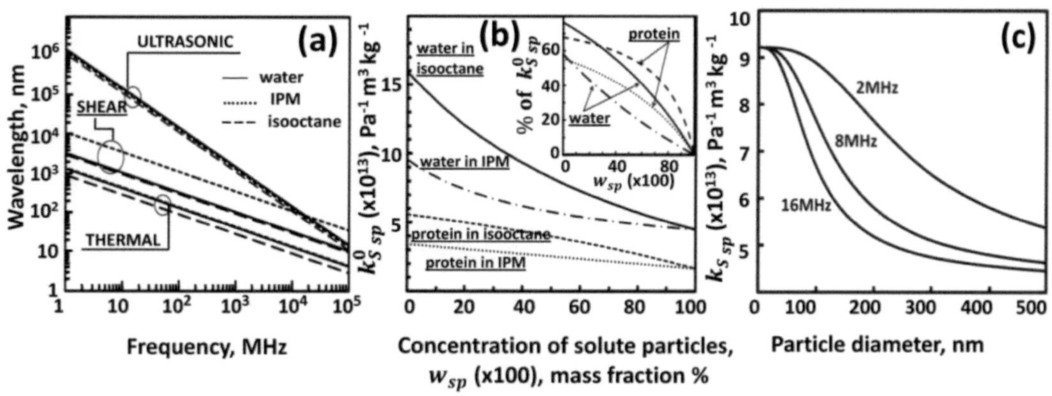

Fig. 1. (a): Wavelength as function of frequency for ultrasonic, λ, thermal, λ_T, and shear, λ_η waves in pure water, IPM system (composition described in Section 6) and in isooctane at 25°C calculated using physical parameters given in Table 1. (High GHz frequency part of the plot for illustrative purposes only as the physical properties of the media at high frequencies are expected to deviate from those at low frequencies). **(b): Main frame.** Specific adiabatic compressibility of protein particles and of droplets of water in IPM system and in isooctane at 25°C calculated for thermal low frequency limit using Equation (14) and physical properties given in Table 1. **Insert.** The 'solvent' part of adiabatic compressibility of protein particles and droplets of water in IPM and in isooctane as % of $k^0_{S\,sp}$ shown in the main frame; calculated using Equations (13) and (16). **(c):** Effect of droplet size on specific adiabatic compressibility of droplets of water in IPM system at different frequencies and at 25°C calculated as described in Section 3.4.

solvent involved in the heat exchange. In liquids λ_T varies between 100nm and 1000nm for the practical frequency range of precision measurements of compressibility, 1 to 100 MHz (Figure 1 (a)). This is within the range of the sizes of particles of dispersions, thus producing a profound effect of the frequency of ultrasound on the measured compressibility of the micro particles in mixtures. With respect to this, two limiting cases can be distinguished. 1) At **thermal low frequency limit** the period of compression is much longer than the time required for the heat exchange within the compressed volume. Under this condition the complete thermal equilibrium within the compressed volume is achieved and all its parts have the same temperature at each time. 2) At **thermal high frequency limit** the period of compression is much shorter than the time required for the heat exchange between the solute particles and the bulk solvent. Under this condition the solute particles and the solvent are thermally insulated. In structural terms the thermal low frequency limit is realised when the size of the solute particles, l, is much smaller than the wavelength of the thermal wave, $l << \lambda_T$. For the thermal high frequency limit $l >> \lambda_T$. In this case the volume (and the thermal mass) of the transition layer around the border between the solute particle and the solvent, where the solute-solvent heat exchange takes place, is negligibly small when compared with the volume of the solute particles and the solvent. Therefore, compressibility of particles, polymers and molecular aggregates with the size about and below 10 nm shall be attributed to the thermal low-frequency limit, while compressibility of particles and aggregates with the size of microns and higher shall be described in terms of thermal high-frequency limit especially at the top of the frequency range. Below the uppercase symbol ^{0} will be used for variables at low frequency limit and $^\infty$ for high frequency limit. An absence of these symbols will presume general cases applicable for both limits, or their irrelevance.

3.3 Apparent compressibility of solutes and compressibility of solute particles at thermal low and thermal high frequency limits

At thermal low frequency limit all parts of the mixture are at the same temperature, and therefore, have the same value of $\frac{\delta T}{\delta P} = \frac{Te}{c_P}$. Thus, the specific adiabatic compressibility of the mixture at this limit, k_S^0, is given by: $k_S^0 = k_T - \frac{Te^2}{c_P}$ where e and c_P are specific thermal expansibility and heat capacity of the mixture given by Equation (9). This and Equations (6) and (7) provide the relationship for specific apparent adiabatic compressibility of solute at thermal low-frequency limit, $\varphi K_S{}^0$, and specific adiabatic compressibility of solute particle at this limit, $k_{S\,sp}^0$:

$$\varphi K_S^0 = \varphi K_T - \frac{e_0^2 T}{c_{P_0}}\left(2\frac{\varphi E}{e_0} - \frac{\varphi C_P}{c_{P_0}} + w\frac{\left(\frac{\varphi E}{e_0} - \frac{\varphi C_P}{c_{P_0}}\right)^2}{1 + w\left(\frac{\varphi C_P}{c_{P_0}} - 1\right)}\right)$$

$$k_{S\,sp}^0 = k_{T\,sp} - \frac{e_{0\,sp}^2 T}{c_{P_0\,sp}}\left(2\frac{e_{sp}}{e_{0\,sp}} - \frac{c_{P\,sp}}{c_{P_0\,sp}} + w_{sp}\frac{\left(\frac{e_{sp}}{e_{0\,sp}} - \frac{c_{P\,sp}}{c_{P_0\,sp}}\right)^2}{1 + w_{sp}\left(\frac{c_{P\,sp}}{c_{P_0\,sp}} - 1\right)}\right) \quad (12)$$

The above equations demonstrate that at thermal low frequency limit the specific apparent adiabatic compressibility of the solute and adiabatic compressibility of solute particles depend on the concentration even in ideal mixtures. The third terms of Equations (12) represent a reduction of the solute–solvent heat exchange with concentrations, w and w_{sp}, due to the lowering of the amount of solvent in the mixture. At infinite dilution ($w, w_{sp} \to 0$) Equations (12) are reduced to:

$$\varphi K_S^0 = \varphi K_T - \frac{e_0^2 T}{c_{P_0}}\left(2\frac{\varphi E}{e_0} - \frac{\varphi C_P}{c_{P_0}}\right); \quad k_{S\,sp}^0 = k_{T\,sp} - \frac{e_{0\,sp}^2 T}{c_{P_0\,sp}}\left(2\frac{e_{sp}}{e_{0\,sp}} - \frac{c_{P\,sp}}{c_{P_0\,sp}}\right) \tag{13}$$

At thermal high frequency limit no heat exchange between solute particles and the surrounding solvent is present. Therefore, the compressions of solute particles and the solvent occur independently on each other, and the adiabatic compressibility of the mixture at this limit, K_S^∞, is an additive thermodynamic characteristic. Thus, the apparent specific adiabatic compressibility at thermal high frequency limit, φK_S^∞, in ideal mixture does not depend on concentration. At $w=1$ the Equations (12) transform to $(\varphi K_S^0)_{w=1} = \varphi K_T - \frac{T\varphi E^2}{\varphi C_P}$ and at $w_{sp}=1$ to $k_{S\,sp}^0 = k_{T\,sp} - \frac{Te_{sp}^2}{c_{P\,sp}}$. Also at $w=1$ the heat exchange between the solute and the solvent vanishes and, therefore, compressions occur at adiabatic condition for the solute and at $w_s=1$ for the solute particle at all frequencies. Therefore, $(\varphi K_S{}^0)_{w=1} = (\varphi K_S{}^\infty)_{w=1}$. As for ideal mixture of solute and solvent the value of φK_S^∞ does not depend on the concentration, the relationship $\varphi K_S^\infty = \varphi K_T - \frac{T(\varphi E)^2}{\varphi C_P}$ shall be valid for any concentration. It is important to note that this relationship is correct only for ideal mixture of solute and solvent. At $w_{sp}=1$, $k_{S\,sp}^0$ represents the adiabatic compressibility of pure solute particle: $k_{S\,sp}^0 = k_{S\,sp}^\infty$. With respect to $k_{S\,sp}^0$ Equations (12) can be rewritten in the following form:

$$k_{S\,sp}^0 = k_{S\,sp}^\infty + k_{HE\,sp}; \qquad k_{S\,sp}^\infty = k_{T\,sp} - \frac{e_{sp}^2 T}{c_{P\,sp}}$$

$$k_{HE\,sp} = \underbrace{\frac{e_{0\,sp}{}^2 T}{c_{P_0\,sp}}\left(\frac{c_{P\,sp}}{c_{P_0\,sp}} - \frac{e_{sp}}{e_{0\,sp}}\right)^2 \frac{c_{P_0\,sp}}{c_{P\,sp}}} - \underbrace{w_{sp}\frac{\frac{e_{0\,sp}{}^2 T}{c_{P_0\,sp}}\left(\frac{c_{P\,sp}}{c_{P_0\,sp}} - \frac{e_{sp}}{e_{0\,sp}}\right)^2}{1 + w_{sp}\left(\frac{c_{P\,sp}}{c_{P_0\,sp}} - 1\right)}} \tag{14}$$

Contribution of heat exchange between the solute particle and infinite amount of solvent

Reduction of contribution of heat exchange caused by substitution of solvent with solute particles

where $k_{S\,sp}^\infty$ is the specific adiabatic compressibility of the solute particle at thermal high frequency limit and k_{HE} is the contribution of the heat exchange between the solute particle and its environment to compressibility. The specific adiabatic compressibility of solute particle at thermal high frequency limit, $k_{S\,sp}^\infty$, represents the 'true' adiabatic compressibility of the particle. In contrast to this, the specific adiabatic compressibility of solute particle measured at thermal low frequency limit has a contribution from the heat exchange between the solute particle and its environment and, therefore, depends on the properties of solvent even in the absence of solvation effects. Figure 1 (b) represents the compressibility $k_{S\,sp}^{0}$ of globular protein and of droplets of water in two solvents, pure isooctane and in IPM system (composition is given in Table 1), calculated according to Equation (14). The physical properties of solute particles ($''_{sp}$) and bulk solvent ($''_{0sp}$) used in the calculations are given in Table 1. As no solvation effects were included in calculations ($w_s = 0$), the concentration w_{sp} represents the concentration of solute, $w_{sp} = w$. Figure 1 (b) shows that the compressibility of solute particles at thermal low frequency limit depends significantly on the properties of solvent. Aqueous mixtures, where water is the solvent, are different in this respect, when compared with others, due to the 'abnormal' physical characteristics of liquid water at room temperatures and below: very low value of thermal expansibility, $e_{0\,sp} << e_{sp}$, and high value of heat capacitance, $c_{P_0\,sp} >> c_{P\,sp}$. For aqueous mixtures $(\frac{c_{P\,sp}}{c_{P_0\,sp}} - \frac{e_{sp}}{e_{0\,sp}})^2 \approx (\frac{e_{sp}}{e_{0\,sp}})^2$ and, according to Equation (14), at low concentrations the adiabatic compressibility of solute particle at thermal low frequency limit, $k_{S\,sp}^{0}$, is close to its isothermal compressibility, $k_{T\,sp}$: $k_{S\,sp}^{0} \approx k_{S\,sp}^{\infty} + \frac{e_{sp}{}^2 T}{c_{P\,sp}} = k_{T\,sp}$.

Equations (12) or (14) can be applied for calculations of adiabatic compressibility of solute particle, or apparent compressibility of solute in a Solvent 2, $k_{S\,sp\,2}^{0}$ or φK_{S2}^{0} from the compressibility measured in a different solvent, Solvent 1, $k_{S\,sp\,1}^{0}$ or φK_{S1}^{0} If the substitution of solvents does not affect the following additive characteristics of solute particles or solute, $k_{S\,sp}^{\infty}$ (or $k_{T\,sp}$), e_{sp} , $c_{P\,sp}$ or φK_T , φE , φC_P , for infinite dilution, $w \to 0$:

$$\varphi K_{S2}^{0} = \varphi K_{S1}^{0} + \frac{e_{01}^{2} T}{c_{P_0 1}} [2\frac{\varphi E}{e_{0\,1}}(1-R) - \frac{\varphi C_P}{c_{P_0 1}}(1-R^2)]; \quad where\ R \equiv \frac{e_{0\,2}\, c_{P_0 1}}{e_{0\,1}\, c_{P_0 2}}$$
$$k_{S2\,sp}^{0} = k_{S1\,sp}^{0} + \frac{e_{0\,1sp}^{2} T}{c_{P_0 1sp}} [2\frac{e_{sp}}{e_{0\,1sp}}(1-R_{sp}) - \frac{c_{P\,sp}}{c_{P_0\,sp1}}(1-R_{sp}{}^2)]; \quad where\ R_{sp} \equiv \frac{e_{0\,2\,sp}\, c_{P_0 1sp}}{e_{0\,1sp}\, c_{P_0\,2\,sp}} \tag{15}$$

3.4 Transition between thermal low and thermal high frequency limits

As frequency increases the compressibility of solute particle changes from its value for thermal low frequency limit to the value at thermal high frequency limit in the frequency range for which λ_T is comparable with the size of solute. The analytical dependence of compressibility on the frequency for a given size of the solute or on the size of the solute for a given frequency can be obtained through analysis of dynamics of the heat exchange between the solute particles and the solvent. The results depend on structural details of the mixture including particle geometry. Algorithms of this analysis were described previously as a part of ultrasonic scattering theories (Isakovich, 1948; Hemar et al., 1998; Allegra & Hawley, 1972; Povey, 1997) although specific calculations normally were limited to the case of spherical particles. Certain aspects of the results can be generalised. An increase of the frequency decreases the volumes of the solute particle and the solvent participating in the heat exchange around the particle-solvent border. These volumes involve the whole particle and the solvent at low frequencies and become to be negligibly small at high frequencies. The dependence of compressibility on lg of frequency (or on lg of size at constant frequency) exhibits an 'S' shape curve. The main transition of compressibility between the thermal low and thermal high frequency limits occurs around frequency at which the size of solute particle is about the wavelength of the thermal wave. The major change of compressibility with frequency occupies within one decade on the frequency range. Figure 1(c) demonstrates the dependence of compressibility of aqueous droplets on their particle size in the IPM system at $w_{sp} = 0.01$ calculated using the PSize software module (v.2.28.01) provided with HR-US 102 spectrometers (Sonas Technologies Ltd.). This software utilises theoretical approaches developed previously for spherical particles in liquid continuous medium (Epstein & Carhart, 1953; Waterman & Truell, 1961; Allegra & Hawley, 1972; Povey, 1997; Austin et al., 1996). The physical properties used in the calculations are given in Table 1. The calculated values of compressibility for zero and infinite diameter agree well with values of adiabatic compressibility for thermal low and thermal high frequency limits, calculated according to Equation (14). An exception to this is a very small deviation at zero limit of diameter of particle. This could be expected as the cited theories do not account well the concentration behaviour of the heat exchange between the particles and the solvent.

3.5 Molecular interpretation of compressibility of solute particles

Molecular interpretation of compressibility requires distinguishing contributions of different origin to compressibility. For an ideal mixture of solute particles and a solvent three major contributions shall be considered: intrinsic compressibility of solute particle, $k_{S\,in}$, ideal contribution to compressibility, $k_{S\,id}$, and relaxation contribution to compressibility, $k_{S\,rel}$. At infinite dilution $k_{S\,sp} = k_{S\,in} + k_{S\,id} + k_{S\,rel}$.

3.5.1 Ideal contribution to compressibility of solute particle

The ideal contribution to compressibility of solute particles represents compressibility of particles with zero intrinsic volume and intrinsic compressibility. It is originated by the heat motion of the solute particles, which produces an 'additional' (osmotic) pressure in the mixture, thus leading to the expansion of solvent. The change of the volume of this expansion with external pressure provides the ideal contribution to compressibility. Due to the origin (heat motion) of ideal compressibility, its molar value does not depend on the molar mass of solute particle, M. Therefore, its contribution to specific compressibility of solute particle is inversely proportional to molar mass, $k_{S\,id} \sim M^{-1}$, and shall vanish as the molar mass increases. Estimations of this contribution in microemulsion systems with typical thermodynamic parameters of oils, water and their mixtures with surfactants, show that for micelles and microdroplets this contribution is small and normally can be neglected.

3.5.2 Relaxation contribution to compressibility of solute particle

The relaxation (also for some systems called 'structural') contribution to the apparent compressibility and the compressibility of solute particles arises if the particles have different states (conformational, protonated, etc.) in the mixture, and the distribution between the states changes following changing pressure and temperature in ultrasonic wave. The relaxation contribution to compressibility represents the change of the volume of the mixture caused by a shift in the distribution of solute particles between different states, per unit of pressure applied. The basic relationships between the molecular characteristics of relaxation processes and their contributions to ultrasonic parameters were described previously (see for example Kaatze et al., 2000). The theories cover the thermal low frequency limit at which solute and solvent have the same temperature at any time. They provide tools for estimation of relaxation contribution from known volume and enthalpy effects of the relaxation process and the equilibrium concentrations of its participants. The frequency dependence of the relaxation compressibility is determined by factor $\frac{1}{1+(2\pi f\tau)^2}$ (τ is the relaxation time of the process). This allows ultrasonic analysis of kinetics of chemical processes. The relaxation compressibility vanishes at high frequencies.

At thermal high frequency limit each solute particle is thermally insulated, thus, oscillations of temperature within the particles of different states and within the solvent are different. Therefore, at this limit relationships between the ultrasonic parameters and molecular characteristics of relaxation processes are complex and depend on the structural details of mixtures.

3.5.3 Intrinsic compressibility of solute particles

Intrinsic adiabatic compressibility of solute particle is the part of compressibility of solute particle associated with intrinsic properties of particles. At thermal high frequency limit it represents the adiabatic compressibility of the material of particles (solutes). For low molecular weight molecules it is determined by the change of Van der Waals radii of atoms and the length of covalent bonds caused by applied pressure. The magnitude of this compressibility can be estimated from compressibilities of relevant crystals. For organic molecules it is much smaller than other contributions to compressibility and normally can be neglected. (see for example Buckin, 1988). Intrinsic compressibility of molecular aggregates and compact polymers at thermal high frequency limit can be qualitatively distinguished by the type of volume changes caused by compression: (a) volume change caused by compression of interatomic contacts (Van der Waals contacts, hydrogen bonding, etc.) and (b) volume change caused by a structural/conformational transitions between the states of different volume (structural contribution). As conformational (structural) transitions normally involve coordinated movement of large number of atomic groups (polymers) or molecules (molecular aggregates) their relaxation time is much longer than the relaxation time for compression of interatomic contacts. This often allows another way to distinguish the (a) and (b) types of contributions to compressibility of molecular aggregates and compact polymers. At high frequencies when the period of oscillation of pressure is much shorter than the relaxation time for a particular conformational transition, this transition will be 'frozen' and its contribution to compressibility vanishes. By their nature the (b) types of contributions to intrinsic compressibility of solutes are the relaxation contributions. For pure liquids the structural contribution normally represents about 30% of their total compressibility (Litovitz & Davis, 1964). Liquid water is an exception as the structural contribution (at room temperature) is about 60% of its compressibility. This is explained by its highly aggregated structure resulted from high level of hydrogen bonding between molecules. The remaining 40% of compressibility is close to the compressibility of ice crystals, which represent type (a) of intrinsic compressibility (Litovitz & Davis, 1964)).

Therefore, hydration processes are often accompanied by a significant decrease of compressibility of water. At thermal low frequency limit the contribution of heat exchange is added to the intrinsic compressibility which is provided by Equations (12) to (14).

3.5.4 Physical location of compressibility of solute particle

Compressibility of solute particle is defined as the difference between the compressibility of the mixture (solute particles in the bulk solvent) and the bulk solvent. At thermal high frequency limit the particle and the bulk solvent are thermally insulated, and, in an absence of ideal and relaxation contributions, compressibility of solute particle represents the compression of the physical volume of solute particle per unit of pressure applied. At thermal low frequency limit, the solute - solvent heat exchange affects the temperature of the particle and the solvent and, therefore, their compressions. Thus, at this frequency limit a part of the measured compressibility of solute particle is physically located in the solvent. It represents the difference between the compression of the solvent in the mixture and the compression of the pure solvent (per unit of pressure applied). A question can be raised on the portions of compressions of solute particle and the solvent contributing to compressibility of solute particle. These portions can be obtained from the following considerations. The specific adiabatic compressibility of the bulk solvent is given as: $k_{S_0\,sp} = k_{T_0\,sp} - e_{0\,sp}\left(\frac{\partial T}{\partial P}\right)_S$. In pure bulk solvent (same composition as in the mixture) $\left(\frac{\partial T}{\partial P}\right)_S = \frac{Te_{0\,sp}}{c_{P_0\,sp}}$. In the mixture at thermal low frequency limit $\left(\frac{\partial T}{\partial P}\right)_S = \frac{Te}{c_P} = T\frac{(1-w_{sp})e_{0\,sp} + w_{sp}e_{sp}}{(1-w_{sp})c_{P_0\,sp} + w_{sp}c_{P\,sp}}$. The difference between the compressibility of the bulk solvent in the mixture and the pure bulk solvent calculated per unit of mass of solute particle provides the portion of compressibility of the solute particle located in the bulk solvent, $k_{S\,sp\,solvent}$: $k_{S\,sp\,solvent} = -e_{0\,sp}(\frac{Te}{c_P} - \frac{Te_{0\,sp}}{c_{P_0\,sp}})$, which can be rewritten as:

$$k_{S\,sp\,solvent} = -(1-w_{sp})\frac{Te_{0\,sp}^{\;2}}{c_{P_0\,sp}}\frac{\frac{e_{sp}}{e_{0\,sp}} - \frac{c_{P\,sp}}{c_{P_0\,sp}}}{1 + w_{sp}(\frac{c_{P\,sp}}{c_{P_0\,sp}} - 1)} \quad (16)$$

It is interesting to note that the 'solvent' part of compressibility of solute particle can be positive or negative, depending on the ratios of thermal expansibility and heat capacity for the particles and the solvent. The insert in the Figure 1(b) illustrates the portion of the 'solvent' part of $k_{S\,sp}^0$ for protein and for droplets of water in isooctane and in IPM system calculated according to the Equations (14) and (16). The physical properties of solute particles ($''_{sp}$) and solvent ($''_{0\,sp}$) used in the calculations are given in Table 1. As no solvation effects were included in calculations ($w_s = 0$), the concentration w_{sp} represented the concentration of solute, $w_{sp} = w$. Figure 1(b) demonstrates that the 'solvent' part of compressibility of protein particles and droplets of water in these solvents is significant and at low concentrations exceeds 50%.

4. Ultrasonic measurements of apparent and partial compressibilities

4.1 Concentration increment of ultrasonic velocity

Equations (4) and (6) allow calculations of apparent adiabatic compressibility of solutes from the measured values of ultrasonic velocity and density. However, for dilute mixtures, where the contribution of solutes to the specific compressibility and the volume of the mixture is much smaller than the contribution of solvent, this way of obtaining of φK_S (and k_{sp}) possesses a problem as it requires very precise measurements of ultrasonic velocity and density in exactly the same mixture and at the same temperature. This is also complicated by the high temperature dependence of ultrasonic velocity and density, which means that during the measurements of velocity and density the temperature must be controlled with high precision. In addition, precision of measurements of the absolute values of ultrasonic velocity and density normally does not exceed 10^{-5} of the absolute value. However, the change in velocity, δu, and density, can be measured much better (down to 10^{-7} of absolute velocity (Buckin & Smith, 1999; Buckin & O'Driscoll, 2002) and 10^{-6} of density). Therefore, calculations of apparent adiabatic compressibility from the contributions of solutes to ultrasonic velocity and to specific volume of the mixture are more precise. These contributions are obtained from the measured change of ultrasonic velocity, δu ($\equiv u - u_0$, where u and u_0 are the velocity in the mixture and in the solvent), and of specific volume, δv, caused by addition of the solute to the solvent.

The contribution of solute to ultrasonic velocity in the mixture can be represented by the concentration increments of ultrasonic velocity of solute, such as a_u or a_{u^2}, defined as:

$$a_u \equiv \frac{\frac{u}{u_0} - 1}{w} = \frac{\delta u}{w u_0}; \qquad a_{u^2} \equiv \frac{(\frac{u}{u_0})^2 - 1}{2w} = \frac{\delta u(2u_0 + \delta u)}{2w{u_0}^2} \tag{17}$$

The two concentration increments of ultrasonic velocity are linked with each other as:

$$a_{u^2} = a_u(1 + \frac{w a_u}{2}) \tag{18}$$

At infinite dilution, $w \to 0$, the increments become to be equal: $a_u = a_{u^2}$. If required, the ultrasonic velocity of liquid can be calculated from the known concentration increment of ultrasonic velocity as: $u = u_0(1 + w a_u)$ or $u = u_0\sqrt{1 + 2w a_{u^2}}$. When analysing the properties of solute particles the concentration increments for solute particle can be introduced, $a_{u\,sp}$ and $a_{u^2\,sp}$, which are defined in the same way as a_u and a_{u^2}:

$$a_{u\,sp} \equiv \frac{\frac{u}{u_{0\,sp}} - 1}{w_{sp}} = \frac{\delta u_{sp}}{w_{sp} u_{0\,sp}}; \qquad a_{u^2\,sp} \equiv \frac{(\frac{u}{u_{0\,sp}})^2 - 1}{2w_{sp}} = \frac{\delta u_{sp}(2u_{0\,sp} + \delta u_{sp})}{2w_{sp}{u_{0\,sp}}^2} \tag{19}$$

where $u_{0\,sp}$ is the ultrasonic velocity for the bulk solvent in the mixture and $\delta u_{sp} = u - u_{0\,sp}$. If the value of ultrasonic velocity in the bulk solvent is the same as in the pure solvent, $u_{0\,sp} = u_0$, the proportionality coefficient $\frac{w_{sp}}{w} = 1 + w_s$ allows recalculations of specific concentration increments of ultrasonic velocity: $a_{u\,sp} = \frac{a_u}{1+w_s}$; $a_{u^2\,sp} = \frac{a_{u^2}}{1+w_s}$.

The concentration increments of ultrasonic velocity characterise the state of the whole solute in the mixture. If the state of the solute depends on its concentration it is often useful to obtain the partial concentration increments of ultrasonic velocity, $\bar{a}_u$ and $\bar{a}_{u^2}$, which represent the effect of addition of a small portion of solute to a solutions at a particular concentration:

$$\bar{a}_u \equiv \left(\frac{\partial}{\partial w} \left(\frac{u}{u_0}\right) \right)_{m_0} = a_u + w \left(\frac{\partial a_u}{\partial w} \right)_{m_0}; \quad \bar{a}_{u^2} \equiv \frac{1}{2} \left(\frac{\partial}{\partial w} \left(\frac{u}{u_0}\right)^2 \right)_{m_0} = a_{u^2} + w \left(\frac{\partial a_{u^2}}{\partial w} \right)_{m_0} \tag{20}$$

Partial concentration increments of solute particles are introduced in the same way.

4.2 Calculations of apparent adiabatic compressibilities from concentration increments of ultrasonic velocity

Combination of Equation (4), (6), and (19) provide the following relationship for calculations of specific apparent adiabatic compressibility of solute and the specific compressibility of solute particle in a liquid from the concentration increments of ultrasonic velocity and the specific apparent volume:

$$\begin{aligned} \varphi K_S &= k_{S_0} \left(2\frac{\varphi V}{v_0} - 2a_{u^2} - 1 + w \frac{\left(\frac{\varphi V}{v_0} - 2a_{u^2} - 1\right)^2}{1 + 2wa_{u^2}} \right) \\ k_{S\,sp} &= k_{S_0\,sp} \left(2\frac{v_{sp}}{v_{0\,sp}} - 2a_{u^2\,sp} - 1 + w_{sp} \frac{\left(\frac{v_{sp}}{v_{0\,sp}} - 2a_{u^2\,sp} - 1\right)^2}{1 + 2w_{sp}a_{u^2\,sp}} \right) \end{aligned} \tag{21}$$

If the mixture contains gel structures, the contribution $\frac{M'_{gel}}{2w\rho u_0^{\,2}}$ shall be subtracted from a_{u^2} or $\frac{M'_{gel}}{2w\rho u_{0\,sp}^{\,2}}$ from $a_{u^2\,sp}$ (Equations (3), (17) and (19)) before applying Equations (21) for calculation of φK_S or $k_{S\,sp}$.

At infinite dilution, $w \to 0$, the Equations (21) are reduced to:

$$\varphi K_S = k_{S_0} \left(2\frac{\varphi V}{v_0} - 2a_u - 1 \right); \quad k_{S\,sp} = k_{S_0\,sp} \left(2\frac{v_{sp}}{v_{0\,sp}} - 2a_{u\,sp} - 1 \right) \tag{22}$$

Equations (21) and (22) allow calculations of the specific apparent adiabatic compressibility of solutes and the specific adiabatic compressibility of solute particles from the measured concentration increment of ultrasonic velocity and apparent specific volume of solutes. When comparing the ultrasonic data with predictions of different models is often more practical to perform the analysis of directly measured concentration increment of ultrasonic velocity rather than of apparent adiabatic compressibility, thus excluding the need of the measurements of apparent volume. For this case Equations (21) can be rewritten in the following way, which allows calculations of concentration increment of ultrasonic velocity (as well as the ultrasonic velocity) from the specific apparent volume and compressibility of solute or the specific apparent volume and compressibility of solute particles:

$$a_{u^2} = \frac{1}{2}\left(2\frac{\varphi V}{v_0} - \frac{\varphi K_S}{k_{S_0}} - 1 + w\frac{\left(\frac{\varphi V}{v_0} - \frac{\varphi K_S}{k_{S_0}}\right)^2}{1 + w\left(\frac{\varphi K_S}{k_{S_0}} - 1\right)}\right);$$

$$a_{u^2\,sp} = \frac{1}{2}\left(2\frac{v_{sp}}{v_{0\,sp}} - \frac{k_{S\,sp}}{k_{S_0\,sp}} - 1 + w_{sp}\frac{\left(\frac{v_{sp}}{v_{0\,sp}} - \frac{k_{S\,sp}}{k_{S_0\,sp}}\right)^2}{1 + w_{sp}\left(\frac{k_{S\,sp}}{k_{S_0\,sp}} - 1\right)}\right) \qquad (23)$$

At infinite dilution, $w \to 0$:

$$a_{u^2} = a_u = \frac{1}{2}\left(2\frac{\varphi V}{v_0} - \frac{\varphi K_S}{k_{S_0}} - 1\right); \qquad a_{u^2\,sp} = a_{u\,sp} = \frac{1}{2}\left(2\frac{v_{sp}}{v_{0\,sp}} - \frac{k_{S\,sp}}{k_{S_0\,sp}} - 1\right) \qquad (24)$$

It is important to note that equations (21) to (24) are valid for the inertial low frequency limit only at which the inertial and gravimetric densities of the mixture are the same.

Equations (15) and (24) can be applied for calculations of a_u and $a_{u\,sp}$ at thermal low frequency limit and at infinite dilution in a Solvent 2, $a_{u\,2}$, from the value of a_u measured in a different solvent, Solvent 1, $a_{u\,1}$, provided that the substitution of solvents does not affect the followingcharacteristics of solute particles or solute, φK_T, φE, φC_P or $k^{\infty}_{S\,sp}$ ($k_{T\,sp}$), e_{sp}, $c_{P\,sp}$:

$$a_{u2} = a_{u1} R_k + \frac{\varphi V}{v_{0\,1}}(R_v - R_k) + TR_h\left[\frac{\varphi E}{e_{0\,1}}(1 - R) - \frac{1}{2}\frac{\varphi C_P}{c_{P_0 1}}(1 - R^2)\right] - \frac{1}{2}(1 - R_k)$$

$$a_{u2\,sp} = a_{u1sp} R_{k\,sp} + \frac{v_{sp}}{v_{0\,1sp}}(R_{v\,sp} - R_{k\,sp}) + TR_{h\,sp}\left[\frac{e_{sp}}{e_{0\,1sp}}(1 - R_{sp}) - \frac{1}{2}\frac{c_{P\,sp}}{c_{P_0 1sp}}(1 - R_{sp}{}^2)\right] - \frac{1}{2}(1 - R_{k\,sp}) \qquad (25)$$

where $R_v \equiv \frac{v_{0\,1}}{v_{0\,2}}$, $R_k \equiv \frac{k_{S_0 1}}{k_{S_0 2}}$, $R_h \equiv \frac{R_k e_{0\,1}^2}{k_{S_0 1}\, c_{P_0 1}}$, $R_{v\,sp} \equiv \frac{v_{0\,1sp}}{v_{0\,2sp}}$, $R_{k\,sp} \equiv \frac{k_{S_0 1sp}}{k_{S_0 2sp}}$ and

$$R_{h\,sp} \equiv \frac{R_k e_{0\,1sp}^2}{k_{S_0 1\,sp}\, c_{P_0 1sp}}.$$

5. Ultrasonic attenuation and particle sizing

Ultrasonic attenuation in emulsions, α, is represented by contributions of two terms: intrinsic absorption, α_I, and scattering losses, α_S: $\alpha = \alpha_I + \alpha_S$ (Povey, 1997)). For ideal mixture of solute particles and a solvent the intrinsic absorption is a sum of attenuations in the solvent, α_{I0}, and in the solute particle, α_{Isp}, weighted according to their volume fractions: $\alpha = \alpha_{I0}(1-x) + \alpha_{Isp}x + \alpha_S$ (x is the volume fraction of solute particles). The scattering contribution to ultrasonic attenuation, α_S, is a function of frequency, particle size and concentration, which is described by ultrasonic scattering theories. In the long wavelength limit, i.e. when the wavelength of ultrasound is much longer than the particle diameter, explicit expressions for the ultrasonic scattering in dispersions of spherical particles have been derived (Epstein & Carhart; 1953, Waterman & Truell, 1961; Allegra & Hawley, 1972; Povey, 1997). The mechanism of interaction of the ultrasonic wave with particles in dispersions in this regime contains two major contributors, thermoelastic and viscoinertial scattering. These contributions result from the 'scattering' of the incident ultrasonic waves into the thermal and the viscous waves propagating from the border between the particle and the continuous medium. The thermoelastic mechanism dominates in emulsions as the difference in density between the continuous and the dispersed phases of emulsions, which determines the magnitude of viscoelastic scattering, is not significant (Powey, 1997). Applications of ultrasonic scattering theories allow calculations of sizes of dispersed particles from the measured values of attenuation and the physical parameters of the continuous medium and the particles. The obtained sizes in microemulsions are in good agreement with the results of other techniques (Hickey et al., 2006, 2010).

6. Materials and methods

Materials. Ethyl oleate (Cat no. 27,074-1), sorbitan mono-laurate (Span 20, Cat no. 1338-39-2), polyoxyethylene 20 sorbitan mono-oleate (Tween 80, Cat no. 9005-65-6), isopropyl myristate (IPM) (cat. no. 110-27-0), n-propanol (cat. no. 71-23-8), doctyl sulfosuccinate sodium salt 98% (AOT) (cat no.577-11-7, lot# STBC 0203V) and 2,2,4-trimethylpentane (isooctane) 99.0% (cat no.540-84-1, lot# 53196MMV) were purchased from Sigma-Aldrich Chemical Co. Ltd. (Dublin, Ireland). Epikuron 200 (E200; soy lecithin, minimum 95 wt.% phosphatidylcholine; fatty acid content: palmitic and stearic, 16-20%; oleic acid, 8-12%; linoleic acid, 62-66%; linolenic acid, 6-8%, was supplied by Lucas Meyer (Germany). All reagents were of the highest purity available and were used as received. The ultrapure water (Millipore Super-Q-System) with the conductivity of 18.2 MΩ · cm^{-1} was used in sample preparation. Details of preparations of emulsions were described previously (Hickey et al., 2006, Hickey et al., 2010).

Microemulsion systems for ultrasonic titrations. **IPM**: isopropyl myristate & [Epikuron 200 : n-propanol (1:1, w/w)]; ratio of oil (isopropyl myristate) to surfactant+cosurfactant (Epikuron 200 + n-propanol (1:1, w/w)) was 60:40 (w/w). **EO**: ethyl oleate & [Tween 80:Span 20 (3:2, w/w)]; ratio of oil (ethyl oleate) to surfactant (Tween 80:Span 20 (3:2, w/w)) was 80:20 (w/w). **AOT**: 0.1M of AOT in isooctane.

Samples for enzyme reactions. D-cellobiose (purity ≥98%, Sigma-Aldrich Corp., Lot. 077K0741) was used for preparation of substrate samples for enzyme reaction. Aqueous solutions of D-cellobiose were prepared as 2.5% (w/w) of D-cellobiose in 10 mM sodium acetate buffer, pH 4.90. Substrate sample of D-cellobiose in microemulsion was prepared by mixing of 2.5

%(w/w) aqueous solution of D-cellobiose and IPM & [Epikuron 200 : n-propanol (1:1, w/w)] giving the following composition of microemuslion: IPM -38% (w/w); [Epikuron 200 : n-propanol (1:1, w/w)] - 38% (w/w); -water – 24% (w/w). Enzyme concentrations were determined by weight as a mass of liquid preparation (Novozyme 188 L (Novo Nordisk A/S); β-glucosidase preparation was obtained from *Aspergillus niger* microorganisms, with a declared minimum activity of 250 CbU per gram of liquid). Amount of protein in the preparation and other details of its composition were described earlier (Resa & Buckin, 2011). Enzyme was added to the substrate (cellobiose) solution (or emulsion) at room temperature and thoroughly mixed for 30 s, then degassed and loaded into the measuring cell of HR-US 102 differential ultrasonic spectrometer (Sonas Technologies Lt.) equilibrated at 25.0°C. The concentration of enzyme preparation in aqueous phase was 0.87 mg/mL. Ultrasonic measurements started immediately after loading (3 min after addition of enzyme to the substrate). Control experiments were performed by adding enzyme to same microemulsion and to the aqueous solution, however, without substrate to verify the stability of ultrasonic parameters within the time interval corresponding to the reaction time.

Titration profiles for ultrasonic velocity and attenuation in microemulsions were measured within 2 to 12 MHz frequency range at 25.0°C using a HR-US 102 SS (IPM, EO, AOT systems) and HR-US 102 (AOT system) differential ultrasonic spectrometers (Sonas Technologies Ltd.). The measuring ultrasonic cell (1.5 mL) was closed with a stopper, which incorporated a line for injection of titrant, drainage line for release of excess of sample and a mechanical mini stirrer. Measuring cell was fully filled with a mixture of oil, surfactant (and cosurfactant) using a calibrated Hamilton syringe. The exact amount of the mixture in the cell was determined by weighing the syringe before and after filling. Hamilton PB600 dispenser with 100uL Hamilton syringe attached to the injection line was used for the titrations. To exclude any possible evaporation the end of the drainage line was immersed in the enclosed vial filled with the mixture. The reference cell was filled with the same amount of the mixture and sealed. Measurements were continuously performed while the sample in the measuring cell was titrated with aliquots of degassed water. The stirring was activated after each addition of water and continued until no further change in the ultrasonic velocity and attenuation occurred, indicating that equilibrium had been reached. Following this stirring was stopped, stability of the measured values and attenuation was confirmed (5 to 10 minutes without stirring approximately) and next portion of water was added. *Density* was measured using Anton Paar 4500 density meter at 25.0°C connected with two flexible pipes to a 2 ml vial with a stirring bar inside filled with the mixture of oil and surfactant (and cosurfactant). Water was added stepwise to the vial in the same way as for ultrasonic measurements. After each addition several cycles of stirring and pumping of the mixture through the measuring cell were performed and stability of density readings were confirmed.

7. Ultrasonic phase diagrams

The ultrasonic phase diagrams can be obtained from titration profiles of ultrasonic velocity and attenuation. Simultaneous measurements of ultrasonic characteristics at various frequencies allow evaluation of their frequency dependence at each titration point. Overall, the titration profiles provide a number of parameters reflecting different levels of microstructural organisation of the system. In most cases transitions between different states of the system are clearly identified by an abrupt change of concentration profile of ultrasonic velocity, attenuation and their frequency dependence. Figure 2 shows the ultrasonic titration profiles measured in three w/o microemulsion systems, IPM, AOT and EO, (described in Section 6) by stepwise additions of water into the mixture of oil & surfactant in the measuring cell of HR-US 102 spectrometer. The IPM and EO systems were studied by us previously (Hickey et al., 2006, 2010), however here we present new data on detailed titration profiles at low concentrations of water. Also, our new titration setup provided higher resolution of measurements. The concentrations of water corresponding to the same transitions for titrations at different oil:surfactant ratios provide the transition lines at ultrasonic pseudo-ternary phase diagrams. The positions of the transitions between the major phases such as microemulsion, coarse emulsion, liquid crystals, biocontinous on the ultrasonic phase diagrams agree well with the phase diagrams produced by other techniques for the above systems (Hickey et al., 2006, 2010). In addition to this, a range of transitions within the microemulsion phase were observed, which allowed upgrading the phase diagrams with additional 'sub phases' within the microemulsion phase. The nature of these 'sub phases' can be related to the state of water and surfactant, including hydration water in dispersed phase, hydration water in swollen reverse micelles and water in aqueous droplets surrounded by surfactant. Although, the details of these interpretations can be further discussed, it is important to recognize that ultrasonic technique allows the identification of these different 'sub-states' on the phase diagram in simple titration measurements. Consideration of these states and their position on the phase diagram is important in design of w/o microemulsions designated for encapsulation of active ingredients (e.g. proteins) functioning of which requires an appropriate level of their hydration. An example of this is discussed in the following chapters.

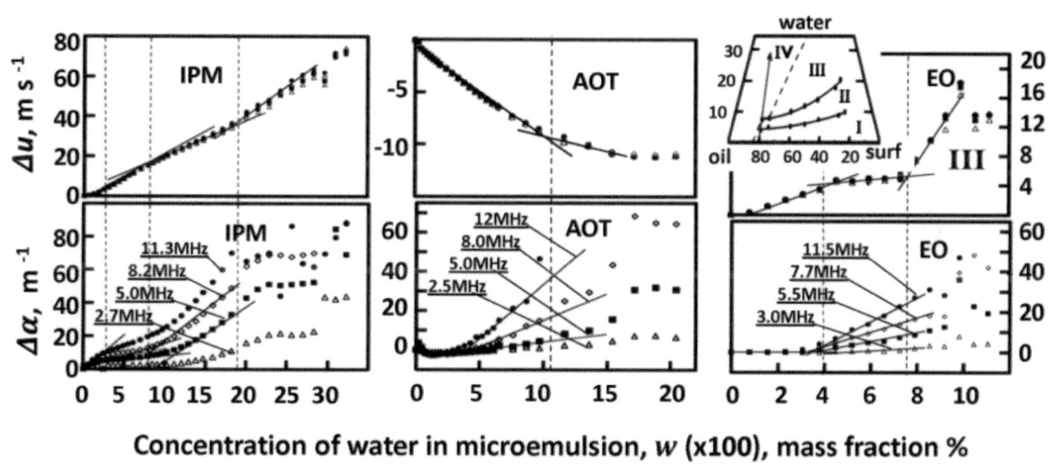

Fig. 2. Change of ultrasonic velocity and attenuation with concentration of water at 25°C in IPM, AOT and EO systems (see Section 6 for composition). The **insert** (EO system) represents the microemulsion part of ultrasonic phase diagram for EO system. The solid line with the arrow bar represents the ultrasonic titration line shown in in the main frame (EO). The points represent positions of the same transitions on ultrasonic titration lines for different oil:surfactant ratios. Transition between phases II and III is accompanied by appearance of UV/VIS light scattering. The dashed line ('- -') represents the data of (Alany et al., 2001) for transition between phases III and IV. I - hydration of surfactant with addition of water; II - formation of microemulsion droplets with a compact pool of water molecules; III - liquid crystals; IV - pseudobicontinuous phase.

In the IPM, AO and EO systems the end of microemulsion phase occurs at concentration of water of about 19%, 10% and 7.5%. Above these concentrations the microemulsions become to be opaque. This transition is marked by the appearance of a dispersion of ultrasonic velocity and abrupt change in the concentration slopes of attenuation and velocity. In addition to this, visible changes in the concentration profile (slope) of ultrasonic velocity and attenuation are observed within the concentration range of the microemulsion phase. For example, in IPM system the attenuation profile shows an initial increase up to the concentration 2%, than attenuation levels off between 2% and 8% approximately (exact concentration depends on the frequency), and begins to rise between 8% and 10% , with a linear part of the rise between 16% and 19%. Same pattern for transitions were observed for other titration lines, however at different concentrations of water. This could be interpreted as an existence of 'sub phases' within the microemulsion phase on the ultrasonic phase diagram, which are realised at concentrations of components between the transition lines. In AOT system the attenuation initially decreases to the concentration about 1.25%, or five molecules of water per one molecule of surfactant. This corresponds to the primary level of surfactant hydration (Spehr, 2010), which shall affect the intrinsic (relaxation) attenuation in the mixture of AOT and isopropanol. At concentrations above 2%, attenuation begins rising, which agrees with formation of droplets with a compact pool of water molecules described earlier (Hasegawa et al., 1994; Vasquez et al., 2011).

Several transition concentrations are also observed on velocity and attenuation profiles of EO system (Figure 2 EO). Formation of microemulsion droplets with a compact pool of water molecules in all systems can be attributed to the concentration points corresponding to a significant rise of ultrasonic attenuation: above 10% in IPM systems, above 2% in AOT system and 4 % for EO system. The amplitudes of this rise of attenuation and their frequency dependence correspond to the nanoscale (up to 20 nm for different parts of phase diagram) sizes of the water pool when ultrasonic scattering theories are applied (Hickey et al., 2006, 2010). This is in agreement with the absence of dispersion (dependence of frequency) of ultrasonic velocity (and compressibility) in the microemulsion part of the phase diagram, thus supporting the thermal low frequency conditions for microemulsion droplets. This is also in agreement with the results of light scattering and other techniques (Hickey et al., 2006, 2010; Hasegawa et al., 1994; Vasquez et al., 2011). The insert in Figure (2) shows the microemulsion part of the ultrasonic phase diagram for the EO system where the phase II corresponds to the existence microemulsion droplets with a compact pool of water molecules.

8. Compressibility of water in microemulsion droplets

The state of water in microemulsion droplets can be characterised by the apparent characteristics of water in microemulsion. As discussed above, the apparent adiabatic compressibility of water at a particular concentration represents the compressibility effect of addition of water to a mixture of oil and surfactant (and cosurfactant) and, therefore, is a sum of contributions of different states of water (water hydrating different atomic groups of surfactant/cosurfactant, 'free' water, if exists), and also contributions of compressibility effect of transferring of surfactant (cosurfactant) from oil to reversed micelles or to the exterior of the droplets of water at higher concentration of water. Therefore, it is also practical to consider the partial adiabatic compressibility of water, which represents the effect of addition of a small amount of water to microemulsion at a particular concentration, and therefore, at a particular microscopic state of microemulsion system. If this addition is not accompanied by a restructuring of microemulsion, it shall reflect the compressibility of water in the state, adopted by the added molecules of water at the particular concentrations of components of microemulsion. If this state is 'free' water in microemulsion droplets, e.g. the properties of water are close to or the same as of the pure water, the specific compressibility of water shall be close to the compressibility of solute particles of 'free' water added to the microemulsion. The same shall be correct for the other partial characteristics of water such as partial volume or partial concentration increment of ultrasonic velocity (Hickey et al., 2006). However, obtaining the partial compressibility of solute as a slope of compressibility vs. concentration is accompanied with higher experimental errors, when compared with apparent compressibility. Thus, the requirements of high precision of ultrasonic measurements are of special importance in this case.

Figure 3 represents the experimental results for apparent and partial compressibilities of water in three microemulsion systems, IPM, EO and AOT, plotted for microemulsion part of the phase diagram. In this part no frequency dependence of compressibility was noticed for IPM, AOT and EO systems in the covered frequency range, 2 to 12MHz. In addition, the apparent and partial compressibility of aqueous solute particles in ideal mixture with oil and surfactant calculated at thermal low frequency limit (Equations (11) and (14)) are also plotted in the Figure 3. The required physical properties of water and the oil/surfactant (and cosurfactant) mixtures are given in the Table 1. Specific apparent volume and specific concentration increment of ultrasonic velocity, a_u, are also given in the Figure together with their partial characteristics and with characteristics of ideal solute particles (Equations (14), (20) and (23)) of pure water in the emulsion are also presented in the Figures. For the plotted range of concentrations the two increments of ultrasonic velocity, a_u and a_{u^2} ,are practically coincide with each other. The data for φV in AOT microemulsion were taken from previous publications (Amararene et al, 1997). The slope of φK_{S_0} and a_u vs. concentration, required for partial characteristics, was taken as a 3 point average on the concentration dependence of φK_{S_0} and a_u .

As can be seen from the Figure 3 at low concentrations of water in microemulsion the apparent compressibilities and volumes and the partial compressibilities and volumes of water in all three microemulsion systems deviate significantly from the values calculated for ideal solute particles ('free' water). This shall be expected as an addition of water at low concentrations is accompanied by hydration of hydrophilic atomic groups of surfactant and co-surfactant and also by restructuring of surfactant aggregates in the mixture. As discussed above, the hydration effects for hydrophilic and charged atomic groups normally produce a significant reduction of compressibility of water. The difference between the measured characteristics of water in the microemulsions and the characteristics expected for ideal droplets of 'free' water decreases with concentration of water, especially in the part of the phase diagram, where a compact pool of water molecules exists. However in all three systems the measured partial compressibilities do not reach the level expected for the droplets of pure water. This indicates that the state of water in the droplets may be different to the state of free water. Some restructuring of surfactant at the water-oil interface, caused by its expanding with concentration of water may also contribute to the difference.

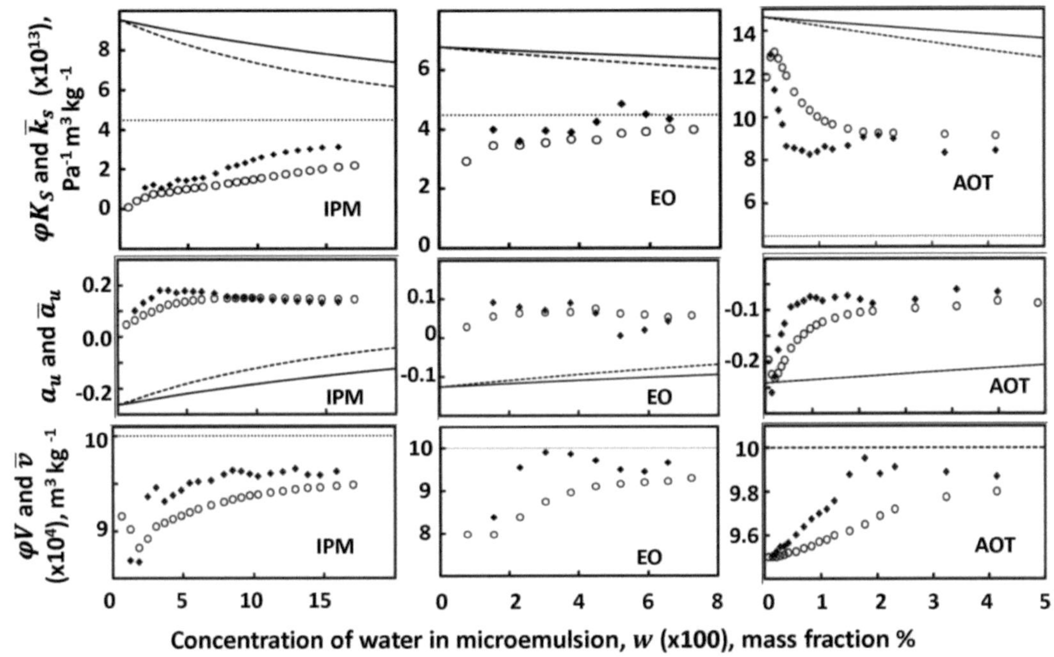

Fig. 3. Apparent (o), and partial (♦) characteristics of water at 25^{0}C in microemulsion part of phase diagrams for three w/o microemulsion systems (described in Section 6): IPM, EO and AOT (data for apparent and partial volume in AOT are taken from Amararene et al., 1997). The continuous ('-') and the dashed ('- -') lines represent the specific and the partial properties of droplets of pure water in microemulsion at thermal low frequency limit calculated according to Equations (14), (11), (23) and (20). The physical properties for these calculations are given in Table 1. The dotted lines ('…') represent the characteristics of pure water.

9. Effects of encapsulation of compressibility of globular proteins

Measurements of compressibility are often applied for characterisation of the state of solute particles in various environments and conditions. This includes characterisation of the state of solutes encapsulated in microemulsion droplets. The state of encapsulated proteins is of particular importance due to the growing interest to applications of microemulsions in various protein related technologies (bio-catalysis, protein extraction etc.). Compressibility of globular proteins, encapsulated on in w/o microemulsions, was studied previously in details (Valdez et al., 2001) for 0.1M AOT isopropanol microemulsion. The measurements were performed at one frequency, 10MHz. Figure 4(a) represents the apparent compressibilities encapsulated in AOT microemulsion for several globular proteins (Valdez et al., 2001) plotted by us as function of weight fraction of water in micoemulsion. The densities of microemulsions required for this plot were calculated using specific apparent volumes of water in 0.1M AOT w/o microemulsions published previously (Yoshimura et al., 2000; Amararene et al., 1997). Figure 4(a) shows that in microemulsion at low water content the apparent adiabatic compressibility of proteins is much higher than the compressibility of proteins in pure water. The increase of concentration of water decreases the apparent compressibility, which approaches the value for pure water at concentration of water above 0.1 (w/w). The high value of the measured apparent adiabatic compressibility of proteins at low water content and its decrease with concentration of water is predicted by the Equations (12) or (14). The contribution of heat exchange to the compressibility of solute particle, $k_{sp\,HE}$, for isooctane solvent is significantly higher than for water. Addition of water to microemulsion reduces the portion of isooctane in the mixture, thus reducing the value of $k_{sp\,HE}$. If the size of microdroplets of water corresponds to the thermal low frequency conditions (all parts of the mixture are at the same temperature at any time of compression period), this substitution shall result in the proportional to the concentration of water, w_{water}, effect on the specific thermal expansibility, $e_{0\,sp}$ and specific heat capacity, $c_{P_0\,sp}$, of the environment of the protein globular:

$$e_{0\,sp} = e_{isooct}(1 - w_{water}) + e_{water}w_{water}; \quad c_{P_0\,sp} = c_{P_0\,isooct}(1 - w_{water}) + c_{P\,water}w_{water},$$

where indexes '*isooct*' and '*water*' mark the corresponding value for isooctane (with 0.1M AOT) and for water. Using these parameters for $e_{0\,sp}$ and $c_{P_0\,sp}$ in Equation (14) we have calculated the compressibility of solute particle having physical properties of globular protein (Table 1) as function of concentration of water in microemulsion, w_{water}. The calculations were performed for infinite dilution of protein ($w_{sp} = 0$), which corresponds to concentrations used in the measurements (Valdez et al., 2001) when concentration dependence of $k^0_{S\,sp}$ is considered (Figure 1(b) or Equation (14)). The dotted line in the Figure 4(a) represents the difference between the values of $k^0_{S\,sp}$ for protein at a particular concentration of water and the value in pure water ($w_{water} = 1$). The values of e_{water} and $c_{P_0\,water}$ were taken as characteristics of pure water, which is not absolutely correct as at least part of the water in the droplets is involved in hydration of polar groups of AOT and its physical properties may deviate from the physical properties of the bulk water. However, these parameters do not affect the limiting value of $k_{HE\,sp}$ at $w_{water} = 0$ and $w_{water} = 1$, as well as the general shape of its dependence on concentration of water. According to Figure 4(a) the above calculations predict reasonably well the apparent compressibility of the encapsulated globular proteins at zero concentration of water, as well as the observed decrease of compressibility with concentration of water. However, the exact functional dependence of the apparent compressibility on the concentration of water is different to the experimental one.

One of the possible explanations of lower than expected compressibility of encapsulated protein could be an effect of confine environment on the mechanical properties of protein globule, which can restrict the mobility of protein segments, thus making it more rigid (F. N. Kolisis & Stamatis, 2010). This will reduce the values of $k_{T\,sp}$ and $k_{S\,sp}$ in Equations (12) and (14) therefore the total compressibility of protein measured in microemulsion, $k_{S\,sp}$.

Another possible explanation can be presented when the structure of microdroplet of water containing a protein molecule is considered. Figure 4(a) shows that the drop of the measured compressibility occurs at much lower than predicted concentrations of water. This indicates that some amount of isooctane could be excluded from thermal exchange with the protein globules. This could be expected when the structure of microemulsion and the physical properties of isooctane are examined. The skin depth, δ_T, for isooctane is shorter than for other common solvents and at 10MHz is about 45 nm. This is comparable with the

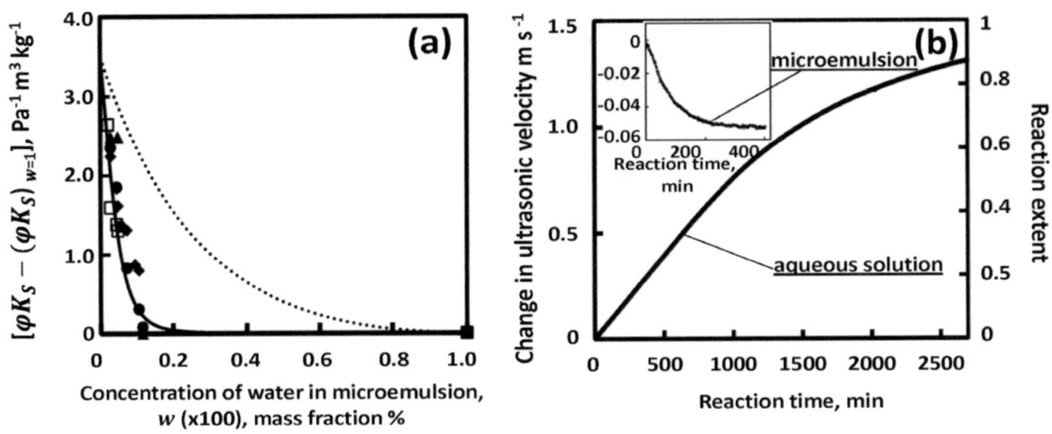

Fig. 4. **(a)**: Effect of concentration of water on specific apparent adiabatic compressibility of proteins (cytochrome c (□), MBP(▲), beta-lactoglobulin (○), lysozyme (♦)) encapsulated in AOT microemulsion. Compressibilities of proteins at 100% of water were subtracted. The lines represent predictions described in Chapter 9. **(b)**: Ultrasonic velocity time profile for hydrolysis of cellobiose by beta-glucosidase in aqueous solution (main chart) and in IPM microemulsion (50:50 w/w ratio of oil to surfactant + cosurfactant) at 25°C as described in Chapter 10.

size of droplets of water formed in 0.1M AOT isooctane microemulsion without protein, 5nm at concentration 0.075 (30 molecules of water per one molecule of surfactant) and larger at higher concentrations of water (van Dijk et al., 1989; Amararene et al., 2000). We could expect that a presence of protein molecules increases the size of the droplets. Thus, the conditions for thermal low frequency limit may not be held for this multilayer system (protein surrounded by water, surrounded by AOT and isooctane). In this case only a part of isopropanol in the mixture will be involved in the heat exchange with the protein globular thus increasing the 'effective' concentration (w/w) of water within the heat exchange volume. The accurate estimations of the effect of the heat exchange on the compressibility of the protein in this multilayer system can be done only when considering dynamics of heat exchange in a course of periodic compression within this system, which is outside of the scope of this paper. However, simplified modelling can be made to relate the effects of the heat exchange with the measured compressibility of proteins in AOT microemulsion. We can suggest that the environment of the protein globule, involved in the heat exchange, is a layer of water around protein with the thickness $r - r_p$ (r is the radius of the droplet and r_p is the radius of molecule of protein) and the layer of isooctane and surfactant of 'effective' thickness δ. The value of δ can be introduced as a thickness of solvent around the droplet which provides same as measured compressibility of protein particle if: (1) the volume comprised of the protein + water + δ layer of solvent is adiabatically insulated from the rest of the mixture and (2) a complete heat exchange between its parts occurs (same temperature for protein water and δ layer of solvent) during the course of compression. For this case the concentration of water in Equation (14) shall be taken as weight fraction of water within the heat exchange volume: protein + droplet + δ

layer of solvent, $w_{HE\,water}$. It can be estimated as: $w_{HE\,water} = \dfrac{(r^3 - r_p^{\,3})\rho_{water}}{((r+\delta)^3 - r^3)\rho_{iso} + (r^3 - r_p^{\,3})\rho_{water}}$.
If we assume that the increase in concentration of water does not affect the number of droplets but results in increase of their size only and that each of the droplets contains one protein molecule, the radius of the droplet can be linked with concentration of water in the mixture as: $r = \sqrt[3]{r_p^{\,3} + (r_0^{\,3} - r_p^{\,3})\dfrac{w_{water}(1 - w_{water\,0})}{w_{water\,0}(1 - w_{water})}}$, where r_0 the size of the droplets at concentration of water $w_{water\,0}$. For a given r_p and r_0 at $w_{water\,0}$ the value of $w_{HE\,water}$ can be calculated and used with Equation (14) to obtain the compressibility of protein as function of concentration of water in the system for any given value of δ. The value of δ can be varied to obtain the best fitting of calculated and experimental compressibilities. For r_p = 2nm (variation of this value by 100% has a minor effect on the results) and r_0 = 7nm (5nm of thickness of layer of water) at w_0 =0.075 the value of δ is 20nm. This is a reasonable estimate for the 'effective' thickness of the layer of solvent around the water droplets participating in the heat exchange. The continuous line in the Figure 4(a) represents the predicted dependence of apparent compressibility of proteins on concentration of water in microemulsion for the above parameters. The predicted dependence describes well the experimental data. An increase in the radius of droplet, r_0, will result in higher values for δ corresponding to the experiments data. Overall, the above estimations demonstrate an importance of examination of the heat exchange between the solvent and the interior of nano-particles when describing the effects of encapsulation on the measured compressibility of the encapsulated material. Detailed analysis of intrinsic compressibility of encapsulated material in AOT w/o microemulsions measured at one frequency requires knowledge of size and, perhaps, other the structural attributes of the droplets. Measurements of compressibility at different frequencies, combined with the measurements of ultrasonic attenuation, have a potential of providing the required information.

10. Ultrasonic monitoring of enzyme reactions in microemulsions

Previously it was shown that high-resolution ultrasonic spectroscopy can be applied for real-time monitoring of enzyme reactions in a wide range of concentrations and substrates. The technique does not require optical transparency, optical markers or secondary reactions and allows real-time non-invasive measurements. The capability of this and other ultrasonic techniques were demonstrated for a variety of reactions catalysed by enzymes, namely, urease, α-amylase and catalase (Kudryashov et. al, 2003), chymosin (Dwyer et al., 2005) proteinase K (Buckin & Craig, 2005), invertase (Resa et al, 2009), and cellobiase (Resa & Buckin, 2011) as well as other types of bioprocesses (McClements, 1995; Povey, 1997). It provided the detailed reaction extent and rate profiles in a broad range of concentrations, allowed analysis of the reverse reaction, determination of equilibrium constant and of the molar Gibbs free energy of reactions as well as analysis of reactions kinetic models (Resa & Buckin, 2011). Figure 4(b) illustrates our results for the real-time ultrasonic monitoring of activity enzyme, β-glucosidase, in aqueous solution (10 mM sodium acetate buffer, pH 4.90) and encapsulated in IPM microemulsion (50:50 w/w ratio of oil to surfactant+cosurfactant) at 24% water in the presence of substrate, cellobiose (disaccharide). The concentration of water in microemulsion corresponds to the part of ultrasonic microemulsion phase diagram, where a presence of water droplets with significant size of the water pool was identified previously (Hickey et al., 2006). This provides some hydration level for encapsulated molecules of enzyme and the substrate. The reaction was started by adding of concentrated solution of enzyme to microemulsion or to the aqueous solution with the substrate. The concentrations of the enzyme and the substrate in aqueous phase of microemulsion were the same as in the aqueous solution. The time profiles for ultrasonic velocity are presented in the Figure 4(b) (main frame - aqueous solution, insert - microemulsion).

In aqueous solutions at our concentrations β-glucosidase provides hydrolysis of cellobiose, during which β-1,4-glycosidic bond is cleaved, one molecule of water is absorbed, and two D-glucose molecules are released. This causes a change in the compressibility and density of the solution, thus resulting in a change of ultrasonic velocity. The ultrasonic reaction time profiles can be recalculated into the reaction extent profile, $\zeta(t) = \frac{w_C^0 - w_C(t)}{w_C^0}$, where w_C^0 and $w_C(t)$ are the concentrations of cellobiose at time zero and at time t. This requires the concentration increment of ultrasonic velocity of the reaction, Δa_u, which represents the relative change of ultrasonic velocity caused by conversion of the reactants into products, calculated per unit of concentration of the reactant (cellobiose) converted (Resa & Buckin, 2011). It is linked with the reaction extent as: $\zeta(t) = \frac{u(t) - u^0}{u_0 \Delta a_u w_C^0}$, where $u(t)$ and u^0 are the ultrasonic velocities at reaction time t and time zero, and u_0 is the ultrasonic velocity in the solvent (solution or microemulsion without the components of the reaction). For hydrolysis of cellobiose $\Delta a_u = a_{uG}\frac{2M_G}{M_C} - a_{uC} - a_{uW}\frac{M_W}{M_C}$, where M represents the molar mass of a particular component and symbols G, W and C stand for glucose, water and cellbiose (Resa & Buckin, 2011). The value of a_{uW} is the concentration increment of ultrasonic velocity of water added to microemulsion in the amount consumed by the reaction. For our conditions in microemulsion this addition corresponds to the change in concentration of water by 0.03% of total 24% and therefore a_{uW} represents the partial concentration increment of water, $\bar{a}_u$. The same is true for the reaction in aqueous solution. We have measured the values of a_{uG}, a_{uW} and a_C in our aqueous solution for concentrations utilised in our reaction. This provided $\Delta a_u = 0.041$, which is close to value previously obtained for 50°C (Resa & Buckin, 2011). The reaction extent, $\zeta(t)$, for reaction in aqueous solution is represented by the right Y scale of the Figure 4(b).

In contrary to the aqueous solution, the reaction in microemulsion is accompanied by a decrease in ultrasonic velocity, however on a significantly smaller scale. Equation (25) can be used to recalculate the value of Δa_u for the aqueous solution to the value of Δa_u in IPM system if the values of φV, φE, φC_P and the physical properties of the continuous medium are known. We have used the data of (Banipal et al., 1997) on the apparent heat capacities at 25°C and the temperature slopes of the apparent volumes of cellobiose and glucose in water at infinite dilution and also the value of a_{uW} in microemulsion obtained in our titration measurements to estimate Δa_u for hydrolysis of cellobiose in IPM microemulsion (the physical properties if IPM system were taken from Hickey et.al., 2006). These estimations provided positive values of Δa_u. Our direct measurements of the difference in ultrasonic

Physical properties at 25°C	Water [(a)]	EO system[(b)] (80:20 w/w oil to surfactant ratio)	IPM system[(c)] (60:40w/w oil to surfactant and cosurfactant ratio)	0.1M AOT in Isooctane	Isooctane	Protein
Specific adiabatic compressibility (x10^{13}), $m^3\,kg^{-1}\,Pa^{-1}$	4.491	6.385	7.655	16.65[(d)]	18.12[(d)]	1.655[(f)]
Specific volume (x10^3), m^3kg^{-1}	1.003	1.109	1.158	1.437[(d)]	1.457[(d)]	0.662[(f)]
Thermal conductivity, $W\,m^{-1}\,K^{-1}$	0.5952	0.12	0.19	0.095[(e)]	0.095[(e)]	-
Specific thermal expansibility (×10^6) $m^3kg^{-1}\,K^{-1}$	0.257	0.827	0.982	1.97[(d)]	2.076	0.07[(f)]
Specific heat capacity, $J\,kg^{-1}\,K^{-1}$	4180	1686	1405	2037[(e)]	2037[(e)]	1403[(g)]
Viscosity, Pa s(×10^3)	0.890	17.2	8.6	0.516 [(d)]	0.479 [(d)]	-

[(a)] Handbook of Chemistry and Physics, 2004-2005; [(b)] Calculated from (Hickey et al., 2010) and (Pratas et al., 2010); [(c)] Calculated from (Hickey et al., 2006); [(d)] (Go´mez-Dı´az et al., 2006); [(e)] value for isooctane Handbook of Chemistry and Physics 2004-2005; [(f)] Taken as intrinsic characteristics of globular proteins (Chalikian, 2003); [g)] Calculated from (Privalov & Dragan, 2007) as average value for three globular proteins: cytochrome C, myoglobin and lysozyme.

Table 1. Physical properties of the media used in calculations.

velocity between microemulsion containing cellobiose (at the its initial concentration in our reaction) and glucose (at concentration corresponding 100% hydrolysis of cellobiose) as well as our data for a_{uW} in microemulsion provided $\Delta a_u \cong 0.005$. This corresponds to 0.04 m/s increase of ultrasonic velocity for 100% of hydrolysis of cellobiose in our microemulsion. The positive sign of Δa_u contradicts to the observed decrease of ultrasonic velocity in the reaction. Two possible explanation of this result can be suggested. 1) The physical state of cellobiose and glucose in microemulsions containing both of these molecules and the enzyme is significantly different than in microemulsions containing glucose and celobiose separately. It may results in the different sign for Δa_u at intermediate degrees of hydrolysis of cellobiose. This option can be verified or excluded through the measurements of ultrasonic velocity in microemulsions containing both components of the reaction. 2) The reaction observed in microemulsion is not the hydrolysis of cellobiose. We can speculate that an alternative reaction is related to transglycosylation activity of some of β-glucosidases resulting in production of tri- and tetrasaccharides. The activity can be enhanced by immobilisation and limited hydration (example: Chen & Ou-Yang, 2004; Marico et al., 2001), which are well known factors affecting functioning of enzymes in microemulsions (Kolisis & Stamatis, 2010). This could explain the observed decrease in ultrasonic velocity. Different energetics (free energy) of reactions of synthesis and hydrolysis of oligosaccharides in microemulsion as compared with aqueous solution can be another contributor. As can be seen from the Figure 4(b) the reaction time in the microemulsion is significantly shorter than in the aqueous solution. The reaction did not produce any substantial change in ultrasonic attenuation; neither a dispersion of velocity in the measured frequency range, 2 to 12 MHz, thus demonstrating that the structure of the microemulsion was not affected by the reaction.

11. Conclusions

High-resolution measurements of ultrasonic velocity and attenuation and their dependence on frequency allow analysis of a range of characteristics of microemulsion systems. This includes determination of phase diagrams, characterisation of size and state of microemulsion droplets and compounds encapsulated in them as well as real time monitoring of chemical reactions curried in microemulsions. Titration measurements with this technique allow identification of different 'sub-states' of microemulsion components, which can be utilised for upgrading the microemulsion phase diagram with additional 'sub phases' corresponding to different states of water and surfactant in microemulsions. Consideration of these 'sub phases' and their position on the phase diagram could be employed in design of w/o microemulsions designated for encapsulation of active ingredients (e.g. proteins), which require an appropriate level of hydration for their functioning. Quantitative analysis of compressibility of microemulsions and its molecular interpretation in most cases can be performed within the thermal low frequency limit approach, which provides direct thermodynamic relationships for concentration dependence of compressibility, and effects of continuous phase on compressibilities of components of microemulsions.

Acknowledgments: This research was supported by the MASAF315 grant from SFI Ireland. We are thankful to Evgeny Kudryashov for collaboration during this project and to Margarida Altas for her assistance in some measurements and preparation of manuscript.

12. References

Alany, R.G., Tucker, I.G., Davies, N.M., Rades, T. (2001). Characterising colloidal structures of pseudoternary phase diagrams formed by oil/water/amphiphile systems. *Drug Dev. Ind. Pharm.*, Vol.27, pp. 31-38.

Allegra, J.R., Hawley, S.A. (1972). Attenuation of sound in suspensions and emulsions: *theory and experiment. J. Acoust. Soc. Am.*, Vol.51, pp. 1545–1564

Amararene, A., Gindre, M., Le Hue´rou, J.Y., Nicot, C., Urbach, W., and Waks, M. (1997). Water Confined in Reverse Micelles: Acoustic and Densimetric Studies. *J. Phys. Chem. B*, Vol. 101, pp.10751-10756

Amararene, A., Gindre, M., Le Hue´rou, J.Y., Urbach, W., Valdez, D., & Waks, M. (2000). Adiabatic compressibility of AOT, sodium bis 2-ethylhexyl sulfosuccinate reverse micelles: Analysis of a simple model based on micellar size and volumetric measurements. *Physical review E.*, Vol. 61, No. 1 pp. 682-689

Austin, J.C., Holmes, A.K., Tebbutt, J.S., Challis, R.E. (1996). Ultrasonic wave propagation in colloid suspensions and emulsions: recent experimental results. *Ultrasonics,* Vol.34, pp. 369–374

Ballaro, S., Mallamace, F., Wanderlingh, F. (1980). Sound velocity and absorption inmicroemulsion. *Phys. Lett.*, Vol.77A, pp. 198–202

Banipal, P. K. , Banipal, T. S., Larkb B. S. and Ahluwaliaa, J. C. (1997). Partial molar heat capacities and volumes of some mono-, di- and tri-saccharides in water at 298.15, 308.15 and 318.15 K. *J. Chem. Soc., Faraday Trans.*, Vol.93, No.1, pp. 81-87

Buckin, V. (1988). Hydration of nucleic bases in dilute aqueous solutions. Apparent molar, adiabatic and isothermal compressibilities, apparent molar volumes and their temperature slopes at 25 ◦C. *Biophys. Chem.*, Vol.29, pp. 283–292

Buckin ,V., Smyth, C. (1999). High-resolution ultrasonic resonator measurements for analysis of liquids, *Sem. Food Anal.*, Vol.4, pp. 113–130

Buckin, V., Kudryashov, E. (2001). Ultrasonic shear wave rheology of weak particle gels, *Advances in Colloid and Interface Science,* Vol.89-90, pp. 401-422

Buckin, V., O'Driscoll B. (2002). Ultrasonic waves and material analysis: recent advances and future trends. *Lab. Plus Int.*, Vol.16 (3), pp. 17–21

Buckin, V. & Craig, E. (2005). New applications of high resolution ultrasonic spectroscopy for real time analysis of enzymatic proteolysis, *European BioPharma Review (Summer),* pp. 50 - 53

Cao Y.N., Diebold G.J., Zimmt M.B. (1997). Transient grating studies of ultrasonic attenuation in reverse micellar solutions, *Chemical Physics Letters,* Vol.276, p.p. 388-392

Chalikian, T. (2003). Volumetric Properties of Proteins. *Ann. Rev. Biophys. Biomol. Struct,* Vol. 32, pp.207–35

Chen, C.W. & Chao-Chih Ou-Yang, C.-C. (2004). Bounded water kinetic model of beta-galactosidase in reverse micelles. *Bioprocess Biosyst. Eng.* Vol. 26, pp. 307–313

Choi, Y. and Okos, M.R. (1986). Effects of Temperature and Composition on the Thermal Properties of Foods . *Journal of Food Process and Applications,*Vol. 1(1). pp. 93–101

Coupland, J.N., McClements, J.D., (2001). Droplet size determination in food emulsions: comparison of ultrasonic and light scattering methods. *J. Food Eng.*, Vol.50, pp.117–120

Dwyer, C., Donnelly, L., Buckin V. (2005). Ultrasonic analysis of rennet-induced pre-gelation and gelation processes in milk, *J. Dairy Res.* Vol.72, pp. 303–310

Epstein, P.S., Carhart, R.J. (1953). The absorption of sound in suspensions and emulsions.I. Water fog in air. *J. Acoust. Soc. Am.* Vol.25, pp.553–562

Go´mez-Dı´az, Mejuto D., and NavazaJ. M. (2006). Density, Viscosity, and Speed of Sound of Solutions of AOT Reverse Micelles in2,2,4-Trimethylpentane, *J. Chem. Eng. Data*, Vol.51, pp. 409-411

Handbook of Chemistry and physics, 85th edition *CRC press*, 2004-2005

Hasegawa, M., Sugimura,T., Suzaki, Y., and Yoichi Shindo, Y. (1994). Microviscosity in Water Pool of Aerosol-OT Reversed Micelle Determined with Viscosity-Sensitive Fluorescenck Probe, Auramine 0, and Fluorescence Depolarization of Xanthene Dyes. *J. Phys. Chem.* Vol. 98, pp. 2120-2124

Hemar, Y., Hocquart, R. And Palierne, J. F. (1998). Frequency-dependent compressibility in emulsions: Probing interfaces using Isakovich's sound absorption, *Europhys. Lett.*, Vol.42 (3), pp. 253-258

Hickey, S., Lawrence, M.J., Hagan, S.A., Buckin, V. (2006). Analysis of the Phase Diagram and Microstructural Transitions in Phospholipid Microemulsion Systems using High-Resolution Ultrasonic Spectroscopy, *Langmuir*, Vol.22, pp. 5575-5583

Hickey, S., Lawrence, M. J., Hagan, S.A., Kudryashov, E., Buckin, V. (2010). Analysis of Phase Diagram and Microstructural Transitions in an Ethyl Oleate/Water/Tween 80/Span 20 Microemulsion System using High-Resolution Ultrasonic Spectroscopy, *International Journal of Pharmaceutics* Vol. 388, pp. 213–222

Isakovich M. A., Zh. (1948). *Eksp. Teor.Fiz.*, Vol. 18, pp. 907-912

Kaatze, U., Hushcha, T. O. & Eggers, F. (2000). Ultrasonic Broadband Spectrometry of Liquids: A Research Tool in Pure and Applied Chemistry and Chemical Physics, *J. Solution Chemistry,* Vol. 29, No. 4, pp. 299-368

Kolisis, F. N. & Stamatis, H. (2010). REVERCE MICELLES, ENZYMES, *Encyclopedia of Industrial Biotechnology: Bioprocess, Bioseparation, and Cell Technology,* ed M. C. Flickinger, John Wiley & Sons, Inc.

Kudryashov E., Smyth C., O'Driscoll B., Buckin, V. (2003). High-Resolution Ultrasonic Spectroscopy for analysis of chemical reactions in real time, *Spectroscopy* Vol.18 (10), pp. 26–32

Lang, J., Djavanbakht, A., Zana, R. (1980). Ultrasonic absorption study of microemulsions in ternary and pseudoternarnary systems. *J. Phys. Chem.* Vol. 84, pp.1541–1547

Letamendia, L., Pru-Lestret, E., Panizza, P., Rough, J., Scioritino, F., Tartaglia, P., Hashimoto, C., Ushiki, H., Risso, D. (2001). Relaxation phenomena in AOT–water–decane critical and dense microemulsions. *Physica A*,Vol.300, pp. 53–81

Litovitz, T.A. and Davis, C.M. (1964). Structural and Shear relaxation in Liquids. In *Physical Acoustic*, Vol.2, Part A, Ed. W.P. Mason, pp. 282-349, Academic Press, NY

Title;Effective Transglycosylation catalyzed by a Lipid-coated Glycosidase in Water-containing Supercritical Fluoroform.

Marico, F., Toshiaki, M., & Yoshio, O. (2001). Effective Transglycosylation catalyzed by a Lipid-coated Glycosidase in Water-containing Supercritical Fluoroform. *Nippon Kagakkai Koen Yokoshu*, Vol.79, No.2, pp.931, ISSN:0285-7626

McClements, (1995). Advances in the application of ultrasound in food analysis and processing. *Trends in Food Science and Technology*, Vol.6 pp. 293-299

Mehta, S.K., Kawaljit. (1998). Isentropic compressibility and transport properties of CTAB–alkanol–hydrocarbon–water microemulsion systems. *Colloids Surf. A*, Vol.136, pp. 35–41

Morse, P.M. & Ingard, K.U. (1986). *Theoretical Acoustic*, Princeton University Press

Pierce, A. D. (1991). *Acoustics*, Acoustical Society of America, NY.

Povey M.J. (1997). Ultrasonic Techniques for Fluid Characterisation, *Academic Press, London*

Pratas, M. J., Freitas S., Oliveira, M. B., Monteiro, S. C., Lima, A.S., and Coutinho, J. A. P. (2010). Densities and Viscosities of Fatty Acid Methyl and Ethyl Esters. *J. Chem. Eng. Data*, Vol.55, pp. 3983–3990

Privalov, P. L. & Dragan, A.I. (2007). Microcalorimetry of biological macromolecules. *Biophysical Chemistry*, Vol.126, pp. 16–24

Resa, P., Elvira, L., Sierra, C., Montero de Espinosa, F. (2009). Ultrasonic velocity assay of extracellular invertase in living yeasts. *Anal.Biochem.*, Vol.384, pp. 68–73

Resa P. & Buckin V. (2011). Ultrasonic analysis of kinetic mechanism of hydrolysis of cellobiose by beta-glucosidase. *Analytical Biochemistry*, Vol.415, pp. 1–11

Smyth, C., Kudryashov, E., Buckin, V. (2001). High-frequency shear and volume viscoelastic moduli of casein particle gel. *Colloids and Surfaces, A: Physicochemical and Engineering Aspects*, Vol.183–185, pp. 517–526

Spehr, T.L. (2010). Water Dynamics in Soft Confinement-Neutron Scattering Investigation of Reverse Micelles. *Technical University of Darmstadt*. PhD. Thesis, Darmstadt, D17

Thurston, R.N. (1964). *Physical Acoustics: Principles and Methods*, Academic Press, New York, Mason, W.P. (Eds.), vol.1, part A, ch. 1

Valdez, D., Le Hue´ rou, J Y., Gindre, M., Urbach, W., and Waks, M. (2001). Hydration and Protein Folding in Water and in Reverse Micelles: Compressibility and Volume Changes *Biophysical Journal* , Vol.80, pp.2751–2760

Van Dijk, M., Joosten, J., Levine ,Y., and Bedeaux, D. (1989). *J. Phys. Chem.* Vol.93, pp. 2506-2512

Vasquez, V. R., Williams, B. C., & Graeve, O. A. (2011). Stability and Comparative Analysis of AOT/Water/Isooctane Reverse Micelle System Using Dynamic Light Scattering and Molecular Dynamics. *J. Phys. Chem. B*. Vol. 115, pp. 2979–2987

Wines, T.H., Dukhin, A., Somasundaran, P. (1999). Acoustic spectroscopy for characterizing heptane/H_2O/AOT reverse microemulsions. *J. Colloid Interface Sci.*, Vol.216, pp.303–308

Waterman, P.C., Truell, R. (1961). Multiple scattering of waves. *J. Math. Phys.*, Vol.2, pp. 512–537

Yoshimura, Y., Abe, I., Ueda, M., Kajiwara K., Hori T., Schelly, Z. (2000). Apparent Molar Volume of Solubilized Water in AOT/Isooctane/Water Reverse Micellar Aggregates. *Langmuir*, Vol.16, pp.3633-3635

Zana, R., Lang, J., Sorba, O., Cazabat , A.M., Langevin, D. (1982). Ultrasonic investigation of critical behaviour and percolation phenomena in microemulsions. *Journal de Physique Lettres*, Vol.43, pp. 829-837

Influence of Linear Aliphatic Alcohols upon the Electric Percolation of AOT-Based Microemulsions

A. Cid[1,2], J.A. Manso[2], J.C. Mejuto[2] and O.A. Moldes[2]
[1]*Chemistry Department, REQUIMTE-CQFB, Faculty of Science and Technology University Nova of Lisbon, Monte de Caparica*
[2]*Department of Physical Chemistry. Faculty of Science. University of Vigo, Ourense*
[1]*Portugal*
[2]*Spain*

1. Introduction

Microemulsions are dynamic structures described as spherical droplets of a dispersed phase (water) into a continuous phase (isooctane) and stabilized by a surfactant (AOT) (see Scheme 1). The components organise themselves in time and space by means of different interactions or collisions, giving rise to coalescence and redispersion processes.

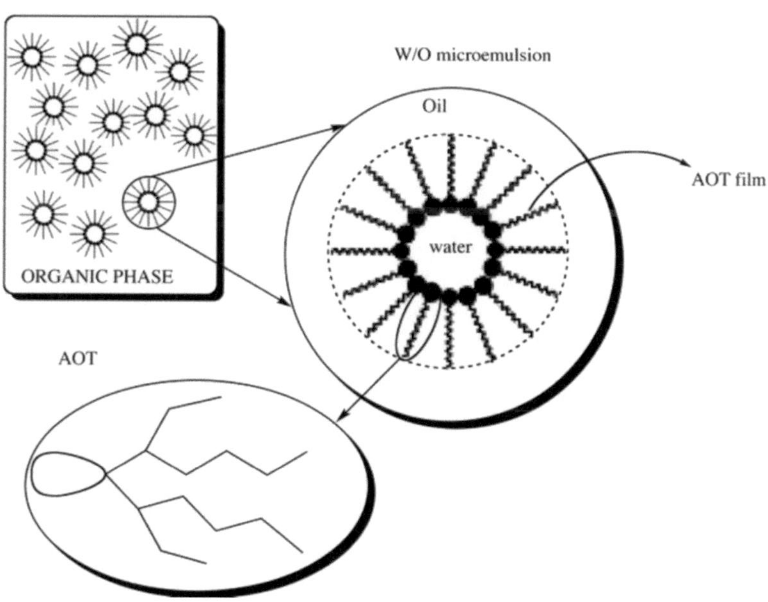

Scheme 1.

In normal conditions, they exhibit a low conductivity (0.001-0.1 $mScm^{-1}$) which is bigger than the corresponding one to the pure isooctane (~10^{-8} $mScm^{-1}$) (Lang et al., 1988; Eastoe et al., 1991). This differential behaviour is justified by the fact that microemulsions are capable of carrying electric charge, and in this sense, the transport capacity is associated with a phenomenon known as electrical percolation, which is an increase in conductivity at a given value of the volume fraction of the dispersed phase, or at a certain temperature, when holding the composition of the microemulsion constant (Mathew et al., 1988; Maitra et al., 1990). This increase is associated with an increase in the electric charge flow between the discrete droplets forming the dispersed phase. Matter exchange rate between droplets increases during percolation phenomena, both ionic and neutral microemulsions; however, in all cases this rate is always lower than the diffusion control limit (Atik & Thomas, 1981; Jada et al., 1989; Jada et al., 1990; Lang et al., 1990).

The mechanism proposed to explain the electric percolation phenomenon is based on the formation of channels exchanging matter between the disperse water droplets in the continuous phase. It is then necessary to have an effective collision between at least two water droplets of the microemulsion, causing the droplets to fuse. Subsequently, an exchange of matter between the water droplets would take place (allowing the charge conduction), which would produce their separation by means of a process of fission (see scheme 2).

Scheme 2.

Moderate or low concentrations of additives have a significant effect on the percolation threshold (García-Río, L. 1994). In the literature there are a large amount of studies about the influence of different additives in general (García-Río et al., 1997; Moulik et al., 1999; Hait et al., 2001; Hait et al., 2002a; Dasilva-Carvalhal et al., 2003; Dasilva-Carvalhal et al., 2005; Dasilva-Carvalhal et al., 2006; Arias-Barros et al., 2010) and, in particular, amphiphilic additives upon the percolation phenomena and the internal dynamics of microemulsions (Nazario et al., 1996; Nazario et al., 2000; García-Río et al., 2000; García-Río et al., 2005; Cid-Samamed et al., 2008). The additives that increase the "rigidity" of the membranes increase the value of the percolation threshold of a microemulsion, (Ray et al., 1993), while those that make them more flexible favour the process (Mathew et al., 1988; Ray et al., 1993). In this sense, the addition of cationic surfactants to AOT-based microemulsions causes an increase in the percolation temperature while the addition of nonconjugated bile hydroxy salts has the opposite effect (Ray et al., 1993).

Regarding the influence of amphiphilic additives upon the electrical percolation of AOT-based microemulsions, previous studies in our laboratory have examined the influence of n-alkyl amines (García-Río et al., 2000) which exhibit a linear dependence of the percolation temperature with the amine chain length, and therefore with the hydrophobicity of the amines. However, the influence of sodium alkyl sulfonates on the electrical percolation of AOT-based microemulsions exhibit a bimodal behaviour according to which a group of alkyl sulfonates increase the percolation temperature (C_3-C_5), only to subsequently decrease it (C_6-C_{18}) (García-Río et al., 2005)

In the present chapter, the analysis of the influence of 1-n-alcohols and 2-n-alcohols upon the electric percolation of AOT/isooctane/water microemulsions is reported.

2. Experimental procedure

2.1 Materials and methods

AOT was supplied by Aldrich (purity 98%), prior to use it was kept in a vacuum desiccator for two days to minimize the water content due to its hygroscopic nature. It was used without any further manipulation. Fluka and Sigma-Aldrich supplied additives, with highest purity commercially available (between 97-99%) were employed.

Microemulsions were made "*in situ*": once microemulsion was prepared by mixing of its three components (water, isooctane, AOT), an adequate amount of organic additive was added (determined by weighing), to obtain the desired concentration of additive. The solution was stirred and heated to achieve a homogenous and isotropic sample. Then the conductivity of samples were measured as a function of temperature. In this study, composition of microemulsions was kept constant, being the values of [AOT] equal to 0.5 M (referred to total microemulsion volume), and W=[AOT]/[H_2O]=22.2. Water used for microemulsions was distilled-deionized, with a conductivity value around 0.10-0.15 $\mu S \cdot cm^{-1}$.

Microemulsions were introduced in 50 mL containers and were properly sealed up for thermostatization. A Teflon bar with magnetic core was used for stirring of the system to maintain a homogenous medium; the system was properly sealed with a lid with two openings, one of them to introduce an electrode and the other one to insert a thermometer to determine the electric conductivity and the temperature of the sample in each moment. The cell with the sample was immersed in a mixed ethanol-water bath, and temperature was measured simultaneously with electrical conductivity. A scheme of the experimental system is shown in the Figure 1.

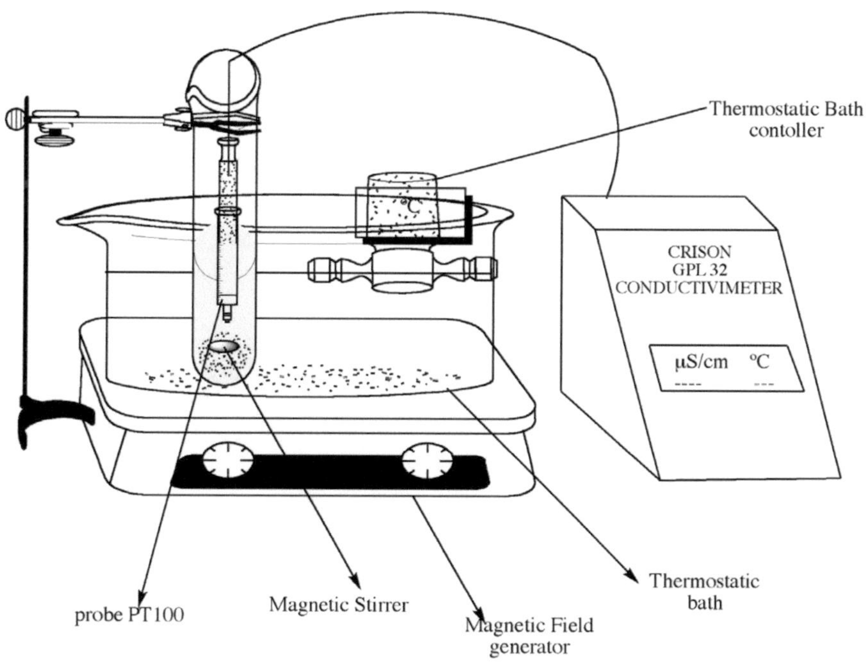

Fig. 1. Experimental equipment

The electrical conductivity was measured using a Crison GPL 32 conductiometer with a cell constant of 0.0109 cm^{-1}. The conductiometer was calibrated using standard solutions of KCl supplied by the apparatus manufacturer with following concentration and specific conductivity: [KCl]=0.0100 mol dm^{-3}, $\kappa=1.143\times10^{-3}\Omega^{-1}cm^{-1}$ (κ=1143μScm^{-1}) at 25 °C and [KCl]=0.100mol dm^{-3}, $\kappa=1.288\times10^{-2}$ $\Omega^{-1}cm^{-1}$(κ=12.88 mScm^{-1}) at 25 °C. Conductivity measurements were carried out using a thermostatic-cryostatic Teche TE-8D RB-5 to control the medium temperature, achieving an accuracy of ±0.1 °C.

Percolation temperature has been obtained experimentally, observing the variation of conductivity with temperature (Figure 2) (Kim & Huang, 1986). Percolation threshold was obtained by analyzing the trace conductivity/temperature using the Boltzmann Sigmoidal Equation (BSE) proposed by Moulik (Hait et al., 2002b).

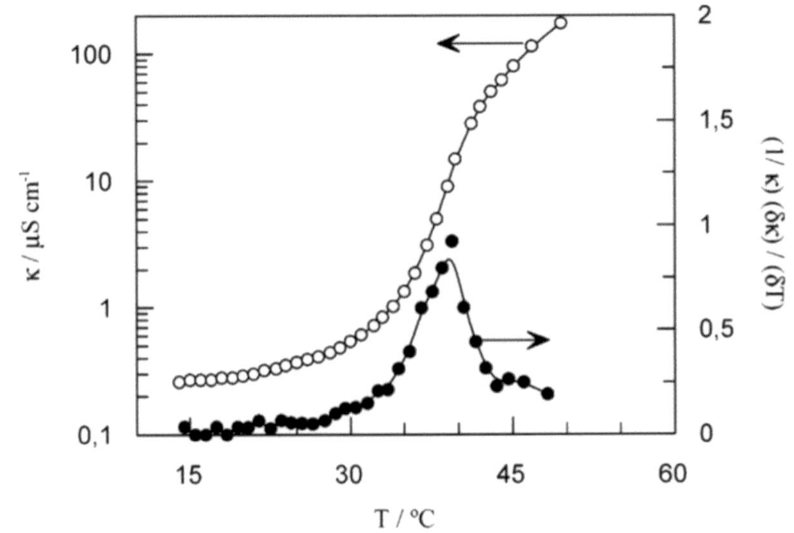

Fig. 2. (O) Determination of percolation temperature, T_p, by Moulik and Papelu procedure (Hait et al., 2002b). Solid line fits the adjustment of experimental data to SBE equation. (●) Determination of percolation temperature, T_p, by Kim and Huang procedure (Kim & Huang, 1986). In both cases adjustments correspond to w/o de AOT/iC_8/H_2O microemulsions ([AOT]=0.5M, W=22).

Determinations of maximum water capacity of microemulsion have been carried out by addition of appropriate volumes of water to sample with well-known AOT and isooctane quantities under agitation, and until permanent turbidity was observed. The samples thus prepared were stored at 25 °C.

3. Results

3.1 Geometry of additives

The geometry of different additives employed in our study were optimized by MM2 (molecular mechanics), using a commercial program, CS ChemBats3D Pro 4.0, supplied by Cambridge Soft Corporation based on QCPE 395 (Burkert & Allinger, 1982; Allinger, 1977).

Chemical structures, of 1-n-alcohols and 2-n-alcohols studied, are shown in Scheme 3. Table 1 shows the values of chain length and head group area for each additive obtained from MM2 calculations.

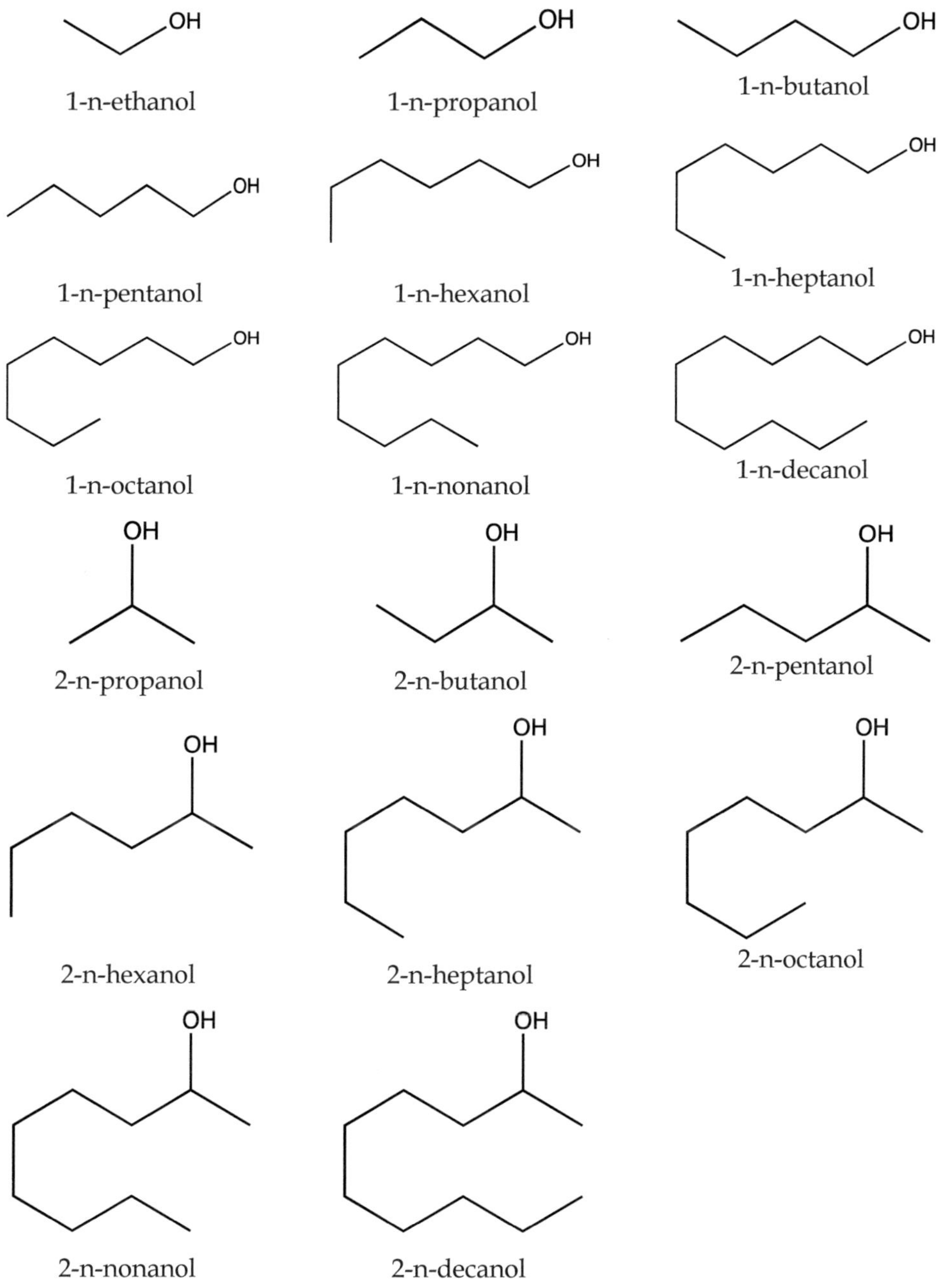

Scheme 3.

Additive	1-n-Alcohol	Number of C atoms	a_0 / Å^2	l_c/ Å
	1-n-Ethanol	$1C_2$		2.16
	1-n-Propanol	$1C_3$		3.48
	1-n-Butanol	$1C_4$		4.84
	1-n-Pentanol	$1C_5$		6.14
	1-n-Hexanol	$1C_6$	1.34	7.30
	1-n-Heptanol	$1C_7$		8.70
	1-n-Octanol	$1C_8$		9.40
	1-n-Nonanol	$1C_9$		11.20
1-n-Alkanols	1-n-Decanol	$1C_{10}$		12.35
	2-n-Propanol	$2C_3$		2.16
	2-n-Butanol	$2C_4$		3.48
	2-n-Pentanol	$2C_5$		4.84
	2-n-Hexanol	$2C_6$	1.86	6.14
	2-n-Heptanol	$2C_7$		7.30
	2-n-Octanol	$2C_8$		8.70
	2-n-Nonanol	$2C_9$		9.40
2-n-Alkanols	2-n-Decanol	$2C_{10}$		11.20

Table 1. Values of critical length of chain and head group area obtained from MM2 calculations for each of the amphiphiles used as additives.

3.2 Influence of 1-n-alcohols upon the electric percolation

The effect of several 1-n-alcohols upon percolation phenomenon has been reported. The addition of these 1-n-alcohols (1-n-ethanol [$1C_2$], 1-n-propanol [$1C_3$], 1-n-butanol [$1C_4$], 1-n-pentanol [$1C_5$], 1-n-hexanol [$1C_6$], 1-n-heptanol [$1C_7$], 1-n-octanol [$1C_8$], 1-n-nonanol [$1C_9$], 1-n-decanol [$1C_{10}$]) to AOT/alkane/water microemulsions has been studied, analyzing the effect of additives concentration on percolation temperature. Chemical structures of these compounds are shown in Scheme 3.

Concentrations used were [C_n]= 0.05, 0.1, 0.25 and 0.5 M referred all of them to the water volume of the microemulsion. In some cases the concentrations range were not completed because of the addition of this additive concentration implies microemulsions destabilization. The percolation temperature values of each specific case were obtained keeping constant the composition of the microemulsion ([AOT]=0.5M and W= 22.2) and varying the additive concentration, as detailed in the experimental procedure section.

In the following figures 3 to 6 the effect of above-mentioned concentrations of 1-n-alcohols can be observed. As we can see in Figures 3 and 4, the observed behaviour of 1-n-alcohols upon the percolation temperature presents a decrease in percolation temperature when increasing the additive concentration studied, it occurs for alcohols which the number of carbon atoms of the alcohol studied is lower than 3. By contrast, the behaviour would be reversed when the alcohol studied has a number of carbon atoms higher than 3, as we quote in Figures 5 and 6. Numerical values of percolation temperature obtained for each alcohol and each concentration are listed in Table 2.

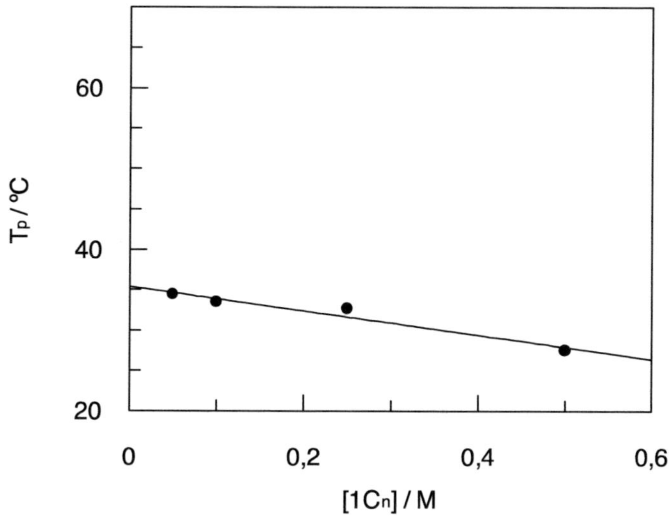

Fig. 3. Influence of $1C_2$ upon electric percolation temperature of AOT/iC_8/H_2O microemulsions

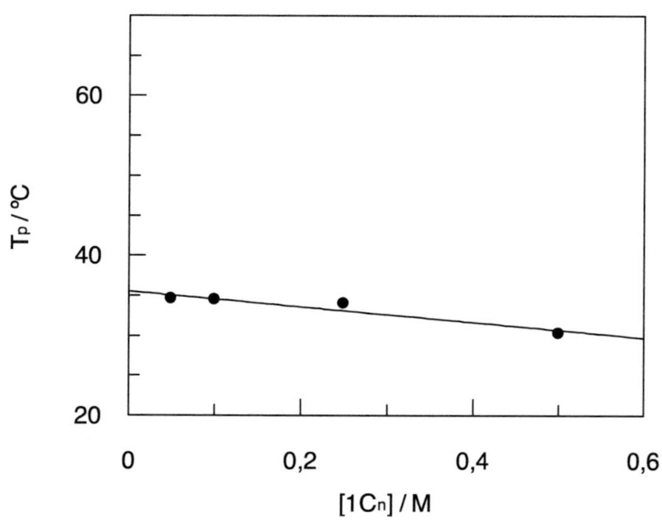

Fig. 4. Influence of $1C_5$ upon electric percolation temperature of AOT/iC_8/H_2O microemulsions with [AOT]= 0.5 M and W= 22.2

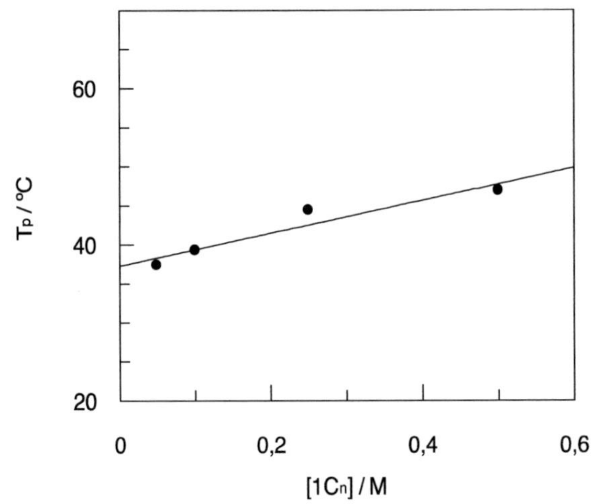

Fig. 5. Influence of $1C_7$ upon electric percolation temperature of AOT/iC_8/H_2O microemulsions with [AOT]= 0.5 M and W= 22.2

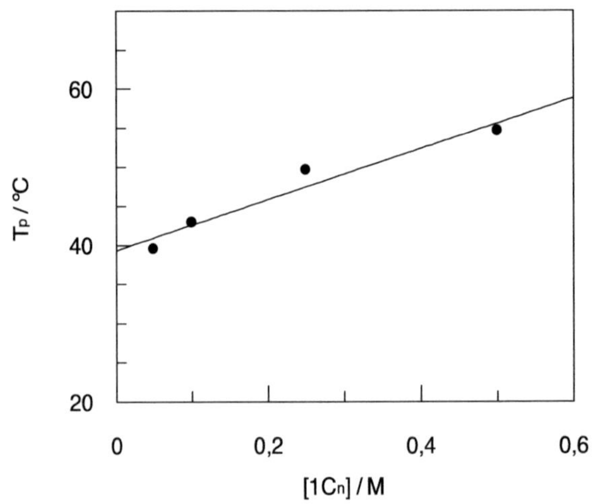

Fig. 6. Influence of $1C_{10}$ upon electric percolation temperature of AOT/iC_8/H_2O microemulsions with [AOT]= 0.5 M and W= 22.2

3.3 Influence of 2-n-alcohols upon the electric percolation

As in the previous section, the effect produced by different 2-n-alcohols (2-n-propanol [$2C_3$], 2-n-butanol [$2C_4$], 2-n-pentanol [$2C_5$], 2-n-hexanol [$2C_6$], 2-n-heptanol [$2C_7$], 2-n-octanol [$2C_8$], 2-n-nonanol [$2C_9$], 2-n-decanol [$2C_{10}$]) to AOT/alkane/water microemulsions upon percolation have been analyzed. Chemical structures of these additives are shown in Scheme 3.

In this case, additive concentrations used were [C_n]= 0.05, 0.1, 0.25 and 0.5 M referred all of them to the water volume of the microemulsion. As quote above, in some cases the concentration ranges were not completed. In this case, an increase in percolation temperature as a function of chain length of the alcohol is observed for all alcohols and all concentration ranges (Table 2).

It must be underlined that this behaviour contrasts with the one that is observed in the case of 1-n-alcohols (vide supra), where there is not a decrease of the percolation temperature with increasing of additive concentration for chain length values lower than 3.

4. Discussion

4.1 Influence of concentration upon electric percolation

In Table 2 we present the values obtained for electric percolation temperature of AOT/iC_8/water microemulsions in the presence of each alcohol. In Figures 7 and 8 the influence of the concentration of $1C_n$ and $2C_n$ on the percolation temperature AOT/iC_8/water microemulsions are depicted. A linear trend can be observed in percolation temperature with the concentration of additives such as 1-n-alcohols and 2-n-alcohols. A comparison of these two graphs allows us to report a different behaviour between 1-n and 2- n-alcohols,

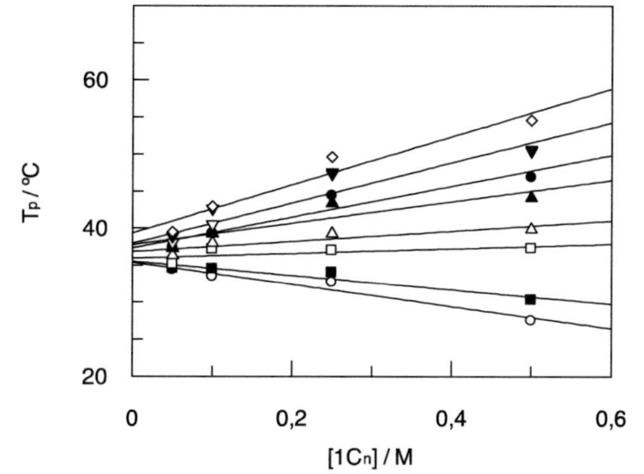

Fig. 7. Influence of the concentration of $1C_n$ on the percolation temperature AOT/iC_8/water microemulsions. [AOT] = 0.5 M, W = 22.2.

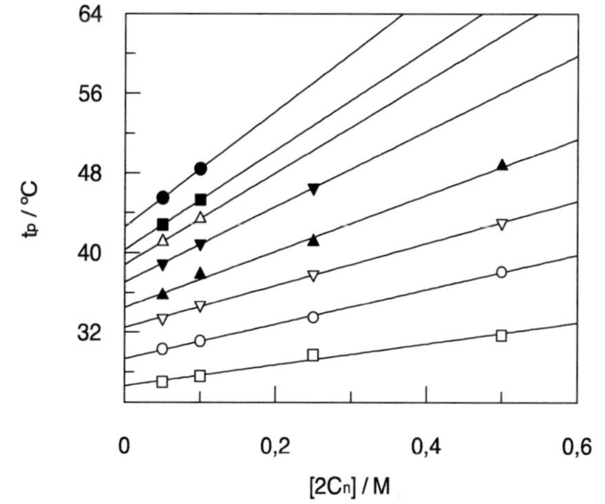

Fig. 8. Influence of the concentration of $2C_n$ on the percolation temperature AOT/iC_8/water microemulsions. [AOT] = 0.5 M, W = 22.2.

although in them, the slope of the growing dependence occurs by increasing the value of n. In both cases, as it was reported, the percolation temperature increases linearly with additives concentration. However, in the case of 1-n-alcohols, and for values of n<3 a negative slope it is observed, i.e. increasing alcohol concentration is a decrease in temperature of percolation (Figure 7). For values of n=3 the percolation temperature is almost independent of the chain length (Figure 7), ultimately for values of n>3 the behaviour is completely analogous between 1-n and 2-n-alcochols (Figure 7). In the case of 2-n-alcohols, an increase in percolation temperature as a function of chain length of the alcohol is observed at a constant additive concentration (Figure 8). This behaviour contrasts with the one that is observed in the case of 1-n-alcohols (Figure 8), where there is not a decrease of the percolation temperature with increasing of additive concentration for chain length values lower than 3. This behaviour would be associated with the polarity of the molecules and hence to the different distributions among the three different domains of the microheterogeneous system and their insertion at the interface as a co-surfactant.

$1C_n$	$[1C_n]$ / M	T_P / °C
$1C_2$	0.05	34.4
	0.10	33.5
	0.25	32.7
	0.50	27.5
$1C_3$	0.05	34.6
	0.10	34.5
	0.25	34.0
	0.50	30.3
$1C_4$	0.05	35.1
	0.10	37.2
	0.25	37.0
	0.50	37.3
$1C_5$	0.05	36.3
	0.10	38.0
	0.25	39.3
	0.50	39.9
$1C_6$	0.05	37.3
	0.10	39.3
	0.25	43.4
	0.50	44.1
$1C_7$	0.05	37.4
	0.10	39.3
	0.25	44.4
	0.50	46.9
$1C_8$	0.05	38.0
	0.10	40.5
	0.25	47.3
	0.50	50.4

$1C_n$	$[1C_n]$ / M	T_p / °C
$1C_9$	0.05	38.9
	0.10	42.6
	0.25	47.5
	0.50	50.6
$1C_{10}$	0.05	39.5
	0.10	42.9
	0.25	49.6
	0.50	54.6
$2C_n$	$[2C_n]$ / M	T_p / °C
$2C_3$	0.05	27.0
	0.10	27.6
	0.25	29.7
	0.50	31.7
$2C_4$	0.05	30.3
	0.10	31.1
	0.25	33.5
	0.50	38.1
$2C_5$	0.05	33.4
	0.10	34.7
	0.25	37.8
	0.50	43.0
$2C_6$	0.05	35.7
	0.10	37.8
	0.25	41.1
	0.50	48.7
$2C_7$	0.05	38.9
	0.10	40.9
	0.25	46.5
	0.50	m/r
$2C_8$	0.05	41.1
	0.10	43.4
	0.25	m/r
	0.50	m/r
$2C_9$	0.05	42.8
	0.10	45.3
	0.25	m/r
	0.50	m/r
$2C_{10}$	0.05	45.5
	0.10	48.4
	0.25	m/r
	0.50	m/r

m/r: no stable microemulsion

Table 2. Continued

4.2 Influence of chain length upon electric percolation

In Figure 9 one can see the effect exerted by the hydrocarbon chain length of the 1-n-alcohols upon percolation temperature at a constant concentration of additive. Similar behaviour has been found for 2-n-alcohol. Taking into account the obtained results we highlight two aspects, the first is the fact that the slope of the experimental curve of t_p *vs* n grows with increasing additive concentration. This result is consistent with the one obtained when studying the effect of additive concentration upon the percolation threshold for each one of the alcohols studied.

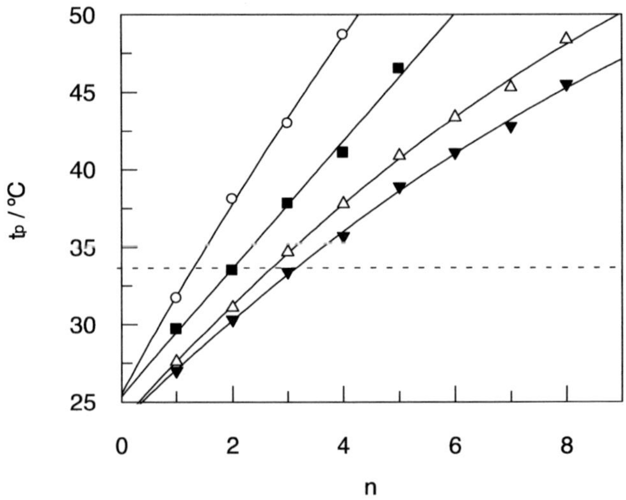

Fig. 9. Influence of *n* (number of carbon atoms) upon t_p in the addition of [$1C_n$] to AOT/iC_8/H_2O microemulsions [AOT]= 0.5 M, W= 22. [1Cn]=0.05 M, [1Cn]=0.1 M, [1Cn]=0.25 M & [1Cn]=0.50M

This behaviour observed would be due to the solubilisation of the alcohol between the interface and the aqueous phase in the case of low concentrations and low chain lengths; the majority solubilisation of the alcohol at the interface in the case of moderate concentrations of additive and moderate and high chain lengths; and the partial solubilisation of the alcohol at the organic phase in the case of high concentrations.

4.3 Comparison of results

As quote above, a different behaviour between both alcohol families was reported. In both cases as it was reported the percolation temperature increases linearly with additives concentration and the growing dependence of percolation temperature with alcohol concentration occurs by increasing the value of n. However, in the case of 1-n-alcohols, there are three different behaviours as a function of alcohol chain length (for n<3 implies a negative slope, for values of n=3 the percolation temperature is almost independent of the chain length and, for n>3 the slope is positive).

This behaviour would be associated with the polarity of the molecules and hence to the different distributions among the three phases of the microemulsion and their insertion at the interface as a co-surfactant. In this sense to rationalize this behaviour, it is necessary to know the polarity of these molecules. The values of the logarithm of the partition coefficient between 1-n-octanol and water of these molecules are shown in Table 5. This parameter is accepted as a plausible way to quantify the polarity of a molecule. This parameter is often accepted as a plausible way to quantify the polarity of a molecule. The observed differences in polarity between 2-n-alcohol and 1-n-alcohols can be noticed for example for a chain length of n=2, 2-n-propanol has a logP=016 while the 1-n-etanol its analogue has a logP=- 0.19.

Other point that must be underlined is the fact that there are two different head-groups with different head-group area, because the 1-n-alcohols would present a head-group area of 5.64 Å^2 , versus 10.86 Å^2 head-group area of the 2-n-alcohols. Thus, the correlation between the logP values and the effect exerted on percolation temperature of the system is shown in Figure 10. Analogous correlation ship can be plotted for 2-n-alcohols.

$2C_n$	Log P
$2C_3$	0.160
$2C_4$	0.691
$2C_5$	1.223
$2C_6$	1.754
$2C_7$	2.285
$2C_8$	2.817
$2C_9$	3.348
$2C_{10}$	3.879

Table 3. logP values for several used 2-n-alcohols

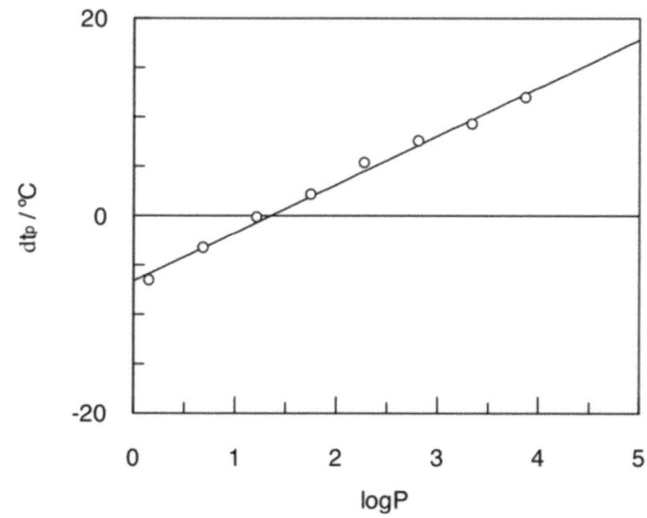

Fig. 10. Influence of hydrocarbon chain length of 2-n-alcohol upon the percolation temperature of AOT/iC_8/H_2O microemulsions. [AOT]= 0.5 M and W= 22.2, [2Cn]= 0.05 $dt_p=t_p^{additive}-t_p$

5. Conclusions

The investigation of the behaviour of 1-n-alcohols on the percolation temperature showed that for the values of n<3 the percolation temperature decreases with increasing the concentration of alcohol added, while, on the contrary, values of n>3 the behaviour would be reversed.

In the same way we have proceeded to study the behaviour of 2-n-alcohols on percolation temperature, obtaining different effects: at low concentrations and short chain lengths we observed an advancement of the percolation temperature, while alcohols with moderate and long-chain lengths showed that there is a delay in the temperature of percolation. This behaviour is clearly different to the results observed for 1-n-alcohols

It must be remarked that the behaviour observed for these two families of additives is quite different to the corresponding one of other amphiphilic molecules studied previously in our research team, as n-alkyl amines (García-Río et al., 2000), sodium alkyl-sulphates (García-Río et al., 2005) or n-alkyl acids (Cid-Samamed et al., 2008)

The behaviour of these studied systems is justified differently depending on the polarity of the molecules and the differences in solubility of the same in the various domains of microemulsions: the solubilisation of the alcohol between the interface and the aqueous phase at low concentrations and low chain lengths; solubilisation of the alcohol takes place mainly at the interface for moderate concentrations of additive and moderate and high chain lengths; and the partial solubilisation of the alcohol at the organic phase in the case of high additive concentrations.

6. Acknowledgment

Antonio Cid Thanks FCT-MCTES (Portugal) to the Postdoctoral grant number SFRH/BPD/78849/2011, and Óscar Moldes thanks the University of Vigo for a research grant P.P. 00VI 131H 64103.

7. References

Allinger, N.L. (1977) Conformational analysis. 130. MM2. A hydrocarbon force field utilizing V1 and V2 torsional terms. *Journal of the American Chemical Society*. Vol.99, pp. 8127-8134.

Arias-Barros, S.I.; Cid, A.; García-Río, L.; Mejuto, J.C. & Morales, J. (2010) Influence of polyethylene glycols on percolative phenomena in AOT microemulsions. *Colloid and Polymer Science*, Vol. 288, pp. 217-221.

Atik,S.S.; Thomas, J.K. (1981) Transport of photoproduced ions water in oil microemulsion: Movement of ions from one water pool to another. *Journal of the American Chemical Society*, Vol.103, pp.3543-3550.

Burkert, U. & Allinger, N.L. (1982) *Molecular Mechanics;* American Chemical Society: Washinton, DC.

Cid-Samamed, A; L. García-Río,L.; D. Fernández-Gándara, D.; Mejuto, J. C.; Morales, J. & M. Pérez-Lorenzo, M. (2008) Influence of n-alkyl acids on the percolative phenomena in AOT-based microemulsions. *Journal of Colloid and Interface Science*, Vol.318, pp. 525–529.

Dasilva-Carvalhal, J.; Carcía-Río, L.; Gómez-Díaz, D.; Mejuto, J.C. & Rodríguez-Dafonte, P. (2003) Influence of Crown ethers on the electrical percolation of AOT/isooctane/water (w/o) microemulsions. *Langmuir*, Vol.19, pp. 5975-5983.

Dasilva-Carvalhal, J.; García-Río, L.; Gómez-Díaz, D.; Mejuto, J.C. & Perez-Lorenzo, M. (2005) Influence of glymes upon percolative phenomena in AOT-based microemulsions. *Journal of Colloid and Interface Science*, Vol.292, pp. 591-594.

Dasilva-Carvalhal, J.; Fernández-Gándara, D.; García-Río, L.; Mejuto, J.C. (2006) Influence of aza crown ethers on the electrical percolation of AOT/isooctane/wáter (w/o) microemulsions. *Journal of Colloid and Inteface Science*, Vol.301, pp. 637-643.

Eastoe,J.; Robinson ,B.H., Steyler, D.C. & Thorn-Leeson, D. (1991) Structural studies of microemulsions stabilised by aerosol-OT. *Advances in Colloid and Interface Sience*, Vol.36, pp. 1-31.

García-Río, L.; Leis, J.R.; Mejuto, J.C. & Peña, M.E. (1994) Effects of additives on the internal dynamics and properties of wáter/AOT/isooctane microemulsions. *Langmuir*, Vol.10, pp. 1676-1683.

García-Río, L.; Hervés, P.; Leis, J.R. & Mejuto, J.C. (1997) Influence of Crown ethers and macrocyclic kryptands upon the percolation phenomena in AOT/isooctane/H_2O microemulsions. *Langmuir*, Vol.13, pp- 6083-6087.

García-Río,L.;Hervés, P.; Mejuto,J. C.; Pérez-Juste, J. & Rodríguez-Dafonte,P. (2000)Effects of Alkylamines on the Percolation Phenomena in Water/AOT/Isooctane Microemulsions. *Journal of Colloid and Interface Science*,Vol.225, pp. 259-26

García-Río, L.; Mejuto, J.C.; Perez-Lorenzo, M.; Rodríguez-Alvarez, A. & Rodríguez-Dafonte, P.(2005) Influence of anionic surfactants on the electric percolation of AOT/isooctane/water microemulsions. *Langmuir*, Vol.21, pp. 6259-6264.

Hait, S.K.; Moulik, S.P.; Rodgers, M.P.; Burke, S.E. & Palepu, R. (2001) Physicochemical studies on microemulsions. 7. Dynamics of percolation and energetics of clustering in water/AOT/isooctane and water/AOT/decane w/o microemulsions in presence of hydrotopes (sodium salicylate, alfa-naphthol, beta-naphthol, resorcinol, catechol, hydroquinone, pyrogallol and urea) and bile salt (sodium cholate). *Journal Physical Chemistry B*, Vol.105, pp. 7145-7154.

Hait, S.K.; Sanyal, A. & Moulik, S.P. (2002a) Physicochemical studies on microemulsions. 8. The effect of aromatic methoxy hydrotropes clustering and understanding of the dynamics of conductance percolation in water/oil microemulsion systems. *Journal of Physical Chemistry B*, Vol.106, pp. 12642-12650.

Hait, S.K.; Moulik, S.P. & Palepu, R. (2002b) Refined method of assessment of parameters of micellization of surfactants and percolation of W/O microemulsions. *Langmuir*. Vol.18, pp 2471-2476.

Jada, A.; Lang, J. & Zana, R. (1989) Relation between electrical percolation and rate constants for exchange of material between droplets in water in oil microemulsions. *Journal of Physical Chemistry*, Vol.93, pp.10-12.

Jada, A.; Lang, J.; Zana, R.; Makhloufi, R.; Hirsch, E. & Candau, S.J. (1990) Ternary water in oil microemulsions made of cationic surfactants, water, and aromatic solvents. 2. Droplet sizes and interactions and exchange of material between droplets. *Journal of Physical Chemistry*, Vol.94, pp. 387-395.

Kim, M.W. & Huang, J.S. (1986) Percolation-like phenomena in oil-continuous microemulsions. *Physical Review A*. Vol.34, pp. 719-722.

Lang, J.; Jada, A. & Malliaris, A. (1988) Structure and dynamics of water-in-oil droplets stabilized by sodium bis(2-ethylhexyl) sulfosuccinate. *Journal of Physical Chemistry*, Vol.92, pp. 1946-1953.

Lang, J.; Mascolo, G.; Zana, R. & Luisi, P.L. (1990) Structure and dynamics of cetyltrimethylammonium bromide water-in-oil microemulsions. *Journal of Physical Chemistry*, Vol.94, pp. 3069-3074.

Maitra, A.; Mathew, C. & Varshney, M. (1990) Closed and open structure aggregates in microemulsions and mechanism of percolative conduction. *Journal of Physical Chemistry*, Vol.94, pp. 5290-5292.

Mathew, C.; Patanjali, P.K., Nabi, A. & Maitra, A. (1988) On the concept of percolative conduction in water-in-oil microemulsions. *Colloids and Surfaces*, Vol.30, pp. 253-263.

Moulik, S.P.; De, G.C.; Bohwmik, B.B. & Panda, A.K. (1999) Physicochemical studies on microemulsions.6. Phase behavior, dynamics of percolation, and energetics of droplet clustering in water/AOT/n-heptane system influenced by additives (Sodium cholate and sodium salicylate) *Journal of Physical Chemistry B*, Vol.103, pp. 7122-7129.

Nazário, L.M.M., Hatton, T.A.& Crespo, J.P.S.G. (1996) Nonionic cosurfactants in AOT reversed micelles: Effect on percolation, size, and solubilization site. *Langmuir*, Vol.12, pp. 6326-6335.

Nazário, L.M.M., Crespo, J.P.S.G., Holzwarth, J.F.& Hatton, T.A. (2000) Dynamics of AOT and AOT/nonionic cosurfactant microemulsions. An iodine-laser temperature jump study. *Langmuir*, Vol.16, pp. 5892-5899.

Ray, S., Bisal, S.R., Moulik, S.P. (1993) Structure and dynamics of microemulsions. Part 1. Effect of additives on percolation of conductance and energetics of clustering in water-AOT-heptane microemulsions. *Journal of the Chemical Society, Faraday Transactions*, Vol.89, pp. 3277-3282.

4

Predictive Modeling of Microemulsion Phase Behaviour and Microstructure Characterisation in the 1-Phase Region

Deeleep K. Rout*, Richa Goyal, Ritesh Sinha, Arun Nagarajan and Pintu Paul
Unilever R&D India, Whitefield, Bangalore, Karnataka India

1. Introduction

Microemulsions are dispersions of water and oil stabilized by one or mixtures of surfactants. The key differentiators from other dispersions are: (i) domain size of the dispersed phase is of the order of 5-100 nm in size (Schulmann et al., 1959); (ii) thermodynamically stable (Zana, 1994), and (iii) optically isotropic (Danielsson and Lindman, 1981). Commonly observed structures are: oil in water (o/w), water in oil (w/o) and bicontinuous (Scriven, 1976). Microemulsion formation has proven to be an effective approach in enhancing oil solubilisation and reducing oil-water interfacial tension (IFT) in many industrial applications. Examples include enhanced-sub surface remediation (Harwell et al., 1992; Brusseau, 1999) , drug delivery (Djordjevic et al., 2004; Kogan & Garti, 2006; Ghosh et al. , 2006), detergency (Rosen, 2004; Srivastava et al., 2006; Solans & Kuneida, 1996; Tongcumpou et al., 2006) and nano particle synthesis (Sjöblom et al., 1996). Forming an efficient microemulsion system requires an equal balance between surfactant-oil and surfactant-water interactions, which has proven to be a very challenging task for certain types of oils (e.g., vegetable oils and triglycerides) (Miller, 1991; Minana-Perez, 1996). In addition, different applications require specific degrees of solubilisation and IFT reduction. It is no surprise, then, that formulating microemulsions for specific industrial use has been referred as both an art and science (Salager, 2006). This is because microemulsion design requires not only a great understanding of molecular interactions, but also a deep understanding of thermodynamic principles (Bourel & Schechter, 1998).

Microemulsions are also known to be effective vehicles for solubilization of certain drugs and as protecting medium for the entrapped drugs from degradation, hydrolysis and oxidation. They can also provide prolonged and controlled release of the drug and prevent irritation despite the toxicity of the drug. Besides, enhanced drug penetration through the skin via microemulsion vehicles is also known (Kogan & Garti, 2006).

In order to investigate the potential of microemulsions as efficient solubilizer as well as delivery vehicles, it is necessary to characterize their microstructures as well as the locus of the

* Corresponding Author

drug or targeted molecule in the loaded microemulsion. Due to the complexity of the microemulsions and the variety of structures and components involved in the construction of the microstructure, as well as the limitation associated with each technique, the characterization of such structures is a rather difficult task. Some of the major methods relevant to the characterization of the microemulsions include viscosity (Bennett et al., 1992; Zu & Neuman, 1998; Angelico, 1998; Ray, 1994; Djordjevic, 2004; Paul, 2000; Feldman et al., 1998) , electrical conductivity measurements (Bennett et al., 1992; Zu & Neuman, 1998; Feldman et al., 1998; D'Angelo et al., 1996; Moulik et al., 2000; Feldman et al., 1996; Mukhopadhyay et al., 1996,1993; Ray et al., 1996,1993; Moulik et al., 1999; Hait et al., 2001, 2002) as well as more advanced methods such as cryo-TEM (Talmon, 1986; Regev et al., 1996; Magdassi et al., 2003), pulsed gradient spin echo (self-diffusion) NMR (Magdassi et al., 2003; Fanun et al., 2001; Spernath et al., 2003), light scattering (Kang et al., 2004; Porras et al., 2004), small angle X-Ray scattering (SAXS) (Bagger et al., 1997; Brunner-Popela et al., 1999; Yaghmur et al., 2004) and small angle neutron scattering (SANS) (Silas & Kaler, 2003; Pedersen, 2004).

Modeling the microemulsion phase transition behavior and the effect of structural parameters of compositions (surfactant structures, e.g., packing parameter, interaction parameters of surfactant mixtures, alkane number of oil); electrolytes (type, lyotropic/hydrotropic number etc.) on phase behavior would considerably reduce experimental effort and time to obtain microemulsion systems. Although a large body of work is available in the area of predictive theoretical models of microemulsion phase behavior, predictive modeling of phase behavior dealing with real (commercial) systems/formulations is still a formidable challenge. Several theories have been proposed based on thermodynamic models (Andelman et al., 1987; Safran et al., 1986; Widom, 1996) ; Winsor-R ratio (Bourrel, M., & Schechter, 1988); packing parameter (Israelachvilli et al., 1976); hydrophilic-lipophilic balance (HLB) (Griffin, 1949); phase inversion temperature (PIT) (Shinoda & Saito, 1969); ultra-low interfacial tension (De Gennes & Taupin,1982); spontaneous curvature and bending elasticity (Hyde et al., 1997; Helfrich, 1973) and hydrophilic-lipophilic-deviation (HLD) model (Salager, 2006). Apart from thermodynamic models, as referred above, all other models are either empirical or semi-empirical models. Although, most molecular models provide an excellent qualitative predictive ability, quantitative differences exist between theory and experiment, which is attributed to a lack of exact consideration of chemical compositions in a surfactant mixture or oil mixture, mostly used for formulation developments.

Once, the condition for microemulsion is known or predicted, experimentalists and applied scientists determine its phase behavior to gain a deeper understanding of the effect of various molecular parameters (surfactant, oil and co-surfactant types and their compositions) as well as field variables (e.g., temperature, electrolyte concentration etc..) on phase changes. Different experimental methods were devised to determine phase diagrams most effectively (Kahlweit Fish diagrams) (Kahlweit, 1988) as well as Shinoda diagrams (Olsson et al., 1986). Kahlweit fish diagram is most widely used in the literature due its simplicity and useful information it provides (minimum concentration of surfactant needed to co-disperse equal amount of water and oil; minimum amount of surfactant needed to form a microemulsion; phase space to be considered for picking of a formulation). Tricritical point (i.e., an unique point in the phase space, signifying the most efficient co-dispersion of oil and water), where three distinct phases (phase separated 2-phases (ϕ), 1-ϕ microemulsion and a 3-ϕ body co-exists) can be obtained from the fish diagram as well. As per the phase rule, there is an unique tricritical point for a three-component system, a line of tricritical points for a four-component system and similarly, a 3-D space of tricritical points in a five component system etc..

2. Experimental techniques and materials

2.1 Electrical conductivity scans to determine the optimum salinity as well as 1-ϕ microstructure

DC electrical conductivity measurements to detect the phase inversion (from oil in water to water in oil) were carried out using a Orion (Model 150) conductivity probe immersed in a well stirred-thermally jacketed vessel at 28 °C. Stirring was facilitated using a Spinot magnetic stirrer and temperature was maintained using a water bath (Haake 3). Phase inversion composition, i.e., the electrolyte concentration (ε = salt/(salt + water)) was thus determined at various R (= oil/(water+oil) values. Electrical conductivity values of 1-ϕ microemulsions at different R values are also measured using the same set up.

AC electrical conductivity measurements were made using a Hewlett Packard 4285A Precision LCR Meter at 75 KHz to 5 MHz frequency. Electrical conductivity data obtained at 75 KHz frequency is used for microstructure characterization. The choice of 75 KHz frequency is based on the fact that this frequency is high enough to prevent space charge contributions and low enough to allow ionic effects contributing to the final conductance values. A home-made liquid cell with a cell constant of 0.73 was used. Temperatures were maintained using the Julabo F25 water circulator.

2.2 Determination of microemulsion phase behavior

Iso-thermal microemulsion phase diagrams (or "fish diagrams") at various R values were determined by visual inspection of samples/mixtures after equilibration. The samples were weighed into centrifuge tubes and sealed. An oil blend of Caprylic Acid and light liquid paraffin oil (LLPO) in the ratio 0.5:9.5 (wt/wt) was used. Commercial linear Lauryl alcohol ethoxylates $C_{12\text{-}15}EO_{<3>}$ and $C_{12\text{-}15}EO_{<7>}$ were taken in the ratio 1:9 (wt/wt) as its cloud point corresponded to near ambient temperature conditions of 30-33°C. The contents in the centrifuge tubes were well mixed using a Spinix vortex mixture and centrifuged in a Remi (R 24) Research Centrifugation facility at 7000 rpm for 30 minutes at a pre-determined temperature. The presence of phase types (2-ϕ, 3-ϕ and 1-ϕ) was ascertained by visual inspection. In addition, the transition between $\underline{2}\phi$ and $\overline{2}\,\phi$ was detected by e-conductivity measurements as and when appropriate. Shinoda diagrams are also determined by visual inspection of the liquid mixture after equilibrating the samples in a graduated Tarsons plastic tube for a few hours in a water bath.

2.3 NMR self-diffusion measurements

NMR measurements was performed on Bruker AV 400 spectrometer with Gradient Amplifier Board (GAB) and a 5 mm BBFO (Broad Band Fluorine Observe) probe equipped with a z-gradient coil, providing a z-gradient strength (G (Gauss)) of up to 55 Gcm^{-1}. The self-diffusion co-efficient of oil (1.3 ppm) and water (4.7 ppm) were determined at required temperatures (28 ± 0.5°C) by monitoring the ^1H signal using bipolar pulsed field gradient stimulated spin echo (BPFG-SSE) technique. Bipolar gradient pulse was used to reduce the eddy current effects (Wu et al., 1995). The pulse sequence consists of two radio frequency pulses, a 90° and 180° pulse separated in time by tau (τ). At the time 2τ, the set up produces an echo. In order to measure the self-diffusion, magnetic field gradients are applied as two short pulses with equal duration, one between 90° and 180° pulse and the another between the 180° and the spin echo. If the molecules under study diffuse between the two field gradient pluses, the spin-echo intensity will decrease compared to when the experiment is performed without a field gradient. Experiments were carried out by varying the gradient strength and keeping the other timing parameters constant (Borkovec et al., 1996). Typically, the values used in the experiment are, Δ = 500 ms, δ = 6.2 ms and G is varied from 1.8 -18 Gcm^{-1} in 32 steps. The self-diffusion co-efficient (D) is given by,

$$I = Io \exp\left(\gamma^2 G^2 \delta^2 \left(\Delta - \frac{\delta}{3}\right) D\right) \quad (1)$$

where, I is the measured signal intensity, I_o is the signal intensity in the absence of field gradient pulses (G = 0), γ is the gyromagnetic ratio for the ^1H nucleus, δ is the gradient pulse length, Δ is the time between the two gradients in the pulse sequence (Diffusion time) and D is the self diffusion coefficient of the component.

2.4 Rheology and viscosity measurements

Viscosity measurements were carried out using a Carrimed CSL^{500} controlled stress rheometer. An acrylic-make cone and plate geometry was used to measure viscosity as a function of shear rate/stress. The viscosity values recorded at a shear rate of 20 S^{-1} are used to compare across different R values.

2.5 Light scattering measurements

The aggregate size of the dispersed phase as well as the correlation length between water and oil domains was measured using light scattering goniometer (Brookhaven Instrument Corporation). The laser (λ=488nm) was produced by an Argon ion laser emitter (Lexel Laser Inc), and the scattered light was detected at 30, 60, 90, 105, 120 and 150° by a photo multiplier (Brookhaven Instrument Corporation). The temperature was maintained at 28 ± 0.1 °C using a temperature controlled water bath (Grant Ltd). The samples were filtered using 0.45 micron filters into pre cleaned vials (washed with methanol) supplied by Brookhaven Instrument Corporation. The viscosity for o/w microemulsions, i.e. from R = 0 to R = 0.4, was taken as that correspond to 15 % by wt. brine (2.62 CP) solution. Similarly, for bicontinuous samples i.e. from R = 0.43 to R = 0.52, a weighted average viscosity of oil and brine was taken and for w/o microemulsion, i.e. from R = 0.6 to R = 0.9, viscosity of oil system, i.e., 19 CP, was considered for estimating correlation length of oil/water domains from light scattering experiments. The refractive index of the particles was chosen as 1.47 while the refractive index of the medium was kept as 1.33 (refractive index of water). The data were analysed by using the CONTIN algorithm.

2.6 Materials

Most surfactants as well as all chemicals, used in the present work, were of LR grade and were used without any further purification. Sodium linear alkyl benzene sulphonates (NaLAS) was prepared in the lab by neutralizing the corresponding acid (97 % purity) with NaOH (99.9 % pure, Sigma-Aldrich). The final concentration of the surfactant was determined by conventional hyamine titration. The surfactant amount in the final product was found to be 82 % of the NaLAS, and the rest has been 16 % by wt. water and 2 % sodium sulphate and other organic impurities. Sodium salt of alkyl ether sulphate ($C_{12}EO_{2.5}SO_4Na$, 70 % aq. paste), was obtained from Galaxy surfactants, Mumbai, India and used as received. Imbentin OA/030 (2-ethyl ethoxylated hexanol, 99 % pure) was obtained from Kolb, Singapore. Linear Lauryl alcohol ethoxylates ($C_{12\text{-}15}\,EO_{<3>}$ and $C_{12\text{-}15}\,EO_{<7>}$, Galaxy Chemicals, India, commercial grade), were used as received, without any further purification. Extended surfactant under the trade name of Alfotera 145-4S (C_{12}-C_{17}, 4 moles propylene oxide (PO)), was purchased from Sasol, Italy. Benzalkonium chloride was obtained from Sigma-Aldrich, India.

Various hydrocarbons such as d-limonene, octane, decane, dodecane were obtained from Sigma Aldrich, India and tetradecane, hexadecane from Fluka, India. Methyl esters of fatty acids were obtained from Proctor & Gamble Chemicals, USA. Light liquid paraffin oil (LLPO, Beta Cosmetics, Silvassa, India, commercial grade) and Caprylic acid (Godrej Industries, India, 80% commercial grade) were used as received. Inorganic salts used in the study such as NaCl, KCl, fused $CaCl_2$, $MgCl_2.6H_2O$, $AlCl_3$, $FeCl_3$ were procured from Merck India and were used without any further purification. In all the experiments, deionised water (pH = 6.5) was used.

3. Hydrophilic lipophilic deviation (HLD) – Equivalent alkane carbon number (EACN) model

A short overview of the history of HLD is given in a paper by Queste et al., 2007. As stated there, the HLD is based on the determination of the so-called optimum formulation with flexible interfacial films. At this optimum formulation composition, the surfactant affinity to the polar and apolar pseudophases of a microemulsion is equal, meaning that the free energy change of a surfactant molecule, when transferred from oil to water and vice versa, is zero. As proposed by Salager et al., 2006, HLD is a dimensionless form of the thermodynamically derived surfactant affinity difference (SAD) equation. For systems with flexible interfacial films, this point is usually found in a tri-phasic system (oil/microemulsion/water, or the Winsor III), when the oil-water interfacial tension is at the lowest. Similarly a negative and positive value of HLD suggest the formation of Winsor Type-I and Type II microemulsions respectively. Under the conditions of this optimum formulation, the HLD value is set to zero. Any changes in the systems such as temperature; type of surfactants; oil molecular weight; salt type and its amount; type and amount of co-surfactants, will then lead to a departure from the reference state, i.e., either to a positive or negative HLD value.

For nonionic surfactants and given aqueous and oil phases, the HLD value of the system is defined from a practical approach as (Salager, 2006),

$$HLD_{ni} = b(S) - k(EACN) - \varphi(a) + \alpha_T \Delta T + \sigma \tag{2}$$

where, S = concentration of electrolyte (g/100 ml); k = characteristic parameter, depending on type of surfactant head group; EACN = equivalent alkane carbon number; ϕ = characteristic parameter, depending on type of co-surfactants; a = mass % of co-surfactant (usually an alcohol); α_T = temperature coefficient; $\Delta T = T - T_{ref}$ where T is temperature of system and T_{ref} is the reference temperature (25 ^oC) at which initially phase inversion is obtained; σ = characteristic curvature of the surfactant, which reflects the hydrophilic/lipophilic nature of the surfactant or surfactant mixtures.

For systems, containing anionic surfactants, the corresponding expression is (Salager, 2006)

$$HLD_i = Ln\,(S) - k(EACN) - \varphi(a) - \alpha_T \Delta T + \sigma \quad (3)$$

The characteristic parameters σ and k can be determined experimentally for new and unknown surfactants by varying, e.g., the salinity and searching for the equivalent temperature shift that compensates the salinity shift. Similarly, the characteristic parameter, b (for electrolyte) and ϕ, for co-surfactants are determined experimentally.

Many a times, practical situations necessitate mixtures of anionic and nonionic surfactants to achieve a desired result, e.g., reduction of the critical micellar concentration (CMC) of a surfactant mixture (Rosen 2004; Shiloach & Blankschtein, 1998; Holland, 1983; Scamehorn, 1986), foam enhancement or reduction; Kraft temperature manipulation and skin irritancy control (Rosen 2004; Shiloach & Blankschtein, 1998; Rodriguez , 1999; Theander et al. ,2003). Attempt has been made to extend the HLD model to surfactant mixtures (Acosta & Bhakta, 2009) with the following expression,

$$HLD_{mix} = X_i . HLD_i + X_{ni} . HLD_{ni} + G_{ex}/RT \quad (4)$$

where the sub index 'i' represents the ionic surfactant and 'ni' the nonionic surfactant. The term G_{ex}/RT represents the excess free energy-non-ideality normalized by RT. Conceptually, G_{ex} reflects the excess free energy of transferring a mole of surfactant (in the mixture) from the oil phase to the aqueous phase. Positive value of G_{ex}/RT indicate that the real mixture is more hydrophobic than the ideal mixture and vice versa.

At the optimum condition (HLD = 0), in the absence of any alcohols (for a microemulsion system with anionic surfactants) or co-surfactants (for the case with nonionic surfactant), and at the reference temperature (ΔT = 0), Equations 3 and 2 are reduced to respectively,

$$\ln S* = k\,(EACN) - \sigma \quad (5)$$

and

$$bS* = k\,(EACN) - \sigma \quad (6)$$

These two equations are incredibly simple and according to Equation (5), both 'k' and 'σ' can be experimentally determined from phase inversion experiments with a number of oils with different EACNs (experimental plot between ln S* and EACN). The slope and intercept respectively provide the above 'k' and 'σ' values. Similarly, a plot of S* and EACN also yields respective surfactant characteristics parameters (for a nonionic surfactant system). This approach has been successfully attempted by Witthayapanyanon et al., 2008 for anionic surfactants.

We have determined S* values with a number of hydrocarbons (starting with *n*-octane to *n*-hexadecane) and using an anionic surfactant, Alfoterra 145-4S (C_{14-15} –$(PO)_4$ SO_4Na), which belongs to a class of extended surfactant because of the presence of propoxylate $(PO)_4$ group near the sulfate head group.

The calculated value of 'k' and 'σ' for Alfoterra 145-4S (C_{14-15} –$(PO)_4$ SO_4Na) respectively are 0.099 and -0.663. It has been reported in the literature (Kiran & Acosta, 2010) that the values of 'k' and 'σ' for a similar propoxylated sulfate (C_{12-13} –$(PO)_8$ SO_4Na) are 0.087 and -0.784

respectively. These parameters have been used to describe surfactant hydrophilicity/ hydrophobicity in a more quantitative manner, e.g., a more negative value of 'σ' suggests the surfactant to be relatively more hydrophilic, while a positive 'σ' values imply a relatively hydrophobic surfactant. In order to describe the surfactant characteristics, both the surfactant parameters 'k' as well as 'σ' needs to be considered. Thus, instead of 'σ' alone, ratio of 'σ/k' is essential to determine the relative hydrophilicity or hydrophobicity of surfactant as both the terms are relative to each other.

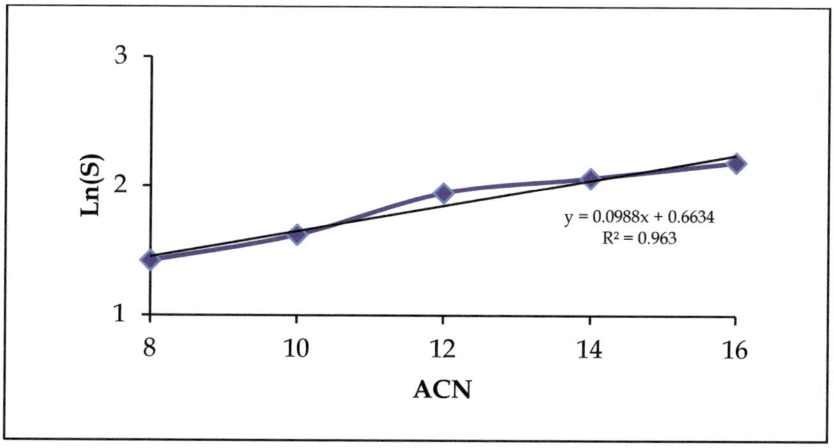

Fig. 1. Determination of k and σ using the HLD equation (4) for anionic surfactant Alfoterra 145-4S at optimal salinity at 25 °C.

Keeping the above in mind, we have obtained the ratio of σ/k for studying the effect of electrolytes on the partitioning of surfactants at the oil-water interface. Using the values of Ln(S) from Fig. 2, (σ/k) is estimated for the surfactant in presence of various electrolytes, used in the present study. Keeping the above in mind, we have tried to obtain the ratio of (σ/k) for studying the effect of electrolytes on the partitioning of surfactants at the oil-water interfaces.

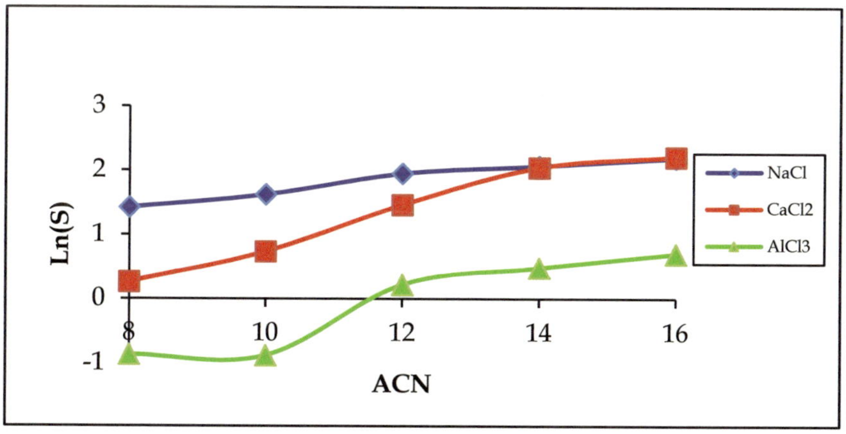

Fig. 2. Ln(S) of electrolytes with different oils (ACN) at 25 °C. The lines are drawn as a guidance to the eye.

Using the procedure outlined in the preceeding section, k, σ, and (σ/k) values are calculated for the surfactant, when various electrolytes (with multi-valent cations) are used. All the data from Fig. 2 is presented in Table 1 below.

Salt	k	σ	R^2	σ/k
NaCl	0.099	-0.66	0.96	-6.71
$CaCl_2$	0.26	1.78	0.96	6.83
$AlCl_3$	0.22	2.77	0.88	12.30

Table 1. k , σ and σ/k values for Alfoterra 145-4S surfactant in the presence of various electrolytes. Oils used for this study are hydrocarbons (octane to hexadecane).

As seen in the table, R^2 values for $AlCl_3$ is 0.88 indicating a very poor fit to the experimental data. Surfactant hydrophobicity 'σ' is very low for mono-valent ions and it increases (from -0.6 to 2.8) as the valency of the cation increases from mono- to tri-valent (Na^+, Ca^{2+}/Mg^{2+} and Al^{3+}) indicating an increase in the hydrophobicity of the surfactant and increasing partitioning to the oil phase. A small but finite 'σ' (~0) would indicate that the surfactant is partitioning equally well to the oil and water phases and the corresponding characteristic curvature of the oil-water interface is also close to zero. It may also be pointed out that the expression (5) is an analytic function of EACN and the intercept will always be a finite σ. The other surfactant characteristics parameter 'k', which is indicative of the surfactant head group characteristics, increases albeit slowly from a value of 0.1 (for NaCl) to 0.22 for $AlCl_3$. In a previous study by Anton et al., 2008, it was shown that 'k' remains almost constant across electrolyte types, especially when anion type is changed. It has also been commented by Salager, 1978 that 'k' values also depends on the type of electrolytes. But for most surfactants, 'k' value was found to range from 0.1 to 0.2 (Anton & Salager, 1990; Acosta& Bhakta, 2009), which is also the case in the present study. The 'k' value is a positive empirical constant determined experimentally, effectively remains the same for most head groups with the identical counterion (sodium) as seen in the past (Anton et al., 2008).

The fact that the 'k' value is changing with the valence of anions/cations, which can be attributed to the change in the head group characteristics of the surfactant due to the presence of multi-valent cations in the microemulsion. There can be two hypotheses supporting this.

The value of 'k' for sodium sulfonates is 0.1, but when a substitution of benzene or xylene moiety near the sulfonate head group takes place, it increases to 0.16. At the same time, if a $(PO)_8$ and $(EO)_2$ group is added near the head group, it decreases to 0.09 and 0.06 respectively (Witthayapanyanon et al., 2008). Benzene and xylene, being aromatic groups, will be more hydrophobic compared to propoxylene (PO) and ethoxylene (EO) groups. In addition to the surfactant head group substitution, counter-ion substitution may also affect 'k' , e.g., if sodium near the surfactant head group is replaced by calcium or aluminium, the head group characteristics may change, as reflected by a change in 'k' values.

Another argument may be made from the CMC (Critical micelle concentration) point of view, where it has been reported that in an aqueous solution of anionic Lauryl sulfates, the CMC decreases in the order, $Li^+ > Na^+ > K^+ > Cs^+ > N(CH_3)_4^+ > N(C_2H_5)_4^+ > Ca^{2+} \sim Mg^{2+}$, which is the same order as the increase in the degree of binding of the cation (i.e., Na^+

counterion in an aqueous medium will have the highest CMC but lower degree of binding with the head group and Ca^{2+} will have lower CMC but higher degree of binding with the head group). So, an increase in the 'k' value of the surfactant, while changing counterions in solution (e.g., from Na^{+} to Ca^{2+}), can be attributed to a higher degree of binding of the cation to surfactant head group. Also, the CMC value would be lower for the latter case. It is also known, for cationic surfactants, that as the degree of binding of the surfactant head group to cation increases, its area decreases (Rosen, 2004).

In the present work as well as past work (Anton et al., 2008; Acosta & Bhakta, 2009) , analysis of σ/k suggests that the value lies between -50 to +20. The lower limit is probably more or less certain as the value for sodium Lauryl ether sulfate (SLES) is -48, which has an HLB value of 40.7 (one of the highest HLB surfactant). The maximum positive value for σ/k is yet to be known as more experiments are required.

Another use of the HLD-EACN model is a practical approach to estimate the EACN value of oil mixtures, modified oils (e.g., partially polymerized oils on kitchen surfaces). We have demonstrated this with an anionic surfactant mixture (a mixture of sodium dodecyl benzene sulfonate (SDBS) and SLES). A characteristic plot of (Ln S - EACN) is determined for the surfactant mixture, which is shown in Fig. 3.

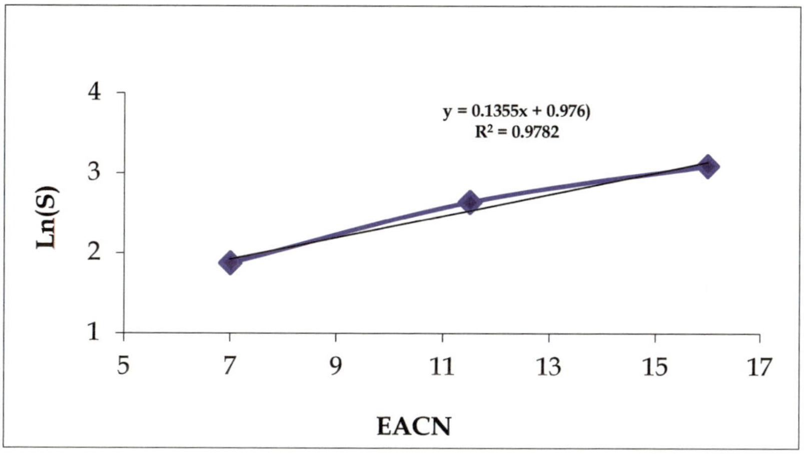

Fig. 3. Ln S- EACN plot for the surfactant mixture of sodium dodecyl benzene sulfonate and sodium Lauryl ether sulfate 1:1 by wt. NaCl is used as the salt. T = 25 °C.

The value of 'k' and 'σ' for the surfactant mixture (0.135 and -0.98 respectively) is in well agreement with reported values in the literature for anionic surfactants (Sabatini et al., 2008). Using these characteristic model parameters for the surfactant mixture, phase inversion for a 1:1 molar mixture of d-limonene (EACN = 6-7) and hexadecane (ACN = 16) is predicted to be at a S* value of 0.12, which matches well with experimentally determined value (0.14). It may be mentioned that a linear mixing rule was used to calculate the EACN value of the oil mixture from individual oil EACN/ACN values. Thus, the predictability of this model indicates that from phase inversion data (Ln S - EACN plot), one can estimate the EACN of an unknown oil by performing a phase inversion experiment with a surfactant or surfactant mixture with known characteristic 'σ/k'.

4. Experimental determination of microemulsion phase diagram and tricritical points

When applying microemulsions in industry or research in general, the temperature T of application, the composition of the oil, amphiphile and that of the electrolyte are specified. To describe a ternary mixture of water (A)–oil (B)–amphiphile (C), it is convenient to introduce a field variable, X (e.g., temperature, co-surfactant concentration, ACN or EACN of oil, electrolyte concentration), the weight percentage of oil in the mixture of oil and water, $R = (B/(A+B))$, and the weight percentage of the amphiphile in the system, $\gamma = C/(A+B+C)$ (Kahlweit, 1995). Each point in the three dimensional phase space is then defined by a certain set of X, R and γ. In order to discuss the phase behavior, an upright phase prism is erected with X as ordinate and the Gibb's ternary triangle of A-B-C as the base. The phase behavior may be discussed along either the horizontal (iso field variable) or vertical (e.g., at fixed R or γ) sections through the phase prism. For example, vertical sections of the phase prism at a fixed 'R' with the field variable X along the Y-axis and γ along the X-axis is termed the "fish diagram" (Kahlweit, 1995; Schwuger et al.,1995; Kunieda & Shinoda,1980; Kahlweit et al, 1987) (see Fig. 4). In the Fish-diagram, there is an unique coordinate, at which the 3-Φ body touches the 1-Φ region, co-existing with the 2Φ ($\underline{2}\Phi$ and $\overline{2}\Phi$), called the tri-critical point (TCP). The TCP has tremendous significance for practical applications requiring the highest mutual dispersibility between oil and water as in the case of supercritical extraction (Schwuger et al.,1995), enhanced oil recovery (Sottmann, 2002; Schwuger et al.,1995) and supercritical pollution oxidation (Schwuger et al.,1995).

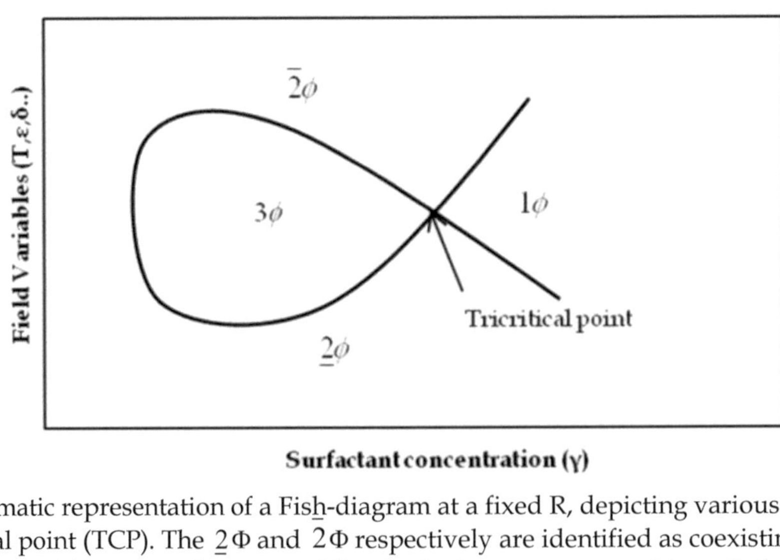

Fig. 4. Schematic representation of a Fish-diagram at a fixed R, depicting various phases and the tricritical point (TCP). The $\underline{2}\Phi$ and $\overline{2}\Phi$ respectively are identified as coexisting aqueous and oil phases with most surfactants residing in the bottom water and top oil phases respectively.

Tri-critical points vary as a function of 'R'. Kahlweit et al., 1995 have performed studies on the (n-decane-C_8EO_3-water) microemulsion system, where variations in the TCP are reported as a function of 'R', with temperature being the field variable. The trajectory of the

TCP when viewed along the temperature (T)-Gamma (γ) plane is parabolic in nature, with each tri-critical point corresponding to a given R. This trajectory shows an interesting trend, where the vertex of the parabola corresponds to the tri-critical point arising at R = 0.5. The tri-critical points corresponding to lower 'R' appear at lower temperatures than $TCP_{R=0.5}$, and tri-critical points at higher 'R' appear at higher temperatures than $TCP_{R=0.5}$, although at lower surfactant concentrations than $TCP_{R=0.5}$.

Thus, the TCP trajectory for field-variables (X) other than temperature (T), can similarly provide opportunities for applications in isothermal conditions. This provides avenues to obtain multiple compositions having ultra-low oil-water interfacial tension (IFT) at isothermal conditions, for a microemulsion that is otherwise strongly responsive to temperature (T). Further, the parabolic trajectory would be significant in predicting the variation of TCP for an unknown oil-water ratio (R). Thereby, an attempt has been made to investigate the TCP variation phenomenon for a commercial grade non-ionic surfactant based microemulsion system with electrolyte as the field variable.

4.1 Phase diagram (tricritical point variation) of the microemulsion system

Fig. 5. shows fish diagrams of the ternary Caprylic Acid/LLPO (0.5:9.5 wt/wt) (oil) - $C_{12-15}EO_{<3>}/C_{12-15}EO_{<7>}$ (1:9 wt/wt) (surfactant)- water system at various 'R' values. The complete fish diagram was determined only for R = 0.5, where it was found that the TCP coordinate at $\varepsilon = 0.12$ and $\gamma = 0.13$. At very high γ however, lamellar liquid crystalline phases are likely to appear which have not been depicted to maintain simplicity and clarity. The TCP depicting the crossover region of $\underline{2}\Phi$, 1Φ, $\bar{2}\Phi$ and 3Φ corresponding to other 'R' are shown along side. Fig. 6 shows the variation of TCP with electrolyte (NaCl) concentration (ε = salt/salt+water) as the field variable.

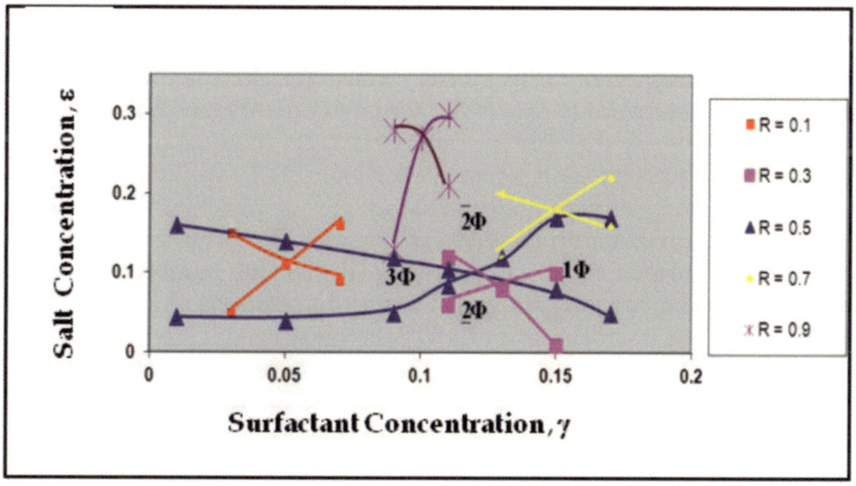

Fig. 5. Fish diagram of the ternary Caprylic Acid/LLPO (0.5:9.5 wt/wt) (oil) - $C_{12-15}EO_{<3>}/C_{12-15}EO_{<7>}$ (1:9 wt/wt) (surfactant) - water system, with epsilon denoting weight fraction of sodium chloride in water and gamma denoting the weight fraction of surfactant blend in oil and water. The complete Fish diagram is determined for R = 0.5, whereas only the TCP were determined for other R values. Temeprature = 28 °C.

The trajectory of the tri-critical points when viewed along the Epsilon (ε)-Gamma (γ) plane is again parabolic in nature, with each tri-critical point corresponding to a unique value of R. The parabola however is inclined to the X-axis and the asymmetric nature may be attributed to the commercial grade of non-ionic surfactants and oil mixtures used in the study. For C_{12-15} $EO_{<3>}$ and C_{12-15} $EO_{<7>}$, there exists a linear carbon chain length distribution ranging from 12-15 units for both compounds and a broad degree of ethoxylation (EO_j, j = 1-10), with a peak at j = 3 for $C_{12-15}EO_{<3>}$ and j = 7 for $C_{12-15}EO_{<7>}$ respectively. These also contribute to the distortions observed in the 3 Φ and 1-Φ regions [86], although the general phase behavior remains the same.

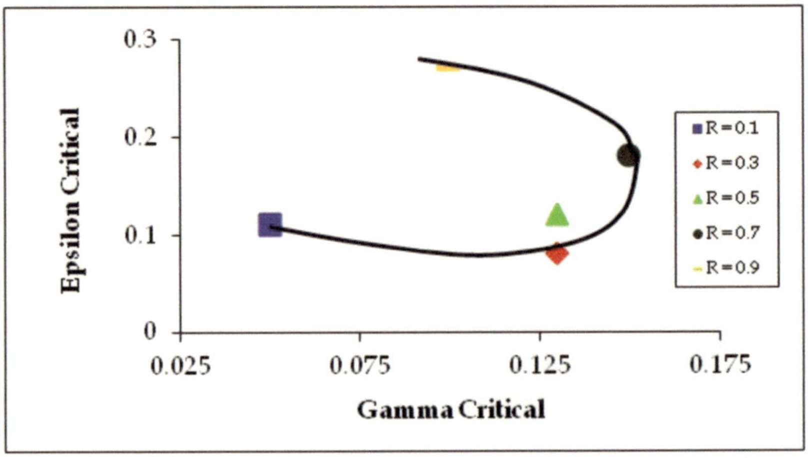

Fig. 6. Variation of TCP coordinates ($\varepsilon_{critical}$) and ($\gamma_{critical}$) as a function of oil in (oil+water) ratio R at a fixed T (= 28 °C). The line drawn though points is only for guidance.

In conclusion, tri-critical points in the present microemulsion system are seen to follow a parabolic profile as a function of R, for a field variable (electrolyte concentration, ε) similar to that observed with temperature in another ternary (C_8E_3-Water-Oil) system (Kahlweit et al., 1995). The similarity in the TCP variation between these two systems indicate the universality of the microemulsion phase behaviour. TCP variation as a function of electrolyte concentration also provides avenues to obtain multiple tri-critical point compositions at isothermal conditions, for a microemulsion forming system, whose phase behavior is otherwise responsive to temperature. Further, the parabolic trajectory offers a guideline to predict the location of tri-critical point compositions of a microemulsion if the oil weight fraction in the microemulsion formulation is altered. Hence, it facilitates a faster process of locating TCP compositions, where mutual dispersibility between oil and water is the maximum and oil-water interfacial tension is at the minimum.

4.2 Shinoda diagram of microemulsion phase behavior

Another method to study microemulsion phase behavior has been to vary the oil amount (R) in a surfactant solution with a fixed surfactant concentration (fixed γ) (also referred as Shinoda-diagram) (Sottmann et al., 2002; Olsson et al,, 2002). The field variable in this case could be one of the following: temperature; electrolyte concentration (ε); co-surfactant concentration (δ = co-surfactant/(cosurfactant + surfactant). This method is quite popular with formulation developers as the phase diagram is determined at a low surfactant concentration. The phase diagram is shown schematically in Fig. 7.

The Shinoda- diagram or the Shinoda-cut as referred by many [86], also depicts a rich array of microstructures as seen in Fig. 7.

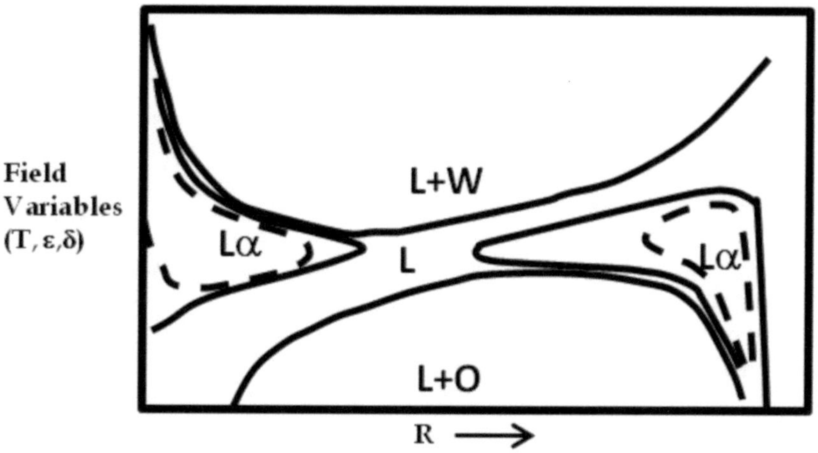

Fig. 7. A schematic Shinoda diagram, representing a rich array of phase behavior (Lamellar L_α; 1-ϕ microemulsion, L; biphasic microemulsion and excess oil, L+O; and biphasic microemulsion and excess water, L+W), is shown at a fixed surfactant concentration (iso-γ). R is the oil to (oil+water) ratio. The phase behavior is generic (i.e., irrespective of the field variables, features of the phase diagram remain the same).

The advantage of the Shinoda diagram is that one can obtain information on the extent of oil incorporation (depicting a 1-ϕ microemulsion) at a given surfactant concentration. The above schematic diagram is a representative behavior, seen for surfactant concentrations in a range of 10-30 % by wt (γ = 0.1 to 0.3) (Sottmann et al., 2002). At lower surfactant concentrations ($< \gamma = 0.1$), the L- region will have a co-existent 1-ϕ, 3-ϕ and 2-ϕ regions, with a tri-critical point (similar to the Fish diagram) (Olsson et al,, 2002). The tricritical point in a Shinoda-diagram also indicate the maximum amount of oil, which can be dispersed in a microemulsion system (a measure of oil incorporation efficiency of a micoemulsion).

We have determined such Shinoda diagrams for a number of surfactant-oil-water systems, e.g., with benzalkonium chloride-methyl esters of *n*-alkyl fatty acid-water. Shinoda diagrams are determined with oils of increasing alkyl chain lengths (e.g., methyl ester of octanoic acid, -decanoic acid and –dodecanoic acid). Calcium chloride is used as the electrolyte. Fig. 8 shows the Shinoda diagram depicting various phases (1-ϕ; 3-ϕ and 2-ϕ regions). The 1-ϕ region is a stable, transparent liquid with an oil-in-water microstructure. The 3-ϕ region is the region, where a microemulsion phase co-exists with the excess oil and water phases. The 2-ϕ region has a microemulsion co-existing with the excess oil phase. For practical applications, the extent of 1-ϕ region is indicative of the ability of the system to provide a stable system, which can incorporate maximum oil possible. Thus, the efficiency is judged from the co-ordinate of the tri-critical point (R and ε), e.g., a higher R and lower ε is indicative of more efficient microemulsion system.

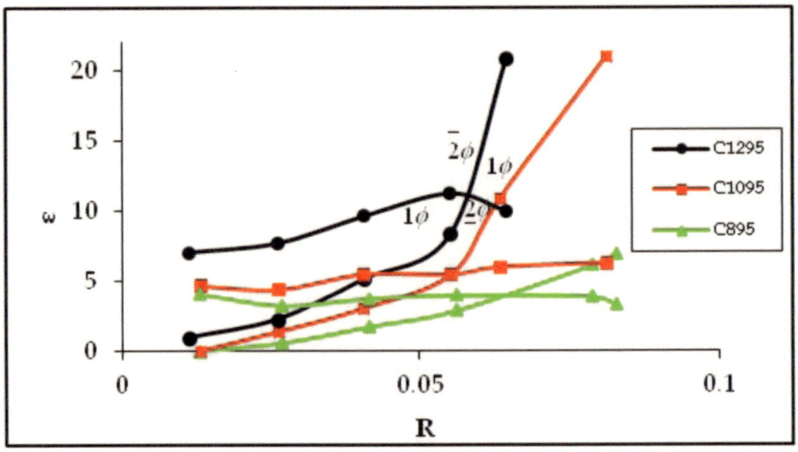

Fig. 8. Shinoda Diagram for a ternary (benzalkonium chloride-water-methyl *n*-alkanoate) with $CaCl_2$ as the electrolyte (field variable in the present case). Benzalkonium chloride concentration is 5 % by wt. (γ = 0.05), Temperature = 25 °C.

From the Fig. 8, it is evident that a higher amount (~ 3 times) of salt is required to emulsify the higher molecular weight (MW) oil (e.g., methyl dodecanoate) compared to low MW oil (methyl octanoate). It is shown in the preceding section that there is a good correlation between electrolyte and temperature requirement for microemulsion formation as well as tri-critical point variation in the phase space. It is also well known (Kahlweit et al., 1987) that the 3-ϕ body shifts to a higher temperature, when the MW of the oil increases. In the present case, the 3-ϕ body shifts to a higher electrolyte concentration when the MW of the oil increases. On the other hand, there seems to be no difference (i.e., no major change in the R value of the tri-critical points, associated with these oils) between the two oils. In terms of microemulsion efficiency, the ratio of oil to surfactant at the tri-critical point is ~1:1, which is comparatively poor compared to some efficient microemulsion systems, known in the past with an oil to surfactant ratio at 4:1 (Shinoda et al., 1984).

5. Micro-structural changes in the 1-phase microemulsion region

5.1 Iso-electrolyte concentration phase diagram

The phase diagrams shown in Fig. 5 have presented us with an opportunity to determine the extent of single-phase domain in the ε-γ space at a fixed temperature. Thus, for a given surfactant concentration (γ= 0.2), and a given electrolyte concentration (ε = 0.15), the extent of 1-ϕ region is determined in the T-R space. As seen in Fig. 9, a wide 1-ϕ channel across R (from 0.1 to 0.9) is seen at 28 °C with the exception for R = 0.7. Thus, it provides an opportunity to study the microstructural changes in the 1-ϕ channel using various experimental techniques. It may be mentioned here that such a 1-ϕ channel at a constant surfactant concentration (γ = 0.2) and constant electrolyte concentration (ε = 0.15) has not been reported before. Most of the past work was concerned with such channels with a varying surfactant concentration (Georges & Chen, 1986). The other advantage of the present work is also that the structural transitions can be monitored by the measurement of electrical conductivity without any unwanted contributions from varying electrolyte concentration across the 1-ϕ channel.

It may also be added here that since the current interest is to study the structural transformations across the 1-ϕ channel as a function of 'R' at a given temperature, such phase channels at other temperatures were not investigated. Subsequent measurements were limited to the 1-ϕ channel at 28 °C only.

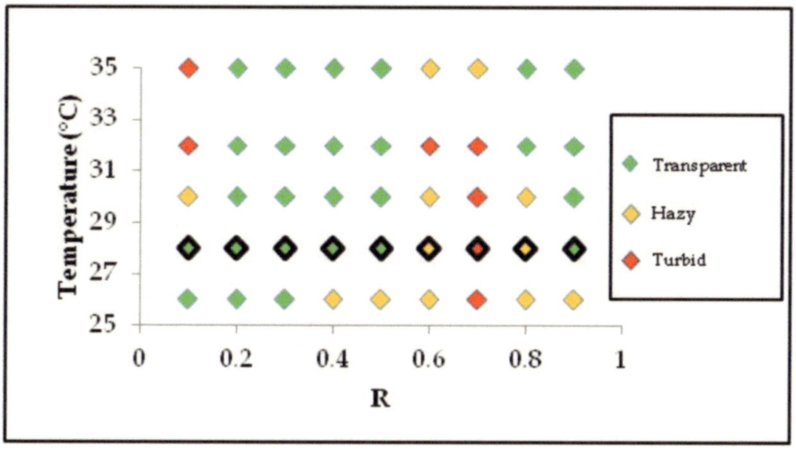

Fig. 9. 1-ϕ channel at various temperatures at a fixed surfactant concentration (γ= 0.2), and electrolyte concentration (ε = 0.15)

5.2 Pulsed gradient spin echo – Nuclear magnetic resonance (PGSE – NMR)

Guering and Lindman (Guering & Lindman, 1985) used the method to study molecular self-diffusion in SDS/ butanol/ toluene/ water/ NaCl microemulsion system. Garti et al (Yaghmur, 2003) and Reimer et al., 2003 used PGSE-NMR for understanding the microstructure of multi component microemulsion systems. Fanun (Fanun, 2007) has used the technique to study a non-ionic surfactant based-microemulsion system with various oils. All these above studies were carried out in the single phase region of the microemulsion. However, the concentration of all the three components was varied while scanning across the 1-ϕ channel.

The microemulsion system under study presents an opportunity to probe the microstructure variation as a function of R (oil / (oil + brine)) in the single-phase microemulsion region, but at an iso-thermal, iso-surfactant, and iso-electrolyte concentration conditions, unlike in the past. The change from oil-in-water to water-in-oil microstructure via a bicontinuous microstructure is probed by measuring the diffusion coefficient of oil and water individually in the system using self-diffusion NMR measurement. Relative self-diffusion coefficient depicted as solvent self-diffusion (D) in the microemulsion to that of the neat solvent (D_0) determines the domains formed by the solvent. The relative self-diffusion coefficient for water (D^{water}/D_0^{water}) is close to unity for an o/w microstructure while that of oil (D^{oil}/D_0^{oil}) is very much less than unity. High values of D/D_0 are observed for both oil and water in a bicontinuous structure. The self-diffusion coefficient of oil is close to D_0^{oil} and that of water is very small for w/o microstructures (Sjöblom et al., 1996).

Fig. 10 shows the relative self diffusion coefficients (D/D_0) of oil and water in the system consisting of brine / C_{12-15} $EO_{<3>}$ - C_{12-15} $EO_{<7>}$ / LLPO - Caprylic acid at 28 °C as a function of oil content (R) in the microemulsion samples where the surfactant mixture equals 20 wt % (γ =0.2) and the salt concentration in water equals 15 wt % (ε =0.15). The experimental points are joined for visual clarification. Increasing the weight fraction of oil in the 1-ϕ microemulsion channel D^{oil}/D_0^{oil} increases from 0.14 at R=0.1 to 0.88 at R=0.8. On the other hand, the values of D^{water}/D_0^{water} decrease from a value of 0.83 at R=0.1 to 0.08 at R=0.9. The relative diffusion coefficients of the oil (D^{oil}/D_0^{oil}) are very low for R values below 0.1 indicating droplet diffusion. These low oil diffusion values can be mainly attributed to the confinement of oils to closed domains. The gradual increase in D^{oil}/D_0^{oil} values until R=0.4 suggests that percolative oil channels are beginning to form from a discrete droplet microstructure. These networked oil channels retard the water diffusion in the continuous medium resulting in a reduction in self diffusion coefficient value. At R=0.4, both values of D^{oil}/D_0^{oil} and D^{water}/D_0^{water} are equal, indicating a clear bicontinuous structure with solvent molecules complementing each other's mobility in their domain. It implies that the diffusion of oil or water molecules in their respective domains is not restricted by the presence of the other phase in the macroscopic distance under consideration. The increase in the D^{oil}/D_0^{oil} values after R=0.4 until R=0.9 depicts a situation, where the molecular diffusion in an oil continuous channel is the dominating process. In a similar manner, the decrease in the D^{water}/D_0^{water} value signifies the gradual change in confinement of water domain from continuous channel to a droplet regime. The data obtained through the NMR self-diffusion measurements is validated and supplemented with the findings from viscosity and dynamic light scattering measurements of the same system, which will be described in subsequent sections.

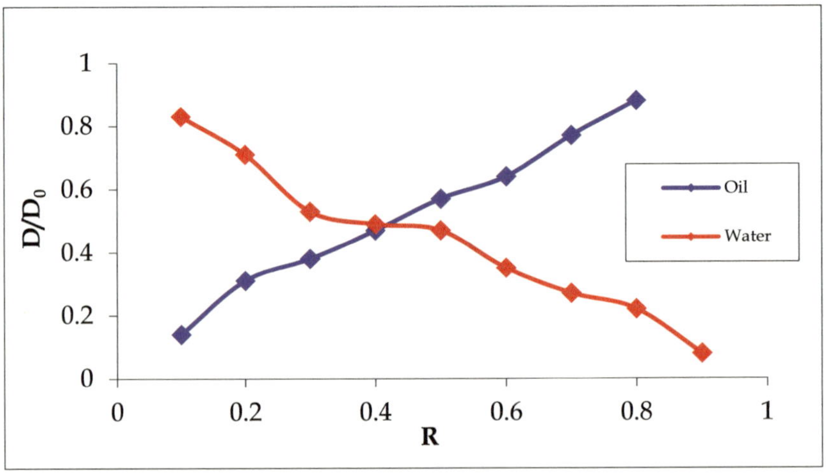

Fig. 10. Relative self diffusion coefficients (D/D_0) of oil and water in the single phase region of microemulsion system at iso-thermal (28 °C), iso-surfactant (20 wt. %) and iso-electrolyte concentration (15 wt. %).

5.3 Viscosity changes across the 1-ϕ channel at 28 °C

Viscosity measurements (viscosity profile as a function of shear rate as well as viscosity measured at a given shear rate for different R) provides information on globular-bicontinuous structural transitions (Georges & Chen 1986; Georges et al., 1987; Podlogar et al, 2004; Ajith et al., 1994; John et al., 1994). Globular (spherical oil droplets in water and vice versa) non-interacting structures are expected to exhibit a Newtonian behaviour ($\eta \approx \dot{\gamma}^0$). A deviation from the Newtonian behaviour is seen when various micro-structures (either in static or under shear flow) interact with each other. Such structures are also manifested in a shear thinning behavior. Such behaviour has been reported in the past for bicontinuous structures as well (John et al. 1994). Viscosity changes as a function of shear rate have been measured at various 'R' at 28 °C. For comparison purpose, viscosity, recorded at 20 S^{-1} , are shown in Fig. 11.

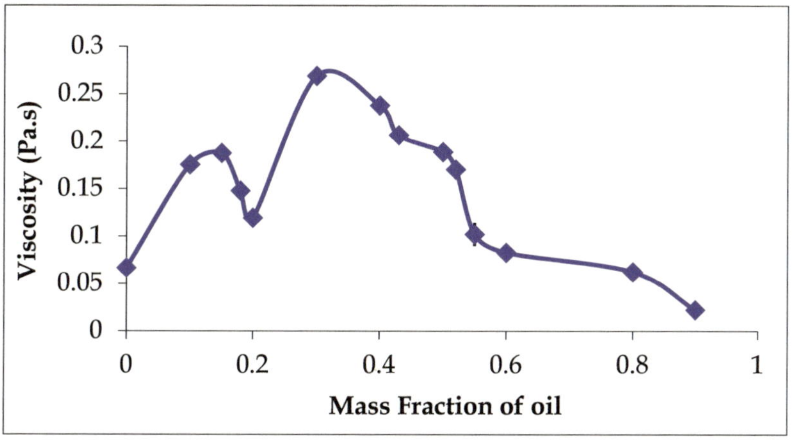

Fig. 11. Viscosity at 20 S^{-1} as a function of 'R' at 28 °C

The viscosity changes in the 1-ϕ channel is far from the bell-shaped curve as reported in the past, indicating the existence of rather more complex micro-structures with increasing oil wt. fraction. To begin with, the surfactant micellar phase (R = 0) exhibited a shear thinning behaviour as shown in Fig. 12.

It may be seen from the Fig. 12 that zero shear rate viscosity is ~1 Pa.S, which decreases to ~0.05 Pa.S at an infinite shear rate ($\dot{\gamma}_\infty$). It is well known that nonionic surfactants form spheroidal micelles at a high surfactant concentration (Lindman, 2001) and therefore, it may be naturally expected to have such structures at γ = 0.2. Under shear, such elongated structures either align (Larson, 1999) with each other or break down (Larson, 1999; Acharya, 2006) to spherical micelles resulting in a reduced viscosity at higher shear rates.

With increasing R, the viscosity increases till R = 0.15 (Fig. 11), following which, it decreases to 0.12 Pa.S at R = 0.18, a value slightly higher than that corresponding to R = 0 (0.08 Pa.S). Such unusual changes in the viscosity in the lower R regions may be explained as follows: With increasing oil wt. fraction, the initial elongated spheroidal micelles swell further to accommodate oil in to its hydrophobic core. Increasing in the size of the structures enhance the propensity to inter-swollen micellar interactions causing the viscosity to increase and also exhibit a stronger shear thinning behaviour (η_0 / η_∞)= 50 in contrast to 20 for R = 0. This observation is also supported by the measurement of correlation length, determined via dynamic light scattering measurements (a mean diameter change from 4.8 nm to 26.2 nm) (Section 5.5). With further addition of oil, there is a rearrangement of swollen micellar structures, causing the elongated swollen, spheroidal micelles to breakdown in to smaller spherical structures with a mean diameter of 4.7 nm. Although, the mean diameter is similar to that observed for R = 0, the structures formed at R = 0.18 are likely to be spherical, non-interacting entities, as manifested in the viscosity profiles recorded at 28 °C.

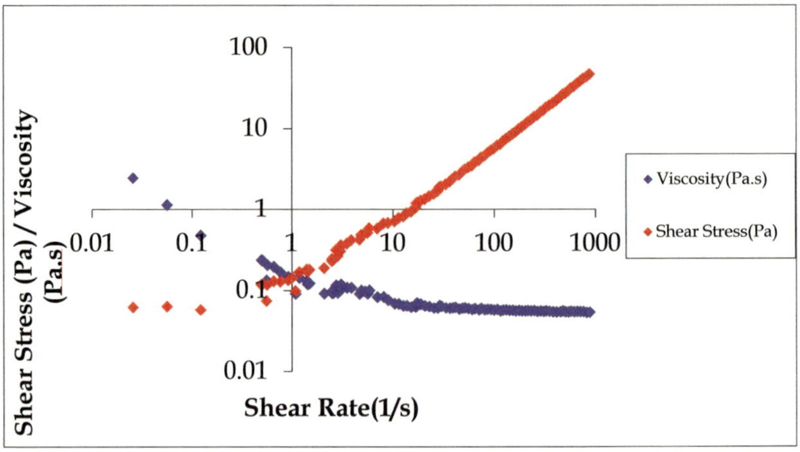

Fig. 12. Shear stress and viscosity variation as a function of shear rate at 28 °C. R = 0.

The viscosity profile for R = 0.18 indicates a Newtonian behaviour with $\eta_0 / \eta_\infty = 1$ and η_0 = 0.1 Pa.S. Subsequently with increasing R, there is a steady increase in the viscosity as seen in Fig. 11. A global maxima in viscosity (0.27 Pa.S) is seen at R = 0.3. It may be said that above R = 0.18, oil droplets exhibit a percolation behaviour, i.e., joining of globular oil droplets leading to the formation of oil channels with high η_0 values and high η_0 / η_∞ values as well. For R = 0.2, these values respectively are, 10 Pa.S and 100 respectively. Thus, a transition from a Newtonian behaviour at R= 0.18 to non-Newtonian behaviour at R= 0.2 signifies the percolation threshold of oil droplets forming oil channels. The global viscosity maxima at R= 0.3 coincides with the formation of a bicontinuous microstructure, characterized by sample spanning, interpenetrating oil and water channels. In the past, both shear thinning (Saidi et al., 1990) and Newtonian flow behavior (Blom & Mellema, 1988; Peyerlasse et al., 1988; Saidi et al., 1990) was observed for bicontinuous structures at R = 0.3-0.5. However, most reports point to a Newtonian behaviour indicating that there could be occurrence of different microstructures other than bicontinuous for the R values indicated above. As observed in the past, it is expected that the bicontinuous microstructure in the 1-ϕ channel exhibits a shear thinning behaviour as well. In the present case, for R= 0.3, η_0 / η_∞ is found to be ~10, which is much smaller than that observed for R = 0.2, indicating that the bicontinuous structures are much more flexible and rearrange rapidly. The mean correlation length (between oil and water channels) is found to be 10 nm, a two-fold increase from that seen at R=0.2. It is also well known (Yaghmur et al., 2003) that the correlation length approaches the maximum near the bicontinuous microstructure, before decreasing further with increasing oil wt. fr.

Following the global maxima in the viscosity at R= 0.3, it decreases subsequently with increasing R, with a shoulder near R = 0.5 (Fig. 11). Interestingly, microemulsions higher than R = 0.3 exhibit Newtonian behavior, which concurs with most past results. One possible explanation for the Newtonian behaviour for the bicontinuous microstructure is the interfacial flexibility as well as the time scale of rearrangement of water, oil channels being much faster than the material response to external shear stimuli. The viscosity changes following the global maxima is very much similar to that seen in the past (Garti et al., 2001; Georges et al., 1987; Langevin, 1992) for 1-ϕ microemulsions with increasing oil wt. fraction. The viscosity values for microemulsions with higher R is lower than its counterpart with low R with a sharp decrease at ~ R = 0.6, which is likely to be the transition to the w/o microemulsion from a bicontinuous structure.

5.4 DC and AC conductivity changes in the 1-ϕ microemulsion region

The microstructural changes from a droplet morphology (e.g., w/o microemulsion) to a bicontinuous morphology via coalescence of water droplets and their percolating networks/channels have been characterized by electrical conductivity and dielectric permittivity measurements (Fanun, 2007; Yaghmur et al., 2003; Capuzzi et al., 1999; Cametti et al., 1995; Peyrelasse & Boned, 1990) . The percolation phenomena as a function of water volume fraction as well as with temperature has been studied in detail and two percolation models, i.e., static and dynamic percolation have been proposed (Cametti et al., 1995). In either of these, there is an order of magnitude change in the DC (ohmic) or low frequency electrical conductivity. The detailed nature of these curves depend on experimental design and parameters, e.g., low frequency conductivity changes at constant surfactant content, at a constant surfactant to oil ratio etc. In the present work, as mentioned earlier in the preceding sections (5.1-3), there is an opportunity to measure e-conductivity changes as a function of oil to water ratio (R) at a constant surfactant concentration (iso-γ); constant electrolyte concentration (iso-ε) and iso-thermal conditions. A plot of DC e-conductivity at various R is shown in Fig. 13.

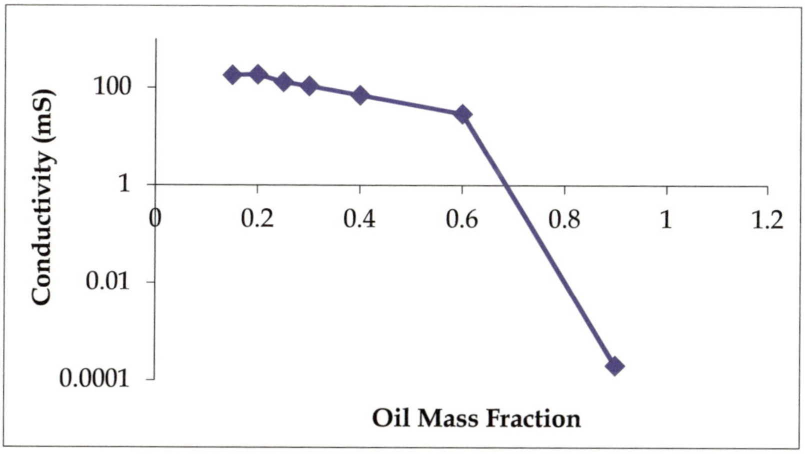

Fig. 13. DC e-conductivity changes at various R. E-conductivity could not be measured below R = 0.15 because of instrumental limitations. The line drawn across points is only for guidance to the eye.

As seen in the conductivity profile, there is a sharp drop in the conductivity values (4-5 orders of magnitude drop) after R =0.6, indicating the transition to a w/o microstructure. The above plot also shows that the percolation threshold at the water-continuous side, i.e., a transition from o/w to bicontinuous structure occurring at R~ 0.2, manifested by a change in the slope of e-conductivity. This is expected as the conduction path is obstructed by the formation of oil channels, resulting in a decrease in the conductivity value, hence a change in the slope. On the other hand, a transition to the oil continuous from bicontinuous microstructure is signified by orders of magnitude change in the conductivity values.

AC e-conductivity scans provide a more accurate picture of electrical conductivity changes across microstructural transitions, because, at high frequency (> 10 kHz frequency), contributions from space charge polarizations can be eliminated. We measured AC conductivity at 75 KHz and the data is shown in Fig. 14.

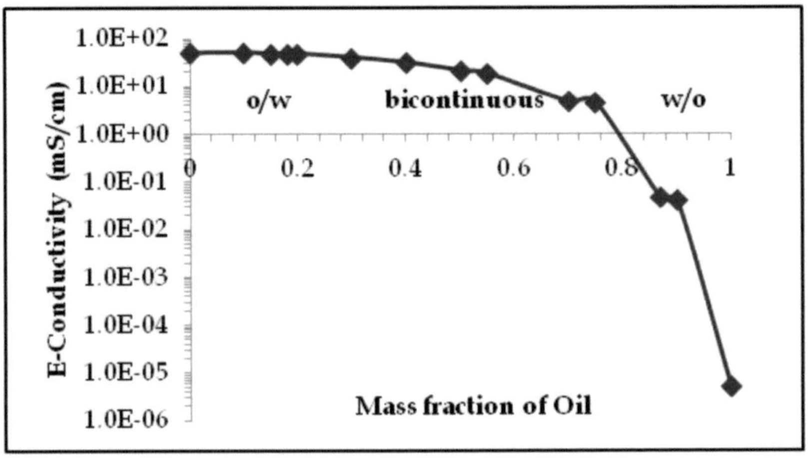

Fig. 14. AC conductivity profile in the 1-ϕ channel. The measurement is carried out at 75 KHz frequency.

The transition from the o/w microstructure to a bicontinuous one is characterized by a change in the slope of conductivity values as is the case in DC conductivity (Fig. 13). However, many a times, because of the high electrolyte concentration in the system, physical measurements become difficult in the case of DC conductivity. Based on the DC- as well as AC-conductivity measurements, the microstructural transitions in the 1-ϕ channel of the current microemulsion system, as a function (R), can be said as, o/w (0.2) bicontinuous (0.75) w/o.

5.5 Diffusion coefficient and correlation length of oil and water micro-domains from light scattering measurements

A plot of diffusion coefficient of various microstructural entities, causing light scattering as a function of R is shown in the Fig. 15.

The diffusion coefficient measurement has reflected various transitions, described so far in the preceding section. The diffusion coefficient peaks ~ R = 0.2, which is the percolation threshold for the oil channels to form. The highest diffusion coefficient is also indicative of lowest viscosity, encountered at R = 0.18. Similarly the transition to the oil-continuous microstructure occurs at R = 0.6, followed by a lowest diffusion coefficient measured at R = 0.5. Using these diffusion coefficient data and viscosity of the continuous phase, correlation length at various R is estimated. The Fig. 16 shows the variation of correlation length as a function of R.

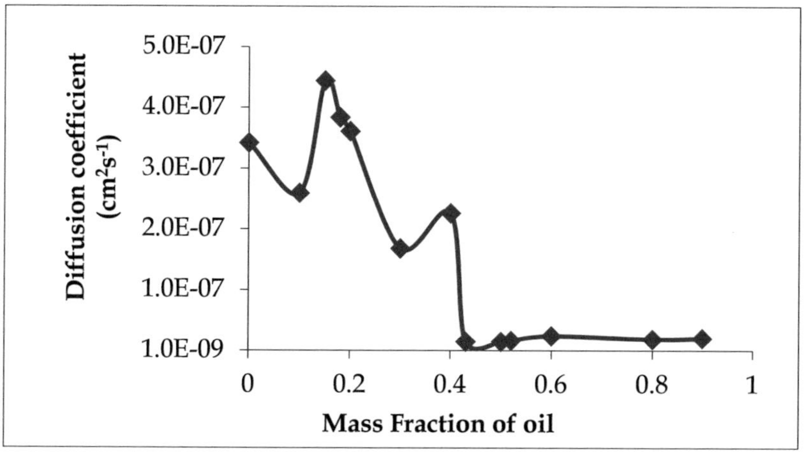

Fig. 15. Diffusion coefficient of oil/water droplets/domains in the 1-ϕ channel as determined from light scattering measurements. The line drawn though all experimental points is only a guidance to the eye.

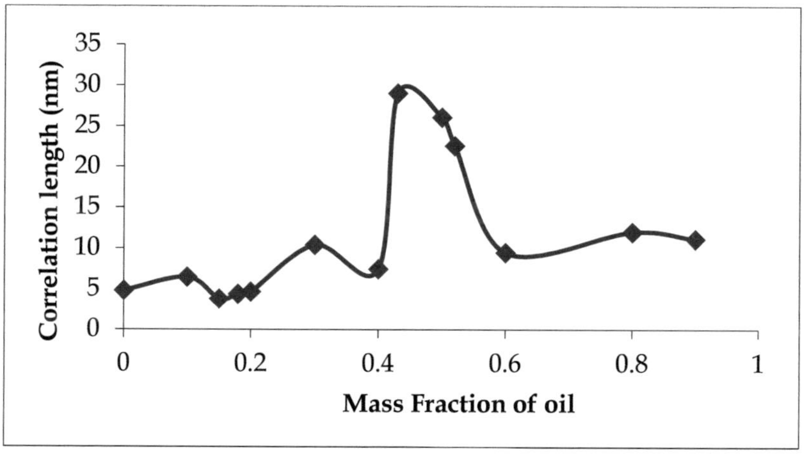

Fig. 16. Correlation length as a function of R in the 1-ϕ channel at $\gamma = 0.2$ and T = 28 °C

The correlation length changes across oil wt. fraction. Initial changes in the oil droplet size follow the changes seen in the viscosity profile (Fig. 11), i.e., an increase in the size of oil domains with increasing oil addition as well as a drop in its size at R = 0.18 (see the discussions in the preceding sections). The global maxima in the correlation length (for the bicontinuous structure) however appears at R = 0.42, which in the viscosity profile corresponds to the appearance of a shoulder. Interestingly, this is the point at which the diffusion coefficient is also the lowest and no structural breakdown of the product under shear takes place. It is common for microemulsion systems, where the correlation length passes through a maximum as per the following relation (de Gennes & Taupin, 1982; Talmon & Prager, 1978)

$$\xi = \frac{6\phi_o \phi_w}{C_s \Sigma} \quad (7)$$

Where, ξ, ϕ_o, ϕ_w, C_s and Σ are respectively, the mean correlation length of oil and water micro-domains, oil volume fraction, water volume fraction, number of surfactant molecules per unit volume and cross-sectional head group area of the surfactant at the oil-water interface. The above relation is derived with the assumption that the oil and water micro-domains form a cubic lattice (Talmon & Prager, 1978). Assuming Σ to be 50 $Å^2$, calculated ξ was found to go through a maxima (13 nm) between R = 0.3 to R = 0.45. The maximum calculated correlation length is roughly half of that estimated from experiments. The source of error could be from the accurate estimation of Σ values as well as the assumption of a close packed cubic lattice structure. Although Σ was taken to be 50 $Å^2$ (Klose et al., 1995), the actual value could be less, especially with a very high electrolyte concentration (ε= 0.15).

A comment on the variation of diffusion coefficient for various structures may be appropriate here. It is seen that diffusion coefficient varies with scattering angle as,

$$D(q) = \frac{D_0 K(q\xi)}{q^2 \xi^2}$$

where K is Kawasaki function,

$$K(x) = \frac{3}{4}[x^2 + 1 + (x^3 - \frac{1}{x})\tan^{-1} x] \quad .$$

Hence, a stronger variation of D(q) with q indicate the contribution of correlation length ξ variations through the 2nd order correction term [$\frac{D_0}{q^2\xi^2}$] to the D_o value obtained at $\theta = 90^\circ$.

Variation in the diffusion coefficient values as a function of scattering vector ($q = \frac{4\pi n Sin(\theta/2)}{\lambda}$) is the least (independent of q) and strongest for R = 0. A strong variation of correlation length with q is also indicative of asymmetric non-spherical structures in the medium, supporting conclusions of spheroidal structures at R = 0 as well as correlation length maxima at R = 0.3.

6. Conclusion

We have shown that the HLD–ACN model can be successfully used to predict microemulsion phase behaviour in systems containing commercial/technical ingredients/mixtures. We also have pointed out various limitations in using the model from a microemulsion formulator point of view. New insights have been obtained by estimating the characteristic surfactant parameters 'k' and 'σ'. It has also been shown that di- and tri-valent cations cause an increase in σ/k values, indicating an increase in the surfactant hydrophobicity, possibily due to an increase in the hydrophobic volume.

Following a brief description on determination of microemulsion phase diagrams (Fish-diagram and Shinoda diagrams), we have shown the generic nature of these phase diagrams, i.e., variation of tricritical points in the phase space and also presented a rare opportunity of 1-ϕ microemulsion channel in the phase space at constant surfactant concentration (iso-γ); constant electrolyte concentration (iso-ε) and iso-thermal conditions. Microstructural transitions, i.e., from an o/w to w/o via a bicontinuous microstructure, as manifested by various experimental parameters (self-diffusion co-efficient ; viscosity; electrical-conductivity and characteristic domain sizes) are demonstrated.

7. Acknowledgements

We would acknowledge our discussions with many co-workers, notably Gautam Kini and Saugata Nad, who have left Unilever now, but contributed immensely during their tenure with Unilever. Similarly, we acknowledge contributions of many students, especially Siddharth Chauhan, Ashwani Aggarwal, Akanksha Agrawal, Rupsa Banerjee, Afreen Anjum, Sneha Kankaria, Ashwini Bhatt and Anushka Pandey during their student internship program with Unilever R&D Bangalore. Finally, we also acknowledge Bangalore R&D management for the encouragement provided to us for this publication.

8. References

Acharya, D.P., Sato, T., Kaneko, M., Singh, Y., Kunieda, H. (2006). Effect of added poly(oxyethylene) dodecyl ether on the phase and rheological behavior of worm like micelles in aqueous SDS solution. *J. Phys. Chem., B* Vol. 110, pp. 754-760.

Acosta, E., & Bhakta, A. (2009). The HLD-NAC model for mixtures of ionic and nonionic surfactants. *Journal of Surfactants and Detergents*, Vol. 12, No. 1, pp. 7-19.

Acosta, E., & Bhakta, A. (2009). The HLD-NAC model for mixtures of ionic and nonionic surfactants. *Journal of Surfactants and Detergents*, Vol. 12, No. 1, pp. 7-19.

Ajith, S., John, A.C., & Rakshit, A.K. (1994). Physicochemical studies of microemulsions. *J.Pure Appl.Chem.* Vol. 66,pp. 509-514

Andelman, D., Cates, M. E., Roux, D., & Safran S. A. (1987). Structure and phase equilibria of microemulsions. *Journal of Chemical Physics*, Vol. 87, No. 12, pp. 7229.

Angelico, R., Palazzo G., Colafemmina, G., Cirkel, P.A., Giustini, M & Ceglie, A. (1998). Water Diffusion and Head group Mobility in Polymer-like Reverse Micelles: Evidence of a Sphere-to-Rod-to-Sphere Transition. *J. Phys. Chem. B*, Vol. 102, No. 16, pp. 2883–2889

Anton, E., Anderez, J., Bracho, C., Vejar, F., & Salager, J. (2008). Practical surfactant mixing rules based on the attainment of microemulsion-oil-water Three-phase behavior systems. *Advances in the Polymer Science*, Vol. 218, pp. 83-113.

Anton, R., & Salager, J. (1990). Effect of the electrolyte anion on the salanity contribution to optimum formulation of anionic surfactant microemulsions. *Journal of Colloid and Interface Science*, Vol. 140, No. 1, pp. 75-81.

Bagger-J, Örgensen, H., Olsson, U., & Mortensen, K. (1997). Microstructure in a ternary microemulsion studied by small angle neutron scattering. *Langmuir*, Vol. 13, No. 6, (March 1997), pp.1413-1421.

Bennett, K.E., Hatfield, J.C., Davis, H.T., Macosko, C.W., & Scriven, L.E. (1982). In: *Microemulsions*, ID Robb, (Ed), pp. (65-84), Plenum Press, ISBN 0306408341, 9780306408342, NewYork.

Blom, C & Mellema, J. (1988). The rheological behavior of a microemulsion in the 3Φ region. *Journal of Dispersion Sciecne and Technology,* Vol. 5, No. 2, pp. 193-218

Borkovec, M., Eicke, H-F, Hammerich, H. & Das Gupta, B. (1988). Two percolation processes in Microemulsions, *J. Phys. Chem.,* Vol. 92, No. 206, pp. 211-109.

Bourel M. & Schechter, R. (1988). In: *Microemulsions and related systems: formulation, solvency, and physical properties,* Marcel Dekker, ISBN S0824779517, 9780824779511 New York

Bourrel, M., & Schechter, R. (1988). *Microemulsion and related systems,* Marcel Dekker, ISBN : 0824779517, New York.

Brunner-Popela, J., Mittelbach, R., Strey, R., Schuebert, K. V., Kaler, E. W., & Glatter, O. (1999). Small- angle scattering of interacting particles. III. D_2O-$C_{12}E_5$ mixtures and microemulsions with n-octane. *Journal of Chemical Physics,* Vol. 110, No. 21, pp.10623.

Cametti, C., Sciortino, F., Tartaglia, P., Rouch, J. & Chen, S. H. (1995). Complex electrical conductivity of water-in-oil microemulsions. *Phys. Rev. Lett.* Vol. 75, pp. 569-572.

Capuzzi, G., Baglioni, P., Gambi, C.M.C. & Shen, E.Y. (1999). Percolation phenomenon of calcium bis(2-ethylhexyl) sulfosuccinate water-in-oil microemulsions by conductivity and dielectric spectroscopy measurements. *Phys. Rev. E.* Vol. 60, pp. 792–798.

D'Angelo M., Fioretto D., Onori G., & Palmieri L. (1996). Dynamics of water-containing sodium bis(2-ethylhexyl)sulfosuccinate (AOT) reverse micelles: A high-frequency dielectric study. *Phys. Rev. E,* Vol.54, pp. 993

Danielsson, I. & Lindman. B. (1981). The definition of a microemulsion. *Colloids Surfaces,* Vol. 3, pp. 391–392.

De Gennes, P. G., & Taupin, C. (1982). Microemulsions and the flexibility of oil/water interfaces. *Journal of Physical chemistry,* Vol. 86, No. 13, pp. 2294-2304.

Djordjevic, L., Primorac M., Stupar M., & Krajisnik D. (2004). Characterization of caprylocaproyl macrogolglycerides based microemulsion drug delivery vehicles for an amphiphilic drug, . *Int. J. Pharm.,* Vol.271, pp. 11–19.

Feldman, Y., Kozlovich N., Nir I. and Garti N., Archipov V., & Idiyatullin Z. (1996). Mechanism of Transport of Charge Carriers in the Sodium Bis(2-ethylhexyl) Sulfosuccinate−Water−Decane Microemulsion near the Percolation Temperature Threshold. *J. Phys. Chem.,* Vol.100, No.9, pp. 3745-3748

Fanun M., J. (2007). Conductivity, viscosity, NMR and diclofenac solubilization capacity studies of mixed nonionic surfactants microemulsions. *Journal of Molecular Liquids,* Vol. 135, Issue 1-3, pp. 5-13.

Fanun, M., Wachtel, E., Antalek, B., Aserin, A., & Garti, N. (2001). A study of the microstructure of four-component sucrose ester microemulsions by SAXS and NMR. Colloids and Surfaces A- Physicochemical and Engineering Aspects, Vol. 180, No. 1-2, (May 2001), pp. 173-186.

Feldman, Y., Kozlovich N, Nir I & Garti N. (1995). Dielectric relaxation in sodium bis(2-ethylhexyl) sulfosuccinate–water–decane microemulsions near the percolation temperature threshold. *Phys Rev E,* Vol. 51, pp.478

Garti, N., Yaghmur, A., Leser, M., Clement, V., & Heribert, J.W. (2001). Improved Oil solubilisation in Oil/Water Food grade microemulsion in the presence of polyols and thanol. *J. Agric. Food Chem.*, 49, pp 2552-2562.

Georges, J., Chen J.W. & Arnaud, N. (1987). Microemulsion structure in the lenticular monophasic areas of Brine/SDS/Pentanol/dodecane or hexane systems: electrochemical and fluorescent studies. *Colloid & Polymer Sci.*, Vol. 265, pp.45-51.

Georges, J., Chen J.W. (1986). Microemulsion studies: Correlation between Viscosity, Electrical Conductivity, Electrochemical and Fluorescent Measurements. *Colloid and Polymer Science*, Vol. 264, pp. 896-907.

Ghosh, P.K., Majithiya, R.J., Umrethia ,M.L., & Murthy, R.S. (2006). Design and Development of Microemulsion Drug Delivery System of Acyclovir for Improvement of Oral Bioavailability, *AAPS Pharm. Sci. Technol.* Vol. 7, No. 3, Article 77

Griffin, W.C. (1949). Classification of surface active agents by HLB. *Journal of the society of Cosmetic chemicals*, Vol. 1, pp. 311.

Guering, P., & Lindman, B. (1985). Droplet and bicontinuous structures in microemulsion from multicomponent self-diffusion measurements. *Langmuir*, Vol. 1, No. 4, pp. 464-468.

Hait, S.K., Moulik, S.P., Rodgors, M.P., Burke, S. E., & Palepu, R. (2001). Physicochemical Studies on Microemulsions. 7. Dynamics of Percolation and Energetics of Clustering in Water/AOT/Isooctane and Water/AOT/Decane w/o Microemulsions in Presence of Hydrotopes (Sodium Salicylate, α- Naphthol, β-Naphthol, Resorcinol, Catechol, Hydroquinone, Pyrogallol and Urea) and Bile Salt(Sodium Cholate). *The Journal of Physical Chemistry B*, Vol.105, No. 29, pp.7145-7154.

Hait, S.K., Sanyal, A., & Moulik, S.P. (2002). Physicochemical studies on microemulsions. 8. The effects of aromatic methoxy hydrotropes on droplet clustering and understanding of the dynamics of conductance percolation in water/oil microemulsion systems. *The Journal of Physical Chemistry B*, Vol. 106, No. 48, pp. 12642-12650.

Harwell J. H., Sabatini D.A., & Knox,(1992). In: Transport and Remediation of sub surface contaminants, Harwell J. H., Sabatini D.A., Knox (Eds), American Chemical Society, ISBN 0841222231, 9780841222236 pp. 124, Washington DC.

Helfrich, W. (1973). Elastic properties of lipid bilayers - Theory and possible experiments. *Z. Naturforsch*, C28, pp. 693-703.

Holland, P., & Rubingh, D. (1983). Nonideal multicomponent mixed micelle model. *Journal of Physical Chemistry*, Vol. 87, No. 11, pp. 1984-1990.

Hyde, S., Blum, Z., Landh, S., Lidin, S., Ninham, B.W., Andersson, K., & Larsson, K. (1997). *The Language of Shape*, Elsevier, ISBN : 978-0-444-81538-5, Amsterdam

Israelachvilli, J.N., Mitchell, D.J., & Ninham, J. (1976). Theory of self-assembly of hydrocarbon amphiphiles into micelles and bilayers, *Journal of the chemical society – faraday transactions II*, Vol. 72, pp. 1525-1568.

John, A.C., & Rakshit, A.K. (1994). Phase Behavior and Properties of a Microemulsion in the Presence of NaCl. *Langmuir*. Vol. *10*, pp.2084-2087.

Kahlweit, M. (1988). Microemulsions. *Science*, Vol. 240, No. 4852, pp. 617-621.

Kahlweit, M. (1995). How to prepare microemulsions at prescribed temperature, oil and brine. *Journal of Physical Chemistry*, Vol. 99, No. 4, pp. 1281-1284.

Kahlweit, M., Strey, R., Haase, D., Kuneida, H., Schmeling, T., Faulhaber, E., Borkovec, M., Eicke, H.F., Busse, G., Eggers, F., Funck, T.H., Richmann, Magid, H., Söderman, L., Stilbs, O., Winkler, P., Dittrich,J. & Jahn, W. (1987). How to study Microemulsion. *J. of Colloid and Interface Science*, Vol. 118, No. 2, pp. 436-453.

Kang, B. K., Chon, S. K., Kim, S. H., Jeong, S. Y., Kim, M. S., Cho, S. H., Lee, H. B., & Khang G. (2004). Controlled release of paclitaxel from microemulsion containing PLGA and evaluation of anti-tumor activity in vitro and in vivo. International Journal of Pharmaceutics, Vol. 286, No. 1-2, pp. 147-156.

Kiran, S., & Acosta, E. (2010). Predicting the morphology and viscosity of microemulsions using HLD-NAC model. *Industrial and Engineering Chemistry Research,* Vol. 49, No. 7, pp. 3424-3432.

Klose,G, Eisenblätter,S. Galle,J., Islamov,A. & Dietrich, U. (1995). Hydration and Structural Properties of a Homologous Series of Nonionic Alkyl Oligo (ethylene oxide) Surfactants. *Langmuir*. Vol. *11,* pp. 2889-2892.

Kogan A. ,& Garti N.(2006). *Adv. Colloid Interface Sci*. Microemulsions as transdermal drug delivery vehicles, Vol. 123 pp. 369

Kogan A., & Garti N, (2006). Microemulsions as transdermal drug delivery vehicles. *Adv. In Colloid and Interf. Sci.*, Vol. 123-12, pp. 369-385

Kogan, A., Shaler, D.E., Raviv, U., Aserin, A., & Garti, N., (1982). Formation and Characterisation of ordered Bicontinuous microemulsions. *J. Phys. Chem.*, Vol. 113, No. 31, pp. 10669-10678.

Kunieda, H., & Shinoda, K. (1980). Solution behavior and hydrophile-lipophile balance temperature in the Aerosol OT-isooctane-brine system - correlation between microemulsion and ultra low interfacial tensions. *J. of Colloid and Interface Science,* Vol. 75, No. 2, pp. 601-606.

Langevin, D., 1992, Chapter 13, In: *Multiphase Microemulsion Systems in Light Scattering by Liquid Surfaces and Complementary Technique., Surfactants Science Series vol. 41* ,D. Langevin (Ed), Marcel Dekkar Inc, ISBN: 0824786076 / ISBN-13: 9780824786076 , New York

Larson R. G., (1999), Chapter 12, In: *The Structure and Rheology of Complex Fluids.* Oxford University Press. ISBN-10: 019512197X, ISBN-13: 978-0195121971,USA

Lindman, B. (2001), Vol. 1, Chapter 19,In: *Handbook of Applied Surface and Colloid Chemistry,*. K. Holmberg, D. O. Shah, & M J Schwuger (Ed), John Wiley & Sons Ltd., ISBN: 978-0-471-49083-8, New Jersey

Magdassi, S., Ben Moshe, M., Talmon, Y., & Danino, D. (2003). Microemulsions based on anionic gemini surfactant. Colloids and Surfaces A- Physicochemical and Engineering Aspects, Vol. 212, No. 1, pp. 1-7.

Mark L. B. (1999), In: *Innovative Subsurface Remediation: Field testing of Physical, Chemical and Characterization Technologies,* M.L. Brusseau (Ed), pp. (49), American Chemical society,ISBN 0841235961, 9780841235960, Washington DC.

Minana-Perez M., Graciaa A., Lachaise J., & Salager J. (1995). *J. Colloid and Interface Sci.*, Vol. 100,pp. 217

Moulik S.P., Digout LG, Aylward W.M., & Palepu R., (2000). Studies on the Interfacial Composition and Thermodynamic Properties of W/O Microemulsions. *Langmuir*, Vol. 16, pp. No. 7, pp (3101–3106)

Moulik, S. P., De, G.C., Bhowmik, B.B., & Panda, A. K. (1999). Physicochemical Studies on Microemulsions. 6. Phase Behavior, Dynamics of Percolation, and Energetics of Droplet Clustering in Water/AOT/n-Heptane System Influenced by Additives (Sodium Cholate and Sodium Salicylate). *The Journal of Physical Chemistry B*, Vol.103, No. 34, pp. 7122-7129.

Mukhopadhyay, L., Bhattacharya, P. K., & Moulik, S. P. (1990). Additive effects on the percolation of water/AOT/decane microemulsion with reference to the mechanism of conduction. *Colloids and Surfaces*, Vol. 50, pp. 295-308.

Mukhopadhyay, L., Bhattacharya, P.K., & Moulik, S.P. (1993). Effect of Butanol and Cholestrol on the Conductance of AOT Aided Water Xylene Microemulsion. *Indian Journal of Chemistry Section A- Inorganic Physical Theoretical & Analytical Chemistry*, Vol. 32, No. 6, pp. 485-490.

Olsson, U. (2002). Specialty surfactants, In: *Handbook of Applied surface and colloid chemistry*, Holmberg, K., pp. 385-405, John Wiley & Sons Ltd., ISBN : 978-0-471-49083-8,

Olsson, U., Shinoda, K., & Lindman, B. (1986). Change of the structure of microemulsions with the hydrophile lipophile balance of nonionic surfactant as revealed by NMR self-diffusion studies. *Journal of Physical chemistry*, Vol. 90, No. 17, pp. 4083-4088.

Paul. B.K., Moulik, S.P. (2000). *Proc. Ind Natl. Sci. Acad.* Vol. 66A, pp. 449

Pedersen, J.S. (1994). Analysis of small-angle scattering data from micelles and microemulsions: free- form approaches and model fitting. *Current Opinion in Colloid & Interface Science*, Vol. 4, No. 3, pp. 190-196.

Peyerlasse, J., Moha-Ouchane, M., & Boned, C., (1988). Viscosity and the phenomenon of percolation in microemulsions *Phys. Rev. A.*, Vol. 38, No. 76, pp. 228.

Peyrelasse, J. & Boned, C. (1990). Conductivity, dielectric relaxation, and viscosity of ternary microemulsions: The role of the experimental path and the point of view of percolation theory *Phys. Rev. A.* Vol. 41, pp. 938–953.

Podlogar, F., Gašperlin, M., Tomšič, M., Jamnik, A., & Rogač, M.B. (2004). Structural characterisation of water–Tween 40®/Imwitor308®–isopropyl myristate microemulsions using different experimental methods. *International Journal of Pharmaceuticals.*, Vol. *276*, pp.115-128.

Porras,M., Solans, C., Gonzalez, C., Martinez, A., Guinart, A., & Gutierrez, J.M. (2004). Studies of formation of W/O nano-emulsions. Colloids and Surfaces A-Physicochemical and Engineering Aspects, Vol. 249, No. 1-3, pp. 1413-1421.

Queste,S., Salager, J.L., Strey, R. & Aubry, J.M. (2007). The EACN scale for oil classification revisted thanks to fish diagrams, J. Colloid & Interface Sci., Vol. 312, pp. 98-107.

Ray S., Bisal S. R., & Moulik S. P. (1994). Thermodynamics of Microemulsion Formation. 2. Enthalpy of Solution of Water in Binary Mixtures of Aerosol OT and Heptane and Heat Capacity of the Resulting Systems. *Langmuir*, Vol.10, No.8, pp. 2507–2510.

Ray, S., Bisal, S. R., & Moulik, S. P. (1993). Structure and dynamics of microemulsions. Part 1.—Effect of additives on percolation of conductance and energetics of clustering in water–AOT–heptane microemulsions. *Journal of Chemical Society, Faraday Transactions*, Vol. 89, No. 17, pp. 3277- 3282.

Ray, S., Paul, S., & Moulik, S. P. (1996). Physicochemical Studies on Microemulsions: V. Additive Effects on the Performance of Scaling Equations and Activation Energy for Percolation of Conductance of Water/AOT/Heptane Microemulsion. *Journal of Colloid and Interface Science*, Vol. 183, No. 1, pp. 6-12.

Regev, O., Ezrahi, S., Aserin, A., Garti, N., Wachtel, E., Kaler, E. W., Khan, A., & Talmon, Y. (1996). A study of the microstructure of a four-omponent nonionic microemulsion by cryo-TEM, NMR, SAXS and SANS. Langmuir, Vol. 12, No. 3, pp. 668-674.

Reimer, J. & Söderman, O., Sottmann, T., Kluge, K. & Strey, R. (2003). Microstructure of Alkyl Glucoside Microemulsions: Control of Curvature by Interfacial Composition. *Langmuir*, Vol. 19, No. 26, pp 10692–10702.

Rodriguez, C., & Scamehorn, J. (1999). Modification of Kraft temperature or solubility of surfactants using surfactant mixtures. *Journal of Surfactants and Detergents*, Vol. 2, No. 1, pp. 17-28.

Rosen M.J.(2004). In: *Surfactants and Interfacial Phenomena*, pp. (353), John Wiley & Sons, ISBN 0471478180, 9780471478188, New Jersey

Safran, S. A., Roux, D., Cates, M.E., & Andelman, D. (1986). Origin of Middle-Phase Microemulsions. *Physical Review Letters*, Vol. 57, No. 4, pp. 491-494.

Saidi, Z., Mathew, C., Peyrelasse, J. & Boned, C. (1990). Percolation and critical exponents for the viscosity of microemulsions. Phy. Rev. A., Vol. 42, No.2, pp. 872-876.

Salager Jean-Louis, Antón Raquel E., Sabatini David A., Harwell Jeffrey H., & Acosta Edgar J., et al. 2005. Enhancing solubilization in microemulsions—State of the art and current trends. J. Surf. Deterg., Vol. 8, No. 1, pp. 3-21.

Salager, J. (1978). PhD Dissertation, University of Texas, Austin

Salager, J., Bourrel, M., Schechter, R., & Wade, W. (1979). Mixing rules for optimum phase-behavior formulations of surfactant-oil-water systems. Society of petroleum Engineers Journal, Vol. 19, No. 5, pp. 271-278.

Scamehorn, J. F. (1986). Phenomena in mixed surfactant system. ACS Symposium series, Vol. 311, NewYork.

Schulman, J. H., Stokenius, W.& Prince, L. M. (1959). Mechanism of Formation and Structure of Micro Emulsions by Electron Microscopy. Journal of Physical Chemistry, Vol. 63,10, pp. 1677–1680.

Schwuger, M., Stickdorn, K., & Schomacker, R. (1995). Microemulsions in technical processes. *Chemical Reviews*, Vol. 95, No. 4, pp. 849-864.

Scriven L.E. (1976). Equilibrium bicontinuous structure. *Nature*, Vol. 263, pp. 123-125.

Shiloach, A., & Blankschtein, D. (1998). Predicting micellar solution properties of binary surfactant mixtures. *Langmuir*, Vol. 14, No. 7, pp. 1618-1636.

Shinoda, K., & Saito, H. (1969). The Stability of O/W type emulsions as functions of temperature and the HLB of emulsifiers: The emulsification by PIT-method. *Journal of Colloid and Interface Science*, Vol. 30, No. 2, pp. 258-263.

Shinoda, K., Kunieda, H., Arai, T., & Saijo, H. (1984). Principles of attaining very large solubilisation (microemulsion): Inclusive understaning of the solubilisation of oil and water in aqueous and hydrocarbon media. *Journal of Physical Chemistry*, Vol. 88, No. 21, pp. 5126-5129.

Silas, J. A., & Kaler, E. W. (2003). Effect of multiple scattering on SANS spectra from bicontinuous microemulsions. *Journal of Colloid and Interfacial Science*, Vol.257, No. 2, (January 2003), pp. 291-298.

Sjöblom, J. (1996). Microemulsions-phase equilibria, characterization, structures, applications and chemical reactions. *Advances in Colloid and Interface Science* Vol. 95, pp. 125.

Sjöblom, J., Lindberg, R., & Friberg, S.E. (1996). Microemulsions – phase equilibria, characterization, structures, applications and chemical reactions. *Advances in Colloid and Interface Science*, Vol. 65, pp. 125-287.

Solans C., & Kunieda H. (1996). In: *Industrial Applications of Microemulsions*, C. Solans, H. Kunieda (Eds), pp. 375, Marcel Dekker, ISBN 0824797957, 9780824797959, New York,

Sottmann, T., Lade, M., Stolz, M. & Schomacker, R. (2002). Phase behavior of non-ionic microemulsions prepared from technical-grade surfactants. *Tenside Surfactants Detergents*, Vol. 39, No. 1, pp. 20-28.

Spernath, A., Yaghmur, A., Aserin, A., Hoffmann, R. E., & Garti, N. (2003). Self-Diffusion Nuclear Magnetic Resonance, Microstructure Transitions, and Solubilization Capacity of Phytosterols and Cholesterol in Winsor IV Food-Grade Microemulsions. Journal of Agriculture and Food Chemistry, Vol. 51, No. 8, pp. 2359-2364.

Srivastava, S., Kini, G. & Rout, D. (2006). Detergency in spontaneously formed emulsions. *J. Colloid & Interface Sci.*, vol. 304, No.1, pp. 214-221.

Talmon, Y. (1986). Imaging Surfactant Dispersions by Electron-Microscopy of Vitrified Specimens. Colloids and Surfaces, Vol. 19, No. 2-3, pp. 237-248.

Talmon,Y. & Prager,S. (1978). Statistical Thermodynamics of Phase Equilibria in Microemulsions. *J.Chem.Phys.*,Vol. *69*, pp. 2984.

Theander, K., & Pugh, R. (2003). Synergism and foaming properties in mixed nonionic/fatty acid soap surfactant systems. *Journal of Colloid and Interface science*, Vol. 267, No. 1, pp. 9-17.

Tongcumpou C., Acosta E. J., Scamehorn J. F., Sabatini D. A. and Yanumet N. (2006), Enhanced triolein removal using microemulsions formulated with mixed surfactants. *J. Surf. Deterg*. Vol. 9, No.2, pp. 181-189.

Widom, B. (1996). Theoritical Modelling: An Introduction. *Physical Chemistry Chemical Physics*, Vol. 100, No. 3, pp. 242-251.

Witthayapanyanon, A., Harwell, J., & Sabitini, D. (2008). Hydrophile-lipophile deviation (HLD) method for characterizing conventional and extended surfactants. *Journal of Colloid and Interface science*, Vol. 325, No. 1, pp. 259-266.

Wu, D., Chen, A., & Johnson, C.S. (1995). An improved Diffusion Ordered Spectroscopy incorporating Bipolar Gradient Pulses. *Journal of Magnetic Resonance, Series A*, Vol. 115, pp. 260-264

Yaghmur, A., Aserin, A., Antalek, B. & Garti, N. (2003). Microstructure Considerations of New Five-Component Winsor IV Food-Grade Microemulsions Studied by Pulsed Gradient Spin-Echo NMR, Conductivity, and Viscosity. *Langmuir*, Vol. *19*, pp. 1063-1068

Yaghmur, A., Aserin, A., Antalek, B., & Garti, N. (2003). Microstructure consideration of new five component winsor IV food grade microemulsions studied by pulsed gradient spin-echo NMR, conductivity and viscosity. *Langmuir*, Vol. 19, No. 4, pp. 1063-1068.

Yaghmur, A., DeCampo, L., Aserin, A., Garti, N.& Glatter, O. (2004). Structural characterization of five-component food grade oil-in-water nonionic microemulsions. *Physical Chemistry Chemical Physics*, Vol. 6, No. 7, pp.1524-1533.

Zana R. (1994), Microemulsions. *HCR Advanced Educational Review*, Vol. 1, pp. 145- 157.

Zu Z.J., & Neuman R.D., (1995). Reversed Micellar Solution-to-Bicontinuous Microemulsion Transition in Sodium Bis (2-Ethylhexyl) Phosphate/n-Heptane/Water System. *Langmuir*, Vol. 11,No. 4, pp1081–1086.

Part 3

Applications in Drug Delivery and Vaccines

5

Nonionic Model Microemulsions to Study Interactions with Active Components and Antioxidant Activity

Joakim Balogh[1,2], Luís Marques[1] and António Lopes[1]
[1]*ITQB Oeiras*
[2]*Physical Chemistry Lund University*
[1]*Portugal*
[2]*Sweden*

1. Introduction

This chapter reviews and presents the latest developments on the work on using nonionic microemulsions stabilized by surfactants of ethylene oxide alkyl ether type, C_mE_n, where *m* is the number of carbons in the alkyl chain and *n* is the number of ethylene oxide groups as models to determine the effect of an added component. We will here investigate if active components can be added to this model drug delivery system without them affecting the system. This is important as there is now several "ready to use" drug delivery systems where one just need to add an active component. We will here see if it is as simple as this or if more work will be required with these formulations.

These model systems stems from research based at Lund University (Olsson & Schurtenberger, 1993) that has been thoroughly investigated and characterized (Balogh, Olsson & Pedersen, 2006, Balogh et al. 2007, Le et al., 1999 Leaver et al., 1994, Leaver et al., 1995, Olsson & Schurtenberger 1993). The effect of changing the alkane (decane changed to hexadecane) has also been studied (Balogh et al., 2006, Balogh, Olsson & Pedersen, 2006, Balogh et al. 2007, Balogh & Olsson 2007). The surfactants and similar systems have been thoroughly investigated trough the massive work by Shinoda and Kuneida (Shinoda & Kunieda, 1973) in Yokohama, Japan and Kahlweit and Strey (Kahlweit et al, 1985) Göttingen, Germany, later Strey and Sottmann (Sottmann & Strey, 1996) with coworkers in Cologne, Germany. More about the early days of these surfactants and systems can be found in these references (Balogh et al., 2010, Lindman 2008). As with most systems the investigations of the phase boundaries are essential. As these systems are temperature sensitive through their surfactant they can change phases by just having the temperature changed. There is also no need for co-surfactants or co-solvents to be able to have a wide variety of phases. It is possible with a small number of samples to study changes into several phases. As the structures are thermodynamically stable one can go up and down in temperature to determine the boundaries more accurately. For the studied system the phase changes are fast and go from turbid to clear, making them easier to investigate. Amongst its many uses (Fanun 2009, Kunieda & Solans 1997) microemulsions have been used as drug delivery systems since long time and have been investigated as such (Garti et al., 2004, Gupta & Moulik, 2008, Heuschkel et al., 2008, Kreilgaard et al., 2001, Lawrence & Rees, 2000, Spernath & Aserin, 2006) especially topical formulations (Grampurohit et al., 2011, Kreilgaard 2002). Nonionic microemulsions of the C_mE_n type has been studied regarding the effect of adding lidocaine (Balogh & Pedersen 2007, Balogh et al 2010). It has also been used as a membrane to study the effect of other analgesics on membranes (Baciu et al., 2007). Microemulsions are usually used to get hydrophobic active components into solution. Microemulsions are also reported to improve drug delivery and as such they are more than just a passive vehicle for the active component. It is an active vehicle that improves the performance of the formulation (Bagwe et al., 2001, Kogan & Garti 2006) especially on topical formulations (Lopes et al., 2010, Santos et al., 2008, Sintov & Levy, 2007). It is therefore important to study the interaction between the active component and the drug delivery system to better know the boundaries of the system. This gives better knowledge on the shelf-life of the formulations. Lidocaine is amongst other used as local anesthetic in dermal creams and is as such, a good candidate as a model drug for a drug delivery system aimed for topical delivery. It is well known that having a hydrophobic compound, which is sensitive to oxidation, in a microemulsion or an emulsion help protect this compound (Gaonkar & Bagwe, 2002). It is also possible to have extra protection by having antioxidant in the system and if formulated well one can have antioxidants in both the oil- and water phase. Antioxidants have now also become a component that is added to many formulations to act as active component and not only to protect other components. This is extra common in skin products where vitamin E is a component that helps sell the products, often labeled as tocopherol acetate, tocopheryl linoleate or tocopheryl nicotinate (Nabi et al., 1998). Some studies have shown that applying vitamin E before UV exposure significantly reduces acute responses such as edema and erythema (Bissett et al., 1992). Topical application is related to decrease of the incidence of ultraviolet UV-induced skin cancer (Krol et al., 2000, McVean & Liebler, 1999). The form of vitamin E that is preferentially absorbed and accumulated in humans is α-tocopherol (Rigotti, 2007). These formulations show many similarities to the model system used here. There is also a trend with "natural" antioxidants in many kinds of products from juices to shampoos and skin products. One fairly available antioxidant is chlorogenic acid, a polyphenol with a dry matter content in coffee beans on 4-14% depending on coffee species (Clifford et al., 1985, De Maria et al., 1994, Ky et al., 2001, Trugo & Macrae, 1984). From the chlorogenic acids family (caffeoylquinic acids), the most reported isomers are, 3-CQA, 4-CQA and 5-CQA. 5-CQA is the most abundant compound found in coffee beans, responsible for about 56-62 % of total chlorogenic acids (Ramirez-Colonel et al., 2004, Ramirez-Martinez 1988). It is also the most abundant antioxidant in coffee foliage (Marques, 2011). It has been getting a lot of attention and the reported health benefits both *in vivo* and *in vitro*, from chlorogenic acids are hepatoprotective, immunoprotective, hypoglycemic and antiviral activities (Basnet et al., 1996, Natella et al., 1998, Scalbert & Williamson, 2000). We are using 5-CQA as our type chlorogenic acid.

Phase studies give a lot of essential information about the system and it has even been done to make sure that the batch of surfactant used is good by constructing a known phase diagram. With three ingoing components and temperature the full spectra of the systems could be illustrated with a phase prism. To optimize the ratio between work and information some cuts (fixing ratios of components) have been used. Three classic ones are the Shinoda cut (Kunieda & Shinoda, 1982) with fixed surfactant ratio varying the water to oil ratio, to investigate the reversal of the system amongst others, the Kahlweit cut (Kahlweit & Strey, 1985) with fixed water to oil ratio often 1:1 on volume base, to investigate the efficiency of the surfactant amongst others, and the Lund cut (Olsson & Schurtenberger, 1993) with fixed ratio of surfactant to oil, often 0.815:1, to study the microemulsion aggregates concentration dependence. The Kahlweit cut is also known as the fish or the fishtail cut due to shape of the 3-phase body and the microemulsion phase. Even though the Lund cut is from the nineties it is only named recently (Balogh, 2010, Balogh et al., 2010). The phase cuts are illustrated in Figure 1.

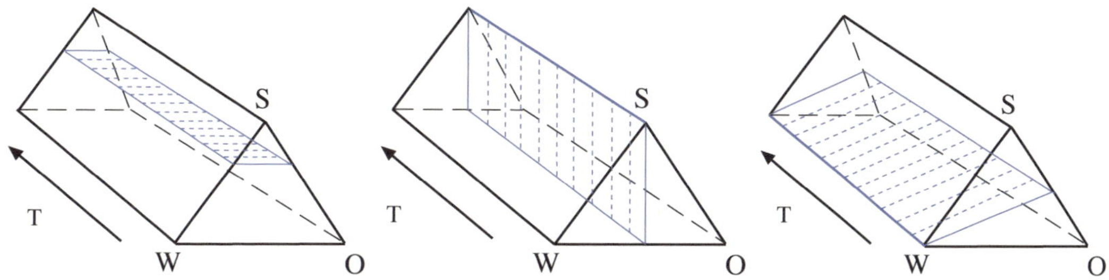

Fig. 1. Phase cuts. Here three different phase cuts are illustrated from the Shinoda cut on the left with the Kahlweit in the middle and the Lund cut to the right. T stands for temperature, W for water, O for Oil and S for Surfactant.

With these systems there are some trends that can be used to help in determining a new system and there is also some scaling. For the Lund system using alkanes and $C_{12}E_5$ the trend is that the lowest temperature where one can have microemulsions, T_{EFB}, increases with the length of oil from octane to octadecane. The microemulsion temperature range, the temperature increase one can do from T_{EFB} and still have 1-phase microemulsion, decreases with increased oil length (Balogh, 2010, Balogh et al., 2010). This is a general finding for all these systems. There have also been reports on the lamellar phase and that it is not only becoming smaller but actually withdrawing gradually to high concentrations only, with aggregates volume fractions above 0.5 (Balogh, Olsson & Pedersen, 2006, Balogh, 2011). We have not seen reports if this is a general finding for these types of systems or not.

2. Result

We will here present the standard system comprised of $C_{12}E_5$, decane and water (surfactant, oil and water) with phase studies and then diffusion studies both at the emulsification boundary and at elevated temperatures. We will then present the systems with an extra added component both phase studies, studies of the zeta potential and diffusion and compare with the standard system. This will be followed by the studies of the antioxidant activity both individually in the microemulsion compared to in a pure solvent and together in the microemulsion.

2.1 Studies of the system $C_{12}E_5$ decane and water

The samples were performed prior to using by diluting a stock-solution of surfactant and oil ($C_{12}E_5$:decane 0.815:1 on volume base) with water in the water rich region. The stock-solution volume fraction, Φ, stretched the 0.005-0.2 region but was mostly investigated in the 0.05-0.2 region. The samples where then mixed above the upper phase boundary (turbid two-phase region) for five minutes and then cooled while stirred into the microemulsion region (clear one-phase solution). They were stored at this temperature until used.

2.1.1 Phase studies of the system $C_{12}E_5$ decane and water

We will here present mostly work using the original type system for the Lund cut, $C_{12}E_5$, decane and water. A phase diagram for the microemulsion phase that we are working in is shown in Figure 2.

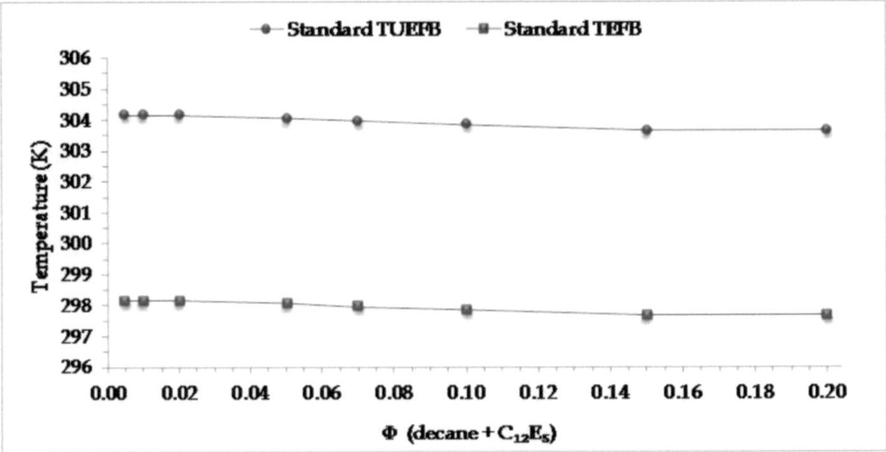

Fig. 2. Partial phase diagram for the standard system ($C_{12}E_5$, decane and water) showing the upper (spheres) and lower phase boundaries (squares) temperatures for the microemulsion phase in the water rich region with the surfactant and oil concentrations from $\Phi = 0.005$ to 0.2.

A comprehensive phase diagram has been determined earlier (Le et al., 1999). This shows that above the microemulsion phase is a lamellar phase and even further up in temperature is a bicontinuous bilayer phase. At higher concentrations just above volume fractions of 0.4 starts liquid crystalline phases like cubic and hexagonal. We have been working in a more limited area of the range mostly in the microemulsion phase with aggregate ratios, Φ, mostly from 0.05 to 0.2. This is done to have as little impact from crystalline structures at higher concentrations and from the edge of the dilution limit as possible. It can be hard to follow the system visually at the highest dilutions with Φ, from 0.005 to 0.02 as they tend to look "clear" even when there is two phases. Initially at the lower phase boundary it starts as discrete aggregates that, at the lower concentrations, are spherical and towards the upper boundary the system becomes bicontinuous. The higher the concentration the lower is the increase in temperature needed to change into bicontinuous system (Balogh & Olsson, 2007, Balogh et al. 2007, Leaver et al., 1994, Leaver et al., 1995). The change happens without macroscopic phase change.

2.1.2 Diffusion studies of the system $C_{12}E_5$, decane and water

With increased temperature the curvature of the surfactant film decreases. As long as one is below the lower emulsification boundary and thus has access to more oil, the aggregates essentially just grow as spheres and become bigger spheres. So at the emulsification boundary the aggregates are spherical at least at lower concentrations. At higher concentrations and especially for longer oils the aggregates become slightly elongated already at the emulsification temperature (Balogh, Olsson & Pedersen, 2006). The growth of these aggregates has been shown to best fit with the one-dimensional ellipsoidal growth (rugby ball) (Leaver et al., 1995). The growth before changing into bicontinuous does not extend axis ratios of 4-6 big axis to small axis (Balogh & Olsson, 2007, Balogh et al. 2007, Leaver et al., 1994, Leaver et al., 1995). Earlier reports of elongated micelles did not investigate if the aggregates had changed into the bicontinuous phase as defined by the self diffusion of the surfactant being different from that of oil (for an oil-in-water system). The behavior in scattering terms of the bicontinuous system is initially such that it easily could be confused with elongated micelles. As there is no macroscopic phase change it is not seen any change when performing phase studies of the microemulsion phase. When performing light scattering experiments one need to remember that for dynamic light scattering one is studying collective diffusion. Collective diffusion has a concentration dependence that makes it important to have in mind when comparing different concentrations. It also means that the size coming from the light scattering machine has not taken this into account as it just using the Stokes Einstein equation to get size from diffusion.

$$D_0 = \frac{K_b T}{6\pi\eta r} \tag{1}$$

Here D_0 is the diffusion at infinite dilution, k_B boltzmann constant, η viscosity and r radius of the aggregate. From this it is easy to see how the size depends on the diffusion in this case at infinite dilution and with spherical aggregates.

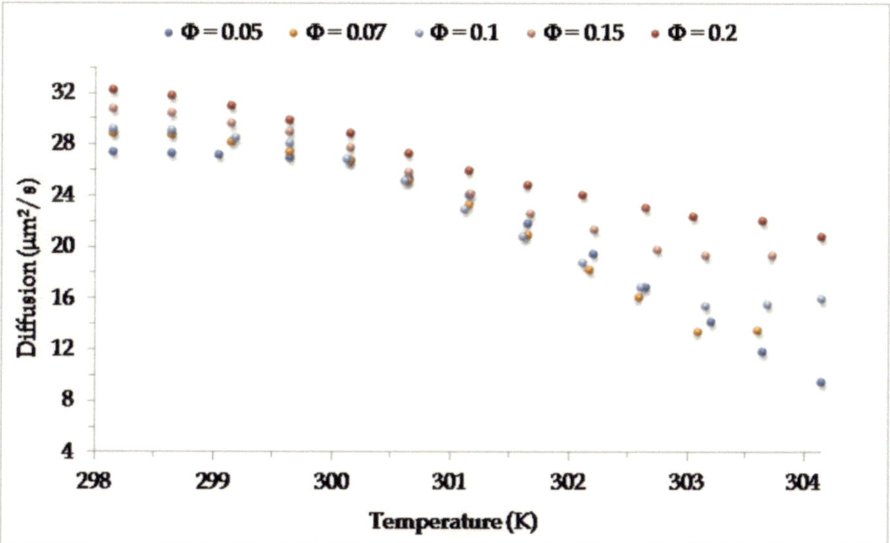

Fig. 3. Diffusion versus temperature for five droplet concentrations from Φ = 0.05 to 0.2.

When looking at Figure 3 for one temperature it seems as if the higher the concentration the higher the diffusion. Just applying Stokes Einstein equation would give smaller size with higher concentration. This is however not the case as we will show. If one want to compare different concentrations it is important that one have the concentration dependence of the collective diffusion in mind. The hard sphere behavior (if there is no growth in the system or repulsion/attraction between the aggregates) is $D_C/D_0 = 1+\Phi_{HS}*1.45$. Where D_C is the collective diffusion, D_0 is the diffusion at infinite dilution and Φ_{HS} the hard sphere volume fraction, for these systems 1.14 times that of the aggregate volume fraction (Olsson & Schurtenberger, 1993). If the result is presented as the diffusion normalized with the diffusion at infinite volume fraction at each temperature (to just have the effect of concentration at each temperature), as in Figure 4, one can see that the difference from the hard sphere behavior is bigger at higher temperatures and concentrations. It has earlier been shown that there is no attraction in these systems that would explain this so the only explanation left is that the aggregates grow. This has been shown to be the case (Balogh & Olsson, 2007, Balogh et al. 2007, Leaver et al., 1994, Leaver et al., 1995). At higher temperatures the aggregates are more elongated and also becomes bicontinuous so that the diffusion do not give a good representation of the aggregate size just some form of repeated unit in the structure and the perceived diffusion may even go up.

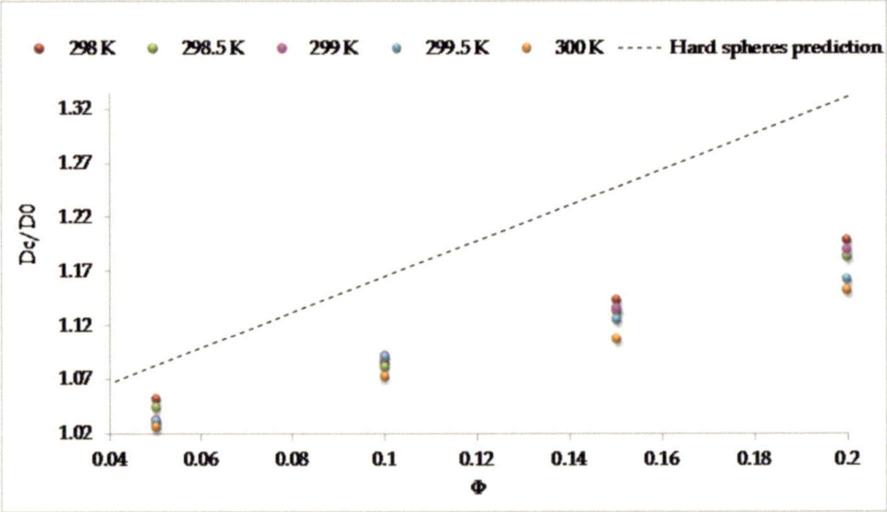

Fig. 4. Diffusion versus droplet concentration from $\Phi = 0.05$ to 0.2 for several temperatures with the collective diffusion behavior for aggregates acting like hard spheres as reference.

Just to once again mention the effect of changing the alkane one would see the same thing as in Figure 4 that the higher temperature the faster they grow with concentration but with increasing chain length. This is seen as the slope of the fitted line would be lower and further away from the hard sphere behavior with increasing chain length or temperature (Balogh, Olsson & Pedersen 2006, Balogh, 2011).

2.2 Studies of the system $C_{12}E_5$, decane and water with an added component

As the model system is well characterized it is easier to study the effect of adding a fourth component to the system especially if one will work with limited number of techniques. Generally the systems used for drug delivery are more complex and often not studied as thoroughly as our model. The findings from an applied system need to be complemented with more techniques to explain what happen on aggregate level. The addition of three different components has been studied. The components are different as one is a local anesthetic, oil soluble and slightly water soluble (20:1 solubility in decane:water) and the other two are natural compounds that act as antioxidants. The antioxidants are different as chlorogenic acid, CQA, is water soluble and α-tocopherol is oil soluble. The molecular structures are shown in Figure 5.

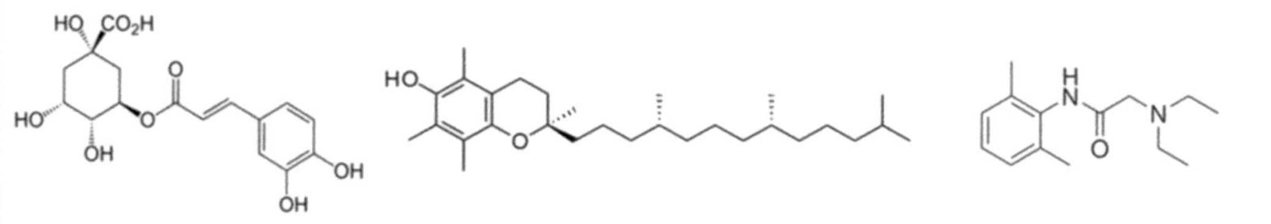

Fig. 5. Molecular structure for the added components. From left to right chlorogenic acid, α-tocopherol and lidocaine.

To try to keep the lipophilic component constant part of the oil was substituted with lidocaine and α-tocopherol respective. The system was close to the limit of how much that could be solved with 10% volume fraction of lidocaine in decane. When adding the surfactant it was no problem to keep it in solution, even in the fridge. The solubility of lidocaine in water was low but there was some solubility estimated to 0.5% giving a ratio of solubility between decane and water to 20:1. There was no problem in solving 10 volumes percent of α-tocopherol in decane. We dissolved a maximum of 2% CQA in water. Tocopherol had very low solubility in water and is often reported as insoluble in water. The solubility of CQA in decane was low. We used the oil with the added substance as "oil" when preparing our stock solution and used water containing CQA to dilute the stock solution when preparing our measurements. From this it is seen that the concentration of CQA was almost constant regardless of the aggregate concentration (giving different ratio). We added 1, 5 and 10 volume percent of lidocaine giving the active compound content of the total system of 1% with a 10% addition and with an aggregate fraction of 0.2. The local analgesic EMLA® from AstraZeneca contains 2.5% lidocaine so the concentrations used are similar to the formulations used today. Concentrations of 2% CQA are much higher than that of coffee and with a maximum of 1% of α-tocopherol on total formulation it seems to be above what is used in formulations today.

2.2.1 Phase studies of the system $C_{12}E_5$, decane and water with an added component

The first thing that needs to be done was to perform phase studies and determine the phase diagram with the additions. In Figure 6 we show the temperature of the lower phase boundaries for different droplet concentrations and different addition of CQA. There is a small but increasing difference in temperature with droplet concentration much more than in the standard and for the other additions.

As the concentrations of the antioxidant components are relatively high compared to many formulations, it was necessary to also study the effect at lower concentrations and for concentrations CQA 0.005-0.02% the change from standard was small. From here it is seen that not only the temperature where the microemulsion form changes, but also the upper boundary.

However the upper boundary does not increase as much as the lower giving a smaller microemulsionrange. The range becomes smaller with higher CQA concentration (and by that the temperature of the emulsification boundary). A similar thing has been seen in system where the oil length has been varied, the lower the temperature the bigger the range. This gives that shorter alkanes has bigger microemulsion temperature range than the longer ones.

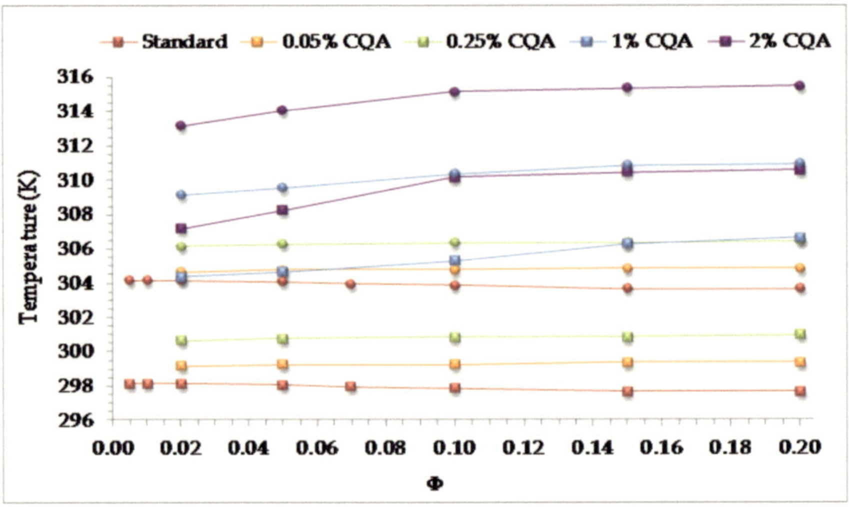

Fig. 6. Partial phase diagram in the water rich region showing the upper (spheres) and lower phase boundaries (squares) for the microemulsion phase with the surfactant and oil concentrations from Φ = 0.005 to 0.2 and with 0.05-2% of water substituted with chlorogenic acid.

Phase investigations were also performed for the system with added lidocaine. As can be seen in Figure 7 there is a significant change when increasing the substitution.

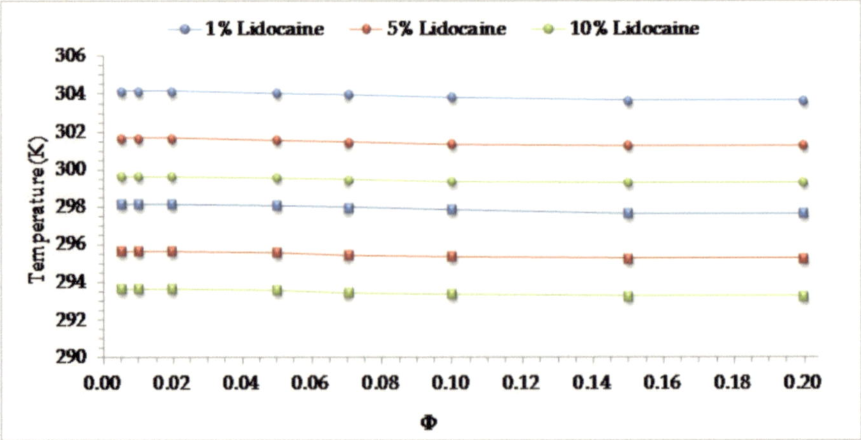

Fig. 7. Partial phase diagram in the water rich region showing the upper (spheres) and lower phase boundaries (squares) for the microemulsion phase with the surfactant and oil concentrations from Φ = 0.005 to 0.2 and with 1, 5 and 10% of the oil substituted with lidocaine.

The phase boundaries of the standard system are very close to that of a 1% substitution with lidocaine. One can also see that with increasing aggregate concentration the phase boundaries are lowered slightly.

Finally phase investigations were performed on the systems with added α-tocopherol and below are shown the resulting phase diagram in Figure 8.

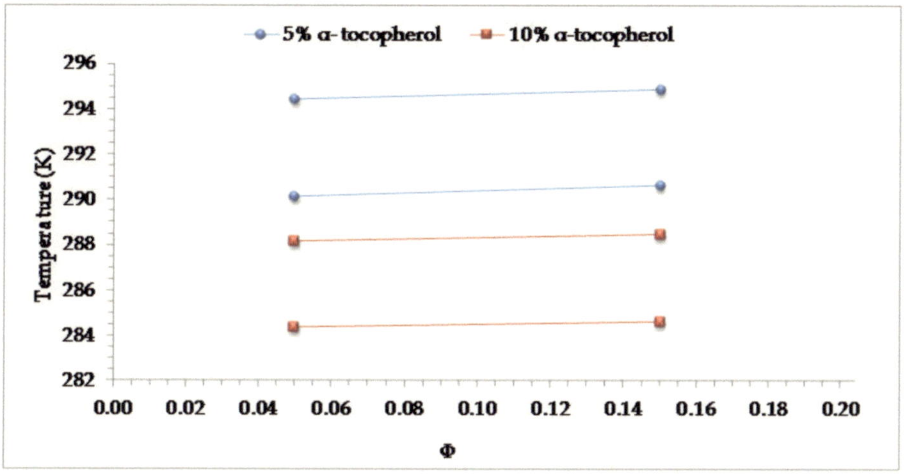

Fig. 8. Partial phase diagram in the water rich region with the surfactant and oil concentrations Φ = 0.05 and 0.15 with 5 and 10% of the oil substituted with α-tocopherol.

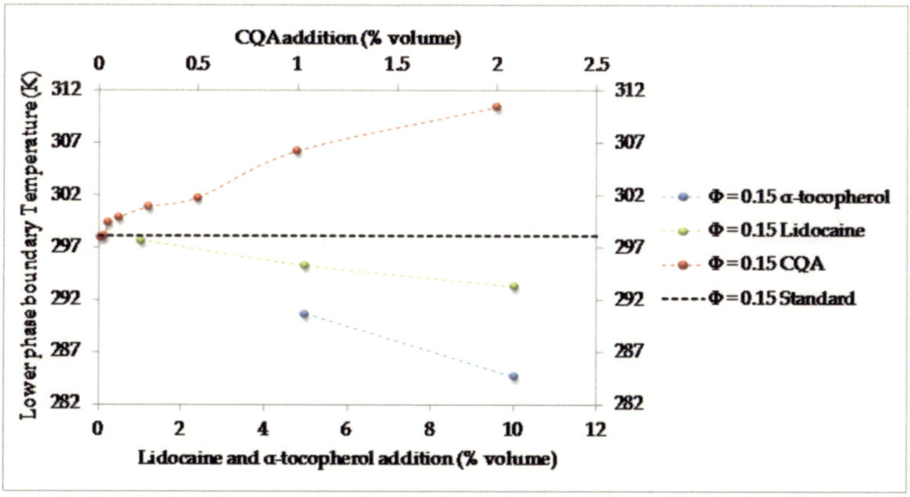

Fig. 9. Partial phase diagram in the water rich region with the surfactant and oil concentrations fixed at Φ = 0.15 with varied amount of the oil substituted with α-tocopherol and lidocaine, and water substituted with CQA.

Here there is a slight increase of phase boundaries with temperature. The substitution with α-tocopherol makes a bigger impact than the same substitution with lidocaine. The temperatures of the phase boundaries are lowered far from the standard system with 298 as lower boundary and for 5% substitution this is above its upper phase boundary. This dramatic phase change needs to be addressed when formulating with this high concentrations. Below Figure 9 is showing that the lower phase boundary of CQA is increasing the temperature and that lidocaine and especially α-tocopherol is lowering the temperature. Here, all of the data is for a droplet ratio of 0.15.

The increase caused by the substitution with CQA is big compared to the others looking at their part of the solvent. As there is more water than decane the change with number of molecules are similar between CQA and α-tocopherol but higher than that of lidocaine.

2.2.2 Zeta potential studies for the system $C_{12}E_5$, decane and water with an added component

The effect of the substitution with CQA was very strong and in order to try to find out why, more investigations were performed. As CQA can be charged in water the possibility of the aggregates also being charged was investigated. Zeta potential measurements were performed and a small Zeta potential was noted. The standard system did not show any zeta potential. To study if it was a "real effect" and not just a coincidence repeated experiments were performed varying the concentrations of the ingoing components. Figure 10 shows the fixed concentration of CQA and the variation of the aggregate concentration.

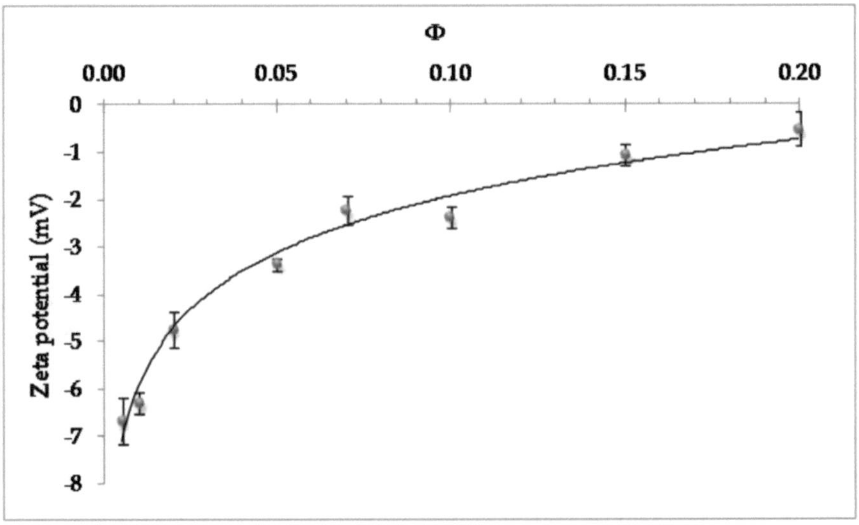

Fig. 10. Zeta potential for different aggregate concentrations with fixed 1 % CQA concentration (triplicate measurements, and standard deviation in the error bars).

As can be clearly seen there is a concentration dependence and the aggregates with most charges are also the ones that has most CQA per aggregate (lower droplet concentrations). To be certain of the effect we also varied the CQA concentration and had fixed aggregate concentration. This is shown in Figure 11.

Here one can also see the relation with charge of the aggregate and the CQA concentration. From here it is evident that the aggregates behave as if they are charged. Up to a bulk concentration of 2% the concentration in the bulk influence the apparent charge of the aggregate. If the charge is connected to the aggregate or if it is several charges that shift and by its close presence to the aggregate make it appear charged is not determined yet. It was not enough to have a charged bulk water to get aggregates that appeared to be charged, as was seen when using sodium chloride with the same charge (Balogh et al., 2012). This had no or minimum effect on the phase boundaries, the zeta potential (had none) and the diffusion compared to the standard system. Lidocaine did not show any effect on the zeta potential even if it had some solubility in water. Tocopherol was not studied as it has so low solubility in water that we did not expect to see any effect.

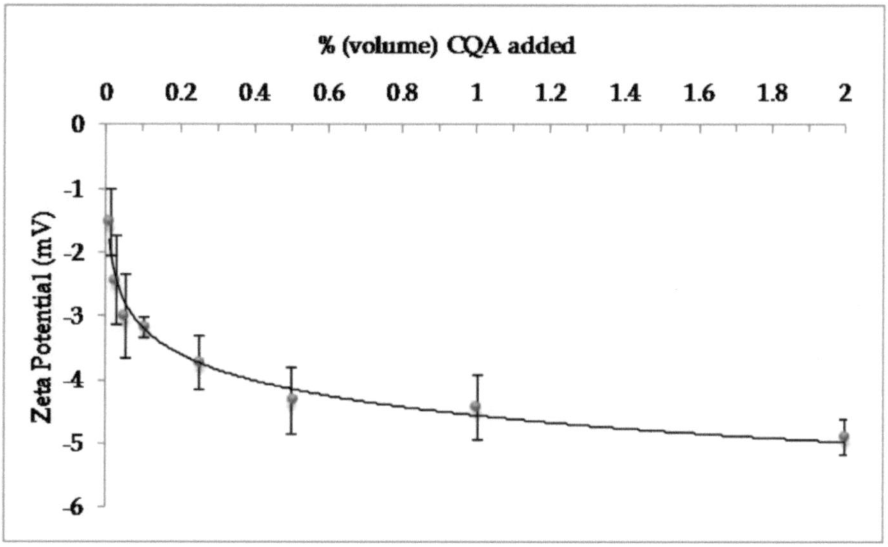

Fig. 11. Zeta potential for different CQA concentrations with fixed aggregate concentrations Φ=0.05 (triplicate measurements, and standard deviation in the error bars).

2.2.3 Diffusion studies for the system $C_{12}E_5$, decane and water with an added component

The diffusion of the system with added compounds did change a bit from standard part of it, which was due to the different temperature range of the microemulsion phase and by that the measuring temperature. To eliminate this temperature effect the curves where normalized with their respective diffusion at the temperature of lower phase boundary. Then one can see that the curves also are slightly different as can be seen from Figure 12. Here one can look at the shape of the curve and compare to see that they have a very similar behavior. For small additions it was hard to see any significant difference but at higher concentrations the profile is slightly different showing that the growth of the aggregates and the temperatures where the systems change into bicontinuous aggregates differs from that of the standard system.

The small differences in how these systems behave with temperature, and hence when they change from discrete aggregates to a bicontinuous system, may influence a lot of properties for a formulation, such as the availability of drug and stability of the formulation, and thus important parameters to be aware of.

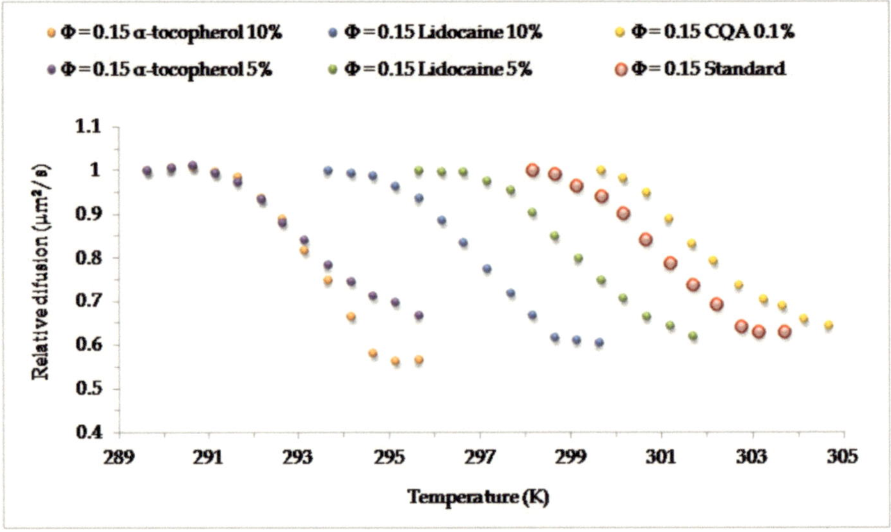

Fig. 12. Relative diffusion the collective diffusion normalized with the diffusion at the respective phase boundary temperature for several systems with added components.

2.3 Studies of the antioxidant properties in the microemulsion

Even though antioxidants can be used to manipulate the phase boundaries, their primary function would be to act as antioxidants either to protect an active component or themselves as main components. It was hence important to confirm that the antioxidants still had the capability to interact with oxidants if these would appear in the microemulsion. We determined the antioxidant activity by using a dye that is a radical 2,2-**dip**henyl-1-**p**icryl**h**ydrazyl (DPPH) hence after called DPPH. The structure of DPPH is shown in Figure 13

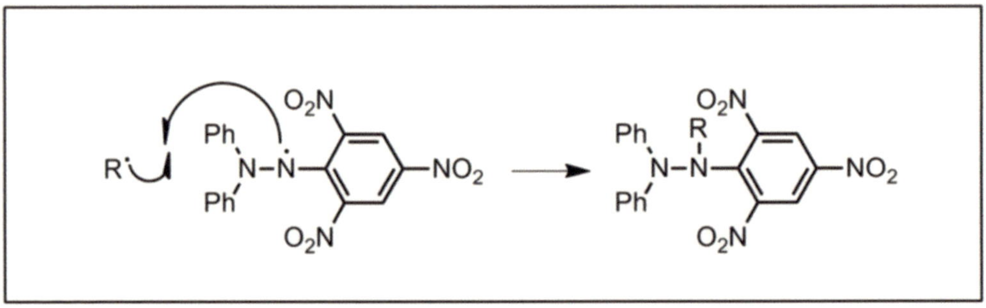

Fig. 13. The molecular structure of 2,2-**dip**henyl-1-**p**icryl**h**ydrazyl (DPPH) and then when it has reacted with another radical.

The problem is to have a system where a dye can be used to test both hydrophobic and hydrophilic compounds. DPPH is soluble in methanol and in decane and it fits our purpose. An added dye may influence the phase boundaries but at the concentrations used here, up to 60 µmol/L, there were no significant changes. First a calibration curve of DPPH was prepared. Then the values after the addition of antioxidant were compared to this to get the concentration. As DPPH is a strong dye, the concentrations that could be used and still be in the region where there is a good linear correlation between absorption and concentration, was quite low as stated earlier (up to 60 µmol/l). This also influenced the maximum concentrations of the antioxidants. It was also possible to use DPPH in decane used to prepare the stock solution for the microemulsion experiments.

2.3.1 Studies of the antioxidant properties of CQA in the microemulsion

We could determine the effect of CQA with DPPH in a methanol solution and then compare that with the effect in the microemulsion. In the microemulsion we dissolved the DPPH in the decane used in our microemulsion and CQA was in the water used to dilute the microemulsion stock solution. This way we did not need to add anything that would influence the microemulsion more than the dye itself. The dye was in such low concentrations that it did not influence the microemulsion phase boundaries. We were somewhat limited in the amount of DPPH we could have in the decane for the microemulsion but not significantly and we were able to have a good concentration interval. Even though DPPH was in decane and CQA in water, CQA had the same effect as if both were in methanol as seen in Figure 14.

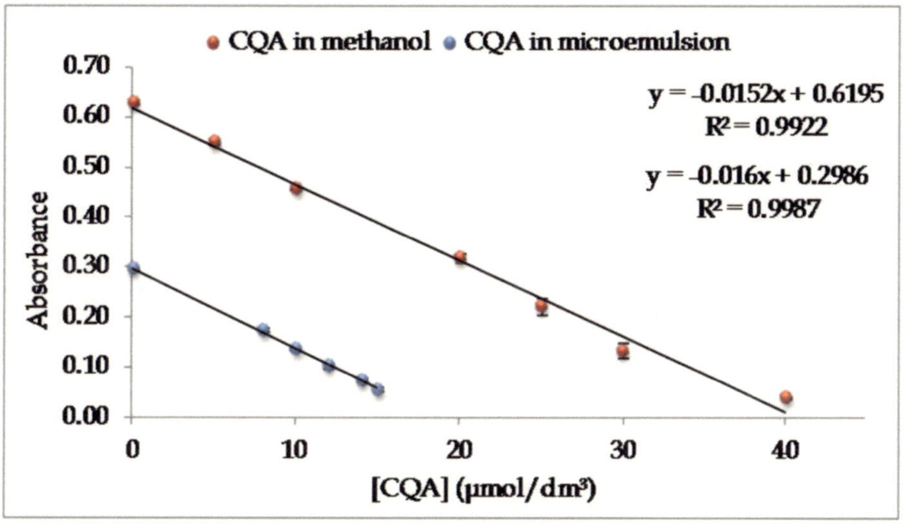

Fig. 14. The decay in DPPH as a function of the CQA concentration both in methanol and in microemulsion.

This shows the exchange between the inside of the microemulsion aggregate and the surrounding water. It also shows that the antioxidant is as effective (if any difference even slightly better) in the microemulsion as in the pure solvent methanol.

2.3.2 Studies of the antioxidant properties of α-tocopherol in the microemulsion

For α-tocopherol we could have it dissolved in the decane used for the microemulsion and had one other microemulsion with DPPH and just mixing these two microemulsions gave us a system with both components. This gave the same effect as if having them mixed in two pure decane solutions as seen in Figure 15. We could not have twice as high concentration of DPPH in one of the microemulsions to get the same initial concentration when mixed so that is why the concentration is lower. The concentration absorbance behavior is the same. The effect in the microemulsion is as good as in the solvent, if anything even slightly better. So this once again shows that there is an interchange between the inside and outside of the aggregates.

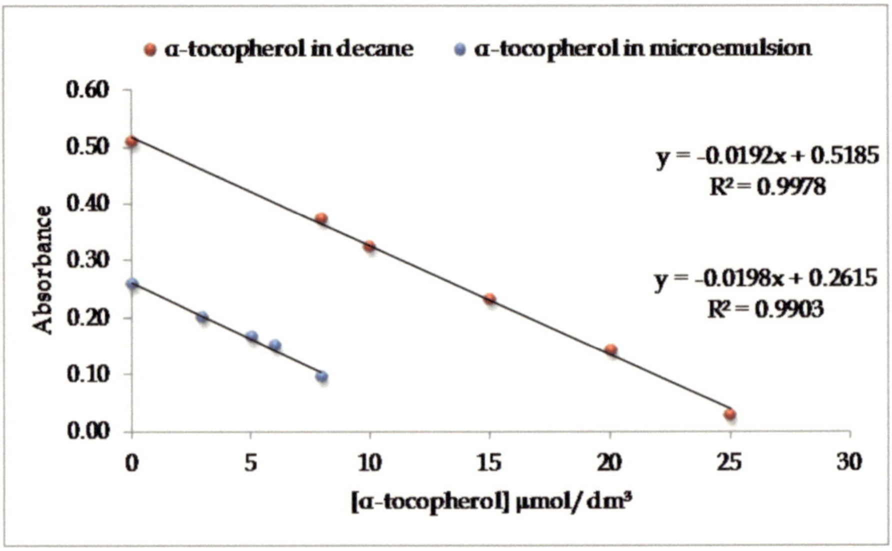

Fig. 15. The decay in DPPH as a function of the α-tocopherol concentration both in decane and in microemulsion.

Once again we showed how there is an exchange between the microemulsion aggregate and the bulk. This is even a good visual way of showing how a purple solution mixed with a clear can become almost clear solution with a brownish color.

2.3.3 Studies of the antioxidant properties of CQA and α-tocopherol in the microemulsion

Having showed that the two antioxidants worked well in the microemulsion when used individually the natural next step is to use them together. Like that there is a bonus in that they have opposite effect on the curvature and hence have smaller effect on the microemulsion boundaries. This was not an issue in the concentration range we used to be able to study the antioxidant effect as the concentrations were so low that they did not have any significant effect. If one would use much higher concentrations in formulations this would however be an advantage. It has earlier been reported that there is an additive effect by having these two antioxidants but that was for a reverse system (Sim et al., 2009). Our microemulsion is as previously stated oil-in-water in this temperature range. To study the antioxidant-effect an α-tocopherol and CQA 1:1 containing microemulsion was mixed with a microemulsion containing DPPH. The effect of total antioxidant versus absorption is shown in Figure 16 and from this it is shown that the antioxidants work well together.

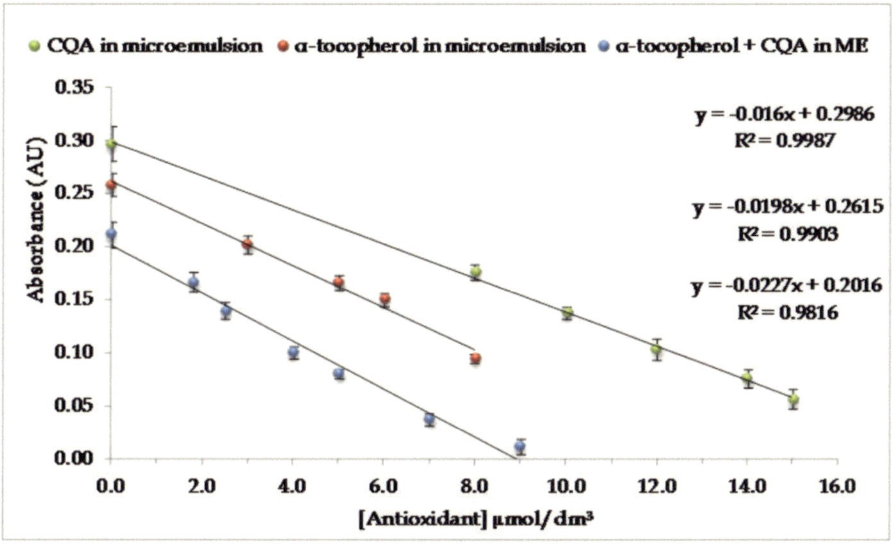

Fig. 16. The decay in DPPH as a function of the antioxidant concentration for CQA, α-tocopherol and the two combined.

By just comparing the decay it is seen how the two antioxidants can work together and by that have the same and even stronger effect with 1/4 extra compared to their individual contribution. This is seen by their decay of -0.027 compared to the average of -0.0160 for CQA and -0.0198 for α-tocopherol. As the two antioxidants has reverse effect on the phase boundaries more can be added and still have the temperature range of the model microemulsion.

3. Conclusion

We have here illustrated how convenient it is to use a model system to determine the effect of some added components to a microemulsion. We have shown both macroscopic and microscopic changes in the systems when adding these compounds. These properties of influencing the microemulsion would severely influence the use of the drug delivery system. Even if the macroscopic temperature effect is taken into consideration, the effect on the microscopic properties are important for properties like drug load and shelf-life. This shows the importance of always performing phase studies of the involved systems when adding components. It also shows the importance to investigate what happens with the system when it is loaded with an active component to see how it differs from a system without the drug. This may be masked to some degree by the more complex systems used today. These systems have much less temperature dependence so therefore the need to sometimes use model systems to better see the changes. We have seen how a hydrophobic compound, α-tocopherol, a hydrophilic compound, CQA, and a compound that is mostly hydrophobic but has a solubility in water as well (20:1 solubility in decane compared to water), lidocaine influence the phase boundaries on the microemulsion. If one uses the ratio of surfactant tail to head as the reason for the curvature of the aggregate one can see the effect of adding these compounds as the effect they have on tails or the heads and in the case of lidocaine tails and to a small extent even the heads. This is illustrated in Figure 17.

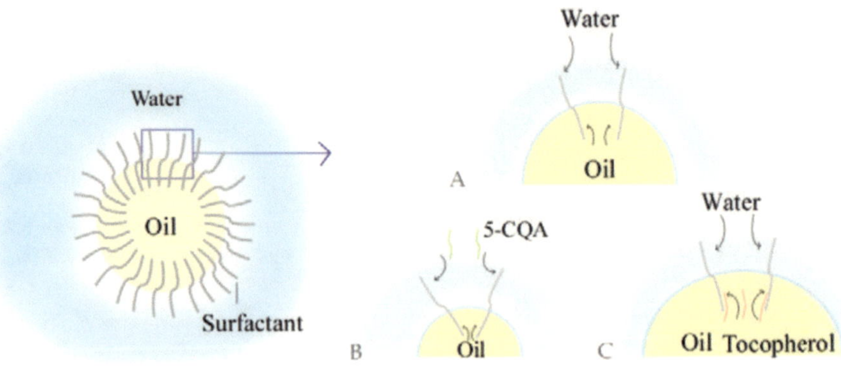

Fig. 17. Effect on the curvature of the surfactant film of the different components. The ingoing components are not set to scale. Here the micelle is shown and the close up of the surfactant and their interaction with water and oil standard system (A), the system with added CQA (B) and system with added tocopherol (C).

It is not determined yet if this is due to the penetration or general proximity to the heads and tails, but we know that it is not caused by the bulk charge in the case of CQA as sodium chloride at the same concentration did not have any significant effect.

Apart from the changes in phase boundary temperatures, the addition of these compounds also influence the temperature increase needed to change into a bicontinuous microemulsion and the growth with temperature and concentration. This influence may be important to be aware of when formulating so that it can be taken into consideration when planning. One can also use antioxidants in the microemulsions and they have the same effect as if they would be used in a pure solvent both for a hydrophobic compound as α-tocopherol and a hydrophilic compound as CQA. We have with these experiments also illustrated that there is an interchange between the inside of the aggregate and the outside. Having a microemulsion or even an emulsion makes it possible to use both hydrophobic and hydrophilic antioxidants and by this have a more complex protection. This also minimizes the effect on the phase boundaries as they have reverse effect. The combination of the two antioxidants showed a synergetic effect of ¼. To further improve the knowledge one would like to study the specific interactions which the components have with the surfactant film. One would also use more substances in order to have a predictive quantitative model and not only a qualitative model. It would also be interesting to do long term tests with antioxidants and components that are sensitive to oxidation. This would eliminate any problem with an added dye.

4. Acknowledgement

This work was supported by Fundação para a Ciência e Tecnologia, (FCT) Portugal thrugh grant PEst-OE/EQB/LA0004/2011. JB would like to thank Fundação para a Ciência e Tecnologia, (FCT) Portugal for financing through post doc fellowship (SFRH/BPD729511/2006).

5. References

Baciu, M., Holmes M. C., & Leaver M. S. (2007). Morphological Transitions in Model Membrane Systems by the Addition of Anesthetics. *The Journal of Physical Chemistry B*, Vol. 111, No 4, (February 2007), pp. 909-917, ISSN 1520-6106.

Bagwe R. P., Kanicky, J. R, Palla, B. J. Patanjali, P. K. Shah, D. O. (2001). Improved drug delivery using microemulsions: rationale, recent progress, and new horizons. *Crit Rev Ther Drug Carrier Syst*, Vol. 18, No 1, pp. 77-140. ISSN: 0743-4863

Balogh, J. (2010). Determining scaling in known phase diagrams of nonionic microemulsions to aid constructing unknown. *Advances in Colloid and Interface Science*, Vol. 159, No 1, (August 2010) pp. 22-31, ISSN 0001-8686.

Balogh, J. Determination of Microemulsion Droplets size and anisotropy using Light Scattering, submitted 2011.

Balogh, J, Olsson U., Pedersen J. S., Kaper H., Wennerström H., Schillén K. & Miguel M. (2010). *Nonionic Microemulsions: Dependence of Oil Chain Length and Active Component (Lidocaine)*, Wiley-VCH Verlag GmbH & Co. KGaA, pp. 59-87, ISBN 9783527632633.

Balogh, J, & Pedersen J. S., Hórvölgyi, Z., & Kiss, É. Eds. (2008). *Investigating the Effect of Adding Drug (Lidocaine) to a Drug Delivery System Using Small-Angle X-Ray Scattering*, Springer Berlin/Heidelberg, Vol. 135, pp. 101-106, ISBN 978-3-540-85133-2.

Balogh, J., & Olsson U. (2007). Dependence on Oil Chain-Length of the Curvature Elastic Properties of Nonionic Surfactant Films: Droplet Growth from Spheres to a Bicontinuous Network. *Journal of Dispersion Science and Technology*, Vol. 28, No 2, (February 2007), pp. 223-230, ISSN 0193-2691.

Balogh, J., Kaper, H., Olsson, U., & Wennerström, H. (2006). Effects of oil on the curvature elastic properties of nonionic surfactant films: Thermodynamics of balanced microemulsions. *Physical Review E*, Vol. 73, No 4, (April 2006) pp. 041506, ISSN 1550-2376.

Balogh, J., Marques, L., & Lopes A. Natural antioxidants in a nonionic microemulsion, interactions and antioxidant activity to be submitted 2012.

Balogh, J., Olsson U., & Pedersen J. S. (2006). Dependence on Oil Chain Length of the Curvature Elastic Properties of Nonionic Surfactant Films: Emulsification Failure and Phase Equilibria. *Journal of Dispersion Science and Technology*, Vol. 27, No 4 pp. 497-510, (February 2007), ISSN 0193-2691.

Balogh, J., Olsson U., & Pedersen J. S. (2007). A SANS Contrast Variation Study of Microemulsion Droplet Growth. *The Journal of Physical Chemistry B*, Vol. 111, No 4, (February 2007) pp. 682-689, ISSN 1520-6106.

Basnet, P., Matushige, K., Hase, K., Kadota, S., & Namba, T., (1996). Four di-O-caffeoyl quinic acid derivatives from propolis. Potent hepatoprotective activity in experimental liver injury models. *Biological & Pharmaceutical Bulletin*, Vol. 19, No 11, (November 1996) pp. 1479-84. ISSN 1347-5215

Bissett, D. L., Chatterjee, R., & Hannon, D. P. (1992). Protective effect of a topically applied anti-oxidant plus an anti-inflammatory agent against ultraviolet radiation-induced chronic skin damage in the hairless mouse. *Journal of the Society of Cosmetic Chemists*, Vol. 43, No 2, (March 1992) pp. 85-92.

Clifford, M., Clarke R. J., &Macrae, R. (1985). *Chlorogenic acids. In: Coffee*, Elsevier Applied Science Publications, Vol 1, London, UK, ISBN 0-85334-368-3.

De Maria, C. A. B., Trugo, L. C., Moreira, R. F. A., & Werneck, C. C. (1994). Composition of green coffee fractions and their contribution to the volatile profile formed during roasting. *Food Chemistry*, Vol. 50, No 2, (October 1994) pp. 141-145, ISSN 0308-8146.

Fanun, M. (2009). *Microemulsions: Properties and Applications*, CRC Press, Surfactant science series, vol. 144, ISBN 9781420089592.

Gaonkar, A. G. & Bagwe R. P. (2002). *Microemulsions in Foods: Challenges and applications in Adsorption and aggregation of surfactants in solution*, editors Mittal K. L. & Shah D. O. 368-388 Marcel Dekker New York USA, ISBN 9780824708436.

Garti, N., Zakhaira, I., Spernath, A. Yaghmur, A., Aserin, A., Hoffman, R. E., & Jacobs, L. Cabuil, V., Leviz, P., & Treiner, C. Eds. (2004). *Solubilization of water-insoluble nutraceuticals in nonionic microemulsions for water-based use*, Springer Berlin/Heidelberg, Vol. 126, pp. 184-189, ISBN 978-3-540-20073-4.

Grampurohit, N., Ravikumar, P., & Mallya, R. (2011). Microemulsions For Topical Use– A Review. *Indian Journal of Pharmaceutical Education and Research*, Vol. 45, No 1, (March 2011) 400-056.

Gupta, S. & Moulik, S. P. (2008). Biocompatible microemulsions and their prospective uses in drug delivery. *Journal of Pharmaceutical Sciences*, Vol. 97, No 1: (January 2008) pp. 22-45, ISSN 1520-6017

Heuschkel, S., Goebel, A., & Neubert, R. H. H. (2008). Microemulsions—modern colloidal carrier for dermal and transdermal drug delivery. *Journal of Pharmaceutical Sciences*, Vol. 97, No 2: (February 2008) pp. 603-631, ISSN 1520-6017.

Kahlweit, M. & Strey, R. (1985). Phase Behavior of Ternary Systems of the Type H2O-Oil-Nonionic Amphiphile (Microemulsions). *Angewandte Chemie International Edition in English*, Vol. 24, No 8, (August 1985) pp. 654-668, ISSN 1521-3773.

Kahlweit, M.; Strey, R., Firman, P., & Haase, D. (1985) Phase behavior of ternary systems: water-oil-nonionic surfactants as near-tricritical phenomenon *Langmuir*, Vol 1, No 3 (March 1985) pp 281-288, ISSN 0743-7463.

Kogan, A. & Garti, N. (2006). Microemulsions as transdermal drug delivery vehicles. *Advances in Colloid and Interface Science*, Vol. 123-126, No. Special Issue in Honor of Dr. K. L. Mittal, (November 2006) pp. 369-385, ISSN 0001-8686.

Kreilgaard, M. (2002). Influence of microemulsions on cutaneous drug delivery. *Advanced Drug Delivery Reviews*, Vol. 54, No Supplement 1, (November 2002) pp. 77-98, ISSN 0169-409.

Kreilgaard, M., Kemme, M. J. B., Burggraaf, J., Schoemaker, R. C., & Cohen, A. F. (2001). Influence of a Microemulsion Vehicle on Cutaneous Bioequivalence of a Lipophilic Model Drug Assessed by Microdialysis and Pharmacodynamics. *Pharmaceutical Research*, Vol. 18, No 5: (May 2001) pp. 593-599, ISSN 0724-8741.

Krol, E. S, Kramer-Stickland K. A., & Liebler D. C. (2000). Photoprotective actions of topically applied vitamin E. *Drug Metabolism Reviews*, Vol. 32, No 3-4, (January 2000) pp. 413-420. ISSN 1097-9883

Kunieda, H. and K. Shinoda. (1982). Phase behaviour in systems of non-ionic surfactant/ water/ oil/ around the hydrophile-lipophile-balance-temperature (HLB-temperature). *Journal of Dispersion Science and Technology*, Vol. 3, No 3, (October 1982) pp. 233-244, ISSN 0193-2691.

Kunieda. H., & Solans C. (1997). *Industrial applications of microemulsions*, New York:Marcel Dekker Inc, vol. 199, pp. 1-17. ISBN 0824797957.

Ky, C.-L, Louarn, J., Dussert, S., Guyot, B., Hamon, S., & Noirot, M. (2001). Caffeine, trigonelline, chlorogenic acids and sucrose diversity in wild Coffea arabica L. and C. canephora P. accessions. *Food Chemistry*, Vol. 75, No 2, (November 2001) pp. 223-230, ISSN 0308-8146.

Lawrence, M. J. & Rees G. D. (2000). Microemulsion-based media as novel drug delivery systems. *Advanced Drug Delivery Reviews*, Vol. 45, No 1: (December 2000) pp. 89-121, ISSN 0169-409.

Le, T. D., Olsson, U., Wennerström, H., & Schurtenberger, P. (1999). Thermodynamics of a nonionic sponge phase. *Physical Review E*, Vol. 60, No 4, (October 1999) pp. 4300-4309 ISSN1550-2376.

Leaver, M. S., Olsson, U., Wennerström, H., Strey, R. Emulsification failure in a ternary microemulsion. *Journal de Physique II*, Vol. 4, No. 3, (March 1994), pp.515-531, ISSN 1155-4312.

Leaver, M., Furo, I. & Olsson, U. (1995). Micellar Growth and Shape Change in an Oil-in-Water Microemulsion. *Langmuir*, Vol. 11, No 5. pp. 1524-1529 (May 1995) ISSN 0743-7463.

Lindman, B. Ed Stubenrauch, C.; (2008). *Microemulsions : background, new concepts, applications, perspectives*, Wiley-Blackwell, pp. XV, ISBN 978-1405-16-7.

Lopes, L. B., VanDeWall, H., Li, H. T., Vengupal, V., Li, H. K., Naydin, S., Hosmer, J., Levendusky, M., Zheng, H., Bentley, M. V. L. B., Levin, R.,& Hass, M. A. (2010). Topical delivery of lycopene using microemulsions: Enhanced skin penetration and tissue antioxidant activity. *Journal of Pharmaceutical Sciences*, Vol. 99, No 3, (March 2010) pp. 1346-1357, ISSN 1520-6017.

Marques, L (2011), *Natural antioxidants extraction and their incorporation into model pharmaceutical systems*, Master Dissertation, FCT-UNL (Lisbon).

McVean, M. & Liebler, D. C. (1999). Prevention of DNA photodamage by vitamin E compounds and sunscreens: Roles of ultraviolet absorbance and cellular uptake. *Molecular Carcinogenesis*, Vol. 24, No 3, (March 1999)pp. 169-176, ISSN 1098-2744.

Nabi, Z., Tavakkol, A., Soliman, N., & Polefka, T. G. (1998). Bioconversion of tocopheryl acetate to tocopherol in human skin: use of human skin organ culture models. *Pathophysiology*, Vol. 5, No 1, (June 1998) pp. 190-190. ISSN 0928-4680.

Natella, F., Nardi, M., Belelli, F., & Scaccini, C. (2007). Coffee drinking induces incorporation of phenolic acids into LDL and increases the resistance of LDL to ex vivo oxidation in humans. *The American Journal of Clinical Nutrition*, Vol. 86, No 3, (September 2007) pp. 604-609. ISSN 0002-9165

Olsson, U. & Schurtenberger, P. (1993). Structure, interactions, and diffusion in a ternary nonionic microemulsion near emulsification failure. *Langmuir*, Vol. 9, No 12, (December 1993) pp. 3389-3394, ISSN 0743-7463

Ramirez-Coronel, M. A., Marnet, N., V. S. Kumar Kolli, V. S. K., Roussos, S., Guyot, S. & Augur, C. (2004). Characterization and Estimation of Proanthocyanidins and Other Phenolics in Coffee Pulp (Coffea arabica) by Thiolysis−High-Performance Liquid Chromatography. *Journal of Agricultural and Food Chemistry*, Vol. 52, No 5, (March 2004) pp. 1344-1349, ISSN 0021-8561.

Ramirez-Martinez, J.R. (1988). Phenolic compounds in coffee pulp: Quantitative determination by HPLC. *Journal of the Science of Food and Agriculture*, Vol. 43, No 2, (March 1988) pp. 135-144, ISSN 1097-0010.

Rigotti, A. (2007). Absorption, transport, and tissue delivery of vitamin E. *Molecular Aspects of Medicine*, Vol. 28, No 5-6, (October 2007), pp. 423-436, ISSN 0098-2997.

Santos, P., Watkinson, A. C., Hadgraft, J., & Lane, M. E. (2008). Application of Microemulsions in Dermal and Transdermal Drug Delivery. *Skin Pharmacology and Physiology*, Vol. 21, No 5, (September 2008) pp. 246-259, ISSN 1660-5527.

Scalbert, A. & Williamson, G. (2000). Dietary Intake and Bioavailability of Polyphenols. *The Journal of Nutrition*, Vol. 130, No 8, (August 2000) pp. 2073S-2085S. ISSN 0022-3166.

Shinoda, K., & Kunieda, H.; (1973) Conditions to produce so-called microemulsions. Factors to increase the mutual solubility of oil and water by solubilizer, *Journal of Colloid Interface Science*, Vol 42; No 2 (February 1973) pp 381–387. ISSN 0021-9797

Sim, W. L. S., Han, M. Y. & Huang, D. (2009). Quantification of Antioxidant Capacity in a Microemulsion System: Synergistic Effects of Chlorogenic Acid with α-Tocopherol. *Journal of Agricultural and Food Chemistry*, Vol. 57, No 9, (May 2009) pp. 3409-3414, ISSN 0021-8561.

Sintov, A. C. & Levy H.V. (2007). A microemulsion-based system for the dermal delivery of therapeutics. *Innovations in pharmaceutical technology*, Vol. 23, (December 2007) pp. 68-72.

Sottmann, T., & Strey, R. (1996) Evidence of corresponding states in ternary microemulsions of water - alkane -C_iE *Journal of Physics: Condensed Matter* Vol, 8,No 25 A (June 1996) pp. A39-A48. ISSN 1361-648X

Spernath, A. & Aserin, A. (2006). Microemulsions as carriers for drugs and nutraceuticals. *Advances in Colloid and Interface Science*, Vol. 128-130, (December 2006): pp. 47-64, ISSN 0001-8686.

Trugo, L.C. & Macrae, R. (1984). Chlorogenic acid composition of instant coffees. *Analyst*, Vol. 109, No 3, (March 1984) pp. 263-266, ISSN 0003-2654.

6

Thermal Reversible Microemulsion for Oral Delivery of Poorly Water-Soluble Drugs

Zhong-Gao Gao
Chinese Academy of Medical Sciences and Peking Union Medical College, Beijing China

1. Introduction

There have been many attempts to improve absorption and bioavailability of poorly water-soluble drugs [1-3]. One of attractive systems is microemulsions which are composed of fine oil-in-water droplets in aqueous medium [4-8]. When such a formulation is released into the lumen of the gut, it disperses to form a fine emulsion, so that the drug remains in liquid state in the gut avoiding dissolution step that frequently limits the rate of absorption of liphophilic drugs [9]. The smaller sizes of droplets will have a larger interfacial surface areas per unit volume and correspondingly large free energy contribution from the liquid-liquid interfacial tension [10].

Self emulsify drug delivery system (SEDDS) is a potency microemulsions for enhancing bioavailability of poorly water soluble drugs after oral administration. Microemulsion is a system of thermodynamically stable and isotropically clear dispersions of two immiscible liquids such as oil and water, stabilized by an interfacial film of surfactant molecules [8,9]. Microemulsion formulations are a good candidate for oral delivery of poorly water-soluble drugs, because of their ability to improve drug solubilization and potential for enhancing absorption in the gastrointestinal tract (GI), caused by surfactant-induced permeability changes [11]. After oral administration, it rapidly disperses in stomach forming small droplets (<5 μm), which promotes wide distribution of the drug throughout the GI tract. In the past decade, we have developed lipophilic drugs microemulsion [6], hydrophobic drug emulsion [12], and thermal sensitive microemulsion[13] for enhancing water insoluble drugs bioavailability. In this chapter, we discuss on the enhancing bioavailability of the poorly water soluble drugs after oral administration by using microemulsion system. Poorly water soluble drug has been designed as a thermal reversible microemulsion system, which can disperse rapidly in the aqueous contents of the stomach and form fine oil in water droplets, and thus leads to improve absorption of the poorly water soluble drugs.

2. Microemulsion system for oral delivery of lipophilic drugs

2.1 Preparation of pseudo-ternary phase diagram

Cyclosporin A as a lipophilic drug model, Caprylic/capric triglyceride (Captex 355®) as a oil, and Polyoxyethylated castor oil (Cremorphor EL®) and Transcutol® as a surfactant and cosurfactant, respectively, were used for preparation of a series of microemulsions. Surfactant

(Cremophor EL®) was mixed with cosurfactant (Transcutol®) in fixed weight ratios (0.5:1, 1:1, 2:1 and 4:1). Aliquots of each surfactant-cosurfactant mixture (S_{mix}) were then mixed with oil (Captex 355®) and finally with aqueous phase (saline or 0.1 N HCl). Mixtures were gently shaken or mixed by vortexing and kept at ambient temperature (25°C) to equilibrate. The equilibrated samples were then assessed visually and determined as being clear and transparent microemulsions, or crude emulsions or gels as shown in Figure 1.

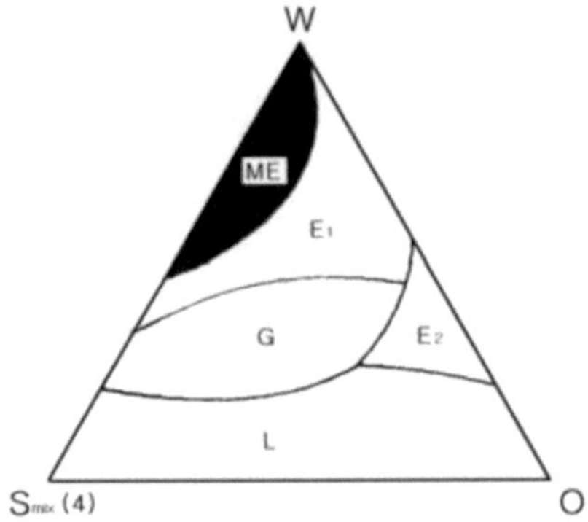

Fig. 1. Pseudo-ternary phase diagrams composed of oil (Captex 355®), Cremophor EL®-Transcutol® mixture (Smix) (Cremophor EL®: Transcutol®= 4:1) and water. Keys: G, gel; L, isotropic region ; ME, single phase o/w microemulsion; E1, crude emulsion; E2, w/o emulsion region.

Phase studies were done to investigate the effect of surfactant/cosurfactant ratio on the extent of stable o/w microemulsion region. The microemulsions in the present study were formed spontaneously at ambient temperature when their components were brought into contact. It is advantageous for developing oral dosage forms of lipophilic drugs, because easy formation at ambient temperature is particularly advantageous for thermo labile drugs such as peptides [14]. Phase study revealed that the addition of Cremophor EL®-Transcutol® mixture in a ratio 4:1 can produce clear and transparent microemulsions in the subsequent study.

2.2 Solubility of drug in the surfactants and oil

The solubility of Cyclosporin A in each component was 98.72 ± 7.75 mg/g in Captex 355®(oil), 56.51 ± 4.90 mg/g in Cremophor EL® (surfactant) and very soluble in Transcutol®(cosurfactant). The solubility of Cyclosporin A in the oil-surfactant mixture was slightly decreased with increasing the content of Cremophor EL® (Fig. 2A). In contrast, the solubility of Cyclosporin A in the oil-cosurfactant mixture was increased linearly with increasing the content of Transcutol® (Fig. 2B). Fig. 2C shows that the solubility of Cyclosporin A in the oil-S_{mix} mixture with varying surfactant-cosurfactant ratio was decreased with increasing the surfactant content. It indicates that the solubilization of Cyclosporin A was greatly affected by the cosurfactant content. After adding the aqueous phase, the solubility of Cyclosporin A in oil-surfactant-water system increased linearly with increasing the content of Cremophor EL® (Fig. 3A). The solubility of Cyclosporin A in oil-cosurfactant-water system also increased dramatically with more than 0.5% of Transcutol®

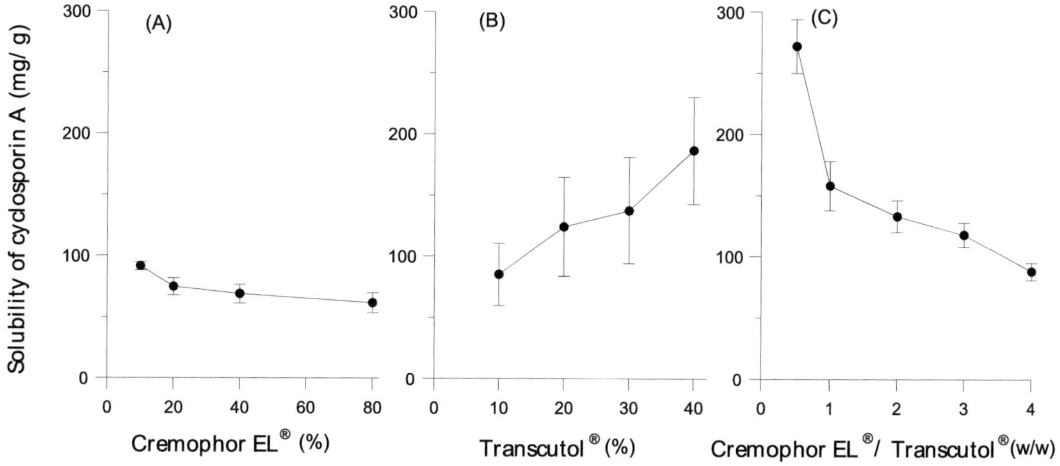

Fig. 2. (A) Effect of the content of Cremophor EL® on the solubility of Cyclosporin A in a mixture of Cremophor EL® and Captex 355®. (B) Effect of the content of Transcutol® on the solubility of Cyclosporin A in a mixture of Transcutol ®and Captex 355®. (C) Effect of the weight ratio of Cremophor EL® to Transcutol® on the solubility of Cyclosporin A in a mixture of Cremophor EL®-Transcutol® (Smix) and Captex 355 (Smix:Captex 355®=15:4).

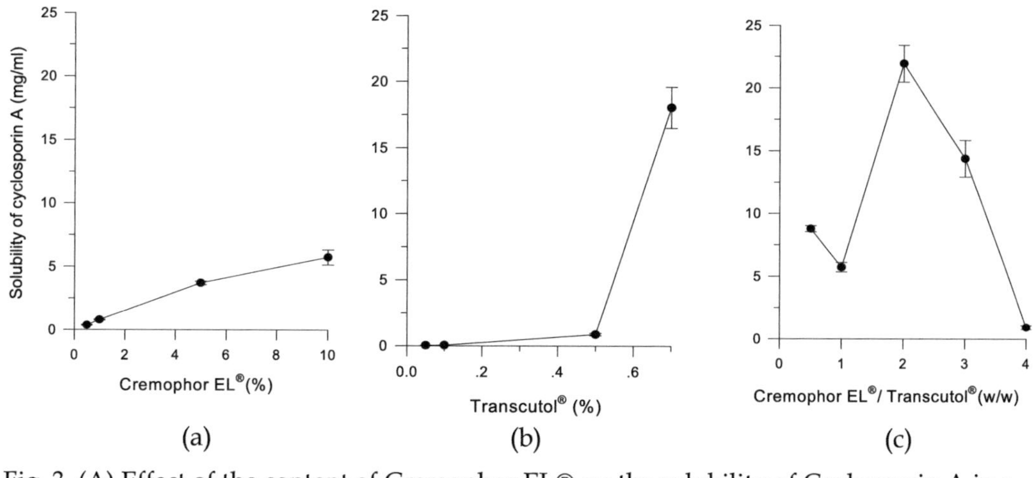

Fig. 3. (A) Effect of the content of Cremophor EL® on the solubility of Cyclosporin A in a mixture of Cremophor EL®, Captex 355® and saline (Cremophor EL®–Captex 355®: saline=19:81). (B) Effect of the content of Transcutol® on the solubility of Cyclosporin A in a mixture of Transcutol®, Captex 355® and saline (Transcutol®–Captex 355®:saline=19:81). (C) Effect of the weight ratio of Cremophor EL® to Transcutol® on the solubility of Cyclosporin A in microemulsion systems obtained by adding saline to a mixture of Cremophor EL®-Transcutol® (Smix) and Captex 355®(Smix:Captex 355®=15:4).

(Fig. 3B). The solubility of Cyclosporin A in a system containing all of the components for producing microemulsions (Fig. 3C) was increased markedly compared to those of systems without surfactant (Fig. 3B) or cosurfactant (Fig. 3A) and it reached the maximum (21.95 ± 1.48 mg/ml) when the ratio of surfactant to cosurfactant of S_{mix} was 2:1 (Fig. 3C). The maximized solubilization is thought to be achieved by the formation of transparent microemulsion with small droplets. At the ratios greater than 2:1, although these mixtures formed microemulsions, the less Transcutol® content in the microemulsion systems was caused the solubilizing capacity of microemulsion to be decreased. In contrast, at the ratio less than 2:1, the clear microemulsion cannot be produced due to the insufficient amount of surfactant. The slight increase of solubility of Cyclosporin A at the ratio of 0.5:1 is thought to be observed due to the fact that, although the clear microemulsion cannot be formed, the excess amount of cosurfactant existed in the water phase and increased the solubility of Cyclosporin A in the aqueous phase. It was also confirmed by determining the partitioning of Cyclosporin A between lipophilic and aqueous phases in the subsequent study (Fig. 4).

2.3 Partitioning of Cyclosporin A between lipophilic and aqueous phases (C_o/C_w)

Partitioning studies using the aqueous and oil phases of corresponding microemulsion are known to be correlated to the observed oral bioavailability and/or *in vitro* permeability [11]. The concentration ratio of Cyclosporin A in lipophilic phase to that in aqueous phase (C_o/C_w) was greatly affected by the ratio of surfactant to cosurfactant of S_{mix} (Fig. 4). C_o/C_w was maximized when the microemulsion system was prepared with S_{mix} at 1:1 ratio of surfactant to cosurfactant. When the ratio was at above or below 1:1, the C_o/C_w was reduced because the excess amount of surfactant or cosurfactant existed in the aqueous phase and contributed to increasing the concentration of Cyclosporin A in aqueous phase. However, if considering the absolute solubility of Cyclosporin A in the lipophilic phase, the maximized concentration of Cyclosporin A could be obtained with microemulsions containing 2:1 mixture of surfactant to cosurfactant.

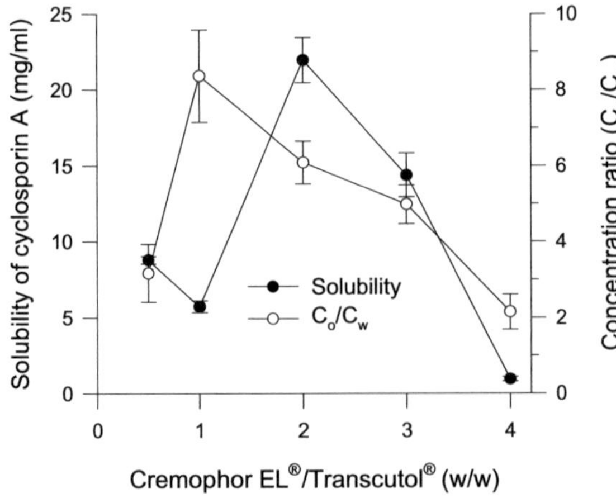

Fig. 4. Effect of the weight ratio of Cremophor EL® to Transcutol® on the concentration ratio of Cyclosporin A in lipophilic phase and that in aqueous phase of microemulsion systems (C_o/C_w).

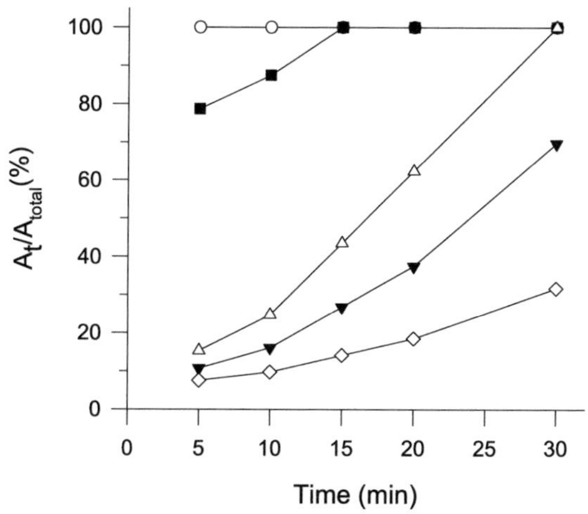

Fig. 5. Disperison rate of mixture of Cremophor/Transcutol(Smix) (Cremophor EL®/Transcutol® as 0.5:1, 1:1, 2:1, 3:1, 4:1 and Captex 355 (Smix:Captex355=15:4) into aqueous media assuming the pH condition of gastric fluid.

2.4 Dispersability and particle size determination

To use the oil-S_{mix} mixture as a pre-microemulsion concentrate, it must be readily dispersed in the stomach to form a microemulsion [10]. Thus, the dispersability of oil-S_{mix} mixture prepared with different weight ratios of surfactant to cosurfactant was compared in aqueous media assuming the pH condition of gastric fluid. The dispersion occurred more slowly with increasing the surfactant/cosurfactant ratio (Fig. 5). Too slow dispersion of pre-microemulsion concentrate prepared with S_{mix} which is greater than 3:1 ratio of surfactant to cosurfactant might retard the absorption of drugs in gastrointestinal tract and it was insisted that the ratio of surfactant to cosurfactant must not exceed 2:1 to allow use of the oil-S_{mix} mixture as a pre-microemulsion concentrate in this study.

It is known that the droplet size distribution is one of the most important characteristics of emulsion for the evaluation of its stability [15] and also *in vivo* fate of emulsion [16]. At first, the effect of each component of microemulsion systems on the resultant droplet size was investigated. The surfactant content at below 20% of mixture did not affect the droplet size significantly (Fig. 6A). However, with increasing the cosurfactant content in oil-cosurfactant-water system, the droplet size decreased linearly (Fig. 6B). The droplet size of microemulsions prepared with S_{mix} was markedly reduced compared with those prepared with surfactant or cosurfactant alone. It was demonstrated that the small and stable microemulsion was formed by the addition of both of them. The droplet size of microemulsion decreased with increasing of the surfactant to cosurfactant ratio and became constant at above 2:1 ratio of surfactant to cosurfactant (Fig. 6C). This result is in accordance with the report that the addition of surfactant on the microemulsion systems causes the interfacial film to condense and to be stable, while the addition of cosurfactant causes the film to expand [14]. The smallest droplet size of microemulsion (22 nm) was obtained from S_{mix} at more than 2:1 ratio of surfactant to cosurfactant, where it can produce clear and transparent microemulsions. However, as mentioned in the proceeding sections, S_{mix} at 3:1 or 4:1 ratio of surfactant to cosurfactant produced microemulsions with reduced drug-solubility capacity and slow dispersion rate, which can be disadvantageous for use as an oral delivery system of Cyclosporin A.

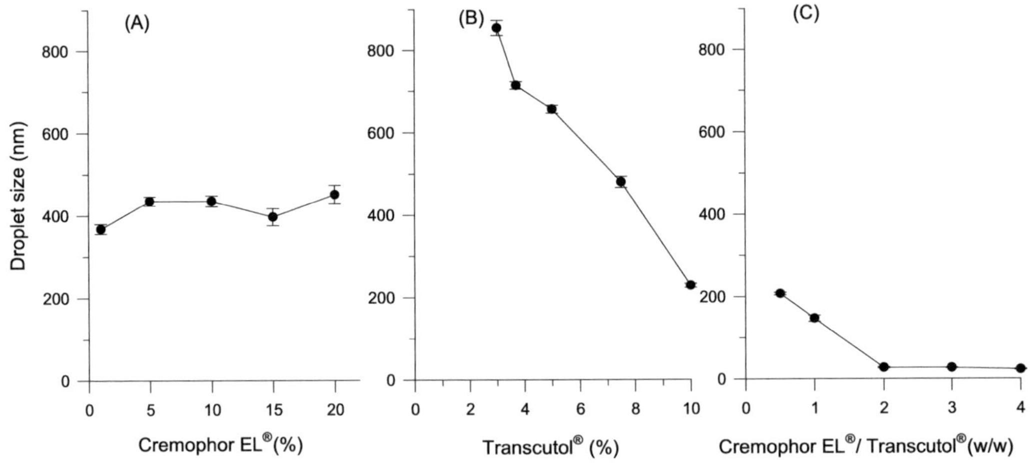

Fig. 6. (A) Effect of the content of Cremophor EL® in a mixture of Cremophor EL®, Captex 355®, and saline (Cremophor EL®–Captex 355®:saline=19:81) on the resultant mean droplet size. (B) Effect of the content of Transcutol® in a mixture of Captex 355®, Transcutol® and saline (Transcutol®–Captex 355®:saline=19:81) on the resultant mean droplet size. (C) Effect of the weight ratio of Cremophor EL® and Transcutol® in empty microemulsion obtained from a mixture of Cremophor®-Transcutol® (Smix), Captex 355® and saline (Smix:Captex 355®:saline=15:4:81) on the resultant mean droplet size.

2.5 Animal studies

The non-compartmental pharmacokinetic parameters in Table 1 were calculated based on the observed blood data. The maximal drug concentration in blood(C_{max}), time at the maximal drug concentration in blood(T_{max}) and Area Under Curve of drug concentration in blood (AUC) of Sandimmun®, Sandimmun Neoral® and microemulsion in this study were 1.285, 2.859, 3.275 (μg· ml^{-1}) and 2.333, 3.000, 3.667 (h) and 12.531, 33.171, 41.332 (μg·h/ml), respectively. The C_{max} of Cyclosporin A loaded in the microemulsion system in the present study was markedly higher compared with Sandimmun® and was not significantly different compared with Sandimmun Neoral®. The AUC of Cyclosporin A via oral administration of microemulsion in this study was significantly increased ($p<0.05$, 3.30 fold), when compared with Sandimmun®. However, no significant difference was found between the AUC of microemulsion and Sandimmun Neoral® ($p>0.05$, 1.25 fold). There was no significant change in T_{max} ($p>0.05$) among the three products. The absolute bioavailability (F) of microemulsion optimized in this study increased about 3.30 and 1.25 fold compared with Sandimmun® and Sandimmun Neoral®. The bioavailability of Cyclosporin A incorporated in microemulsion optimized did not show significant difference compared with Sandimmun Neoral® (0.416 versus 0.518) and 3.30 fold increased compared with Sandimmun® (0.157 versus 0.518). It is thought that this result support the report that for lipophilic drugs and peptides, where absorption is dissolution rate limited, a strong correlation exists between the particle size of emulsion and bioavailability because the particle size of Sandimmun®,

Sandimmun Neoral® and microemulsion (with Smix 2:1) in this study was 864, 104 nm and 22 nm, respectively, measured with Zetasizer Nano ZS90 (Malvern Instruments Ltd., UK) after diluting with water 50 times.

Parameters	Intravenous	Oral		
		Sandimmun®	Sandimmun Neoral®	Microemulsion
C_{max} (μg· ml-1)	–	285± 0.088	2.859± 0.322	3.275± 0.367 [a]
T_{max} (h)	–	333± 0.441	3.000± 0.354	3.667± 0.333
AUC (μg·h/ml)	390± 0.193	12.531± 0.088	33.171±5.534 [a]	41.332± 4.532 [a]
Absolute bioavailability (F) [b]		0.157	0.416	0.518

[a] $p<0.05$ by the student T-test when compared to Sandimmun®.

Table 1. Analysis of noncompartmental pharmacokinetic parameters after oral administration of cyclosporin A products to rats.

2.6 Conclusion

The surfactant to cosurfactant ratio greatly affected on the physicochemical characteristics of resultant microemulsion systems obtained by using polyoxyethylated castor oil (Cremophor EL®) as a surfactant, Transcutol® as a cosurfactant and caprylic/capric triglyceride (Captex 355®) as an oil. The stable microemulsion with high solubility of Cyclosporin A, small droplet size and fast dispersion rate was obtained from mixture composed of 10/5/4 mixture of Cremophor EL®/ Transcutol®/ Captex 355®. The enhancement of bioavailability of Cyclosporin A by using o/w microemulsion optimized in this study is thought to be due to the combination of factors including the drug solubilization effect and the increase of drug permeability to membrane. In other words, the bioavailability of drugs loaded in microemulsions was dependent on the physicochemical properties of drug and o/w microemulsions. This system might be applicable to formulate the liquid and solid dosage forms of Cyclosporin A for enhancing its bioavailability after oral administration. This formulation approach can also help to improve the oral bioavailability of other poorly soluble peptide drugs as is the case for Cyclosporin A.

3. Emulsion for hydrophobic drug oral delivery

Biphenyl dimethyl dicarboxylate (BDD) is an agent against virally induced hepatic injury and has been found to be effective in improving liver function and symptoms of patients with chronic viral hepatitis [17, 18, 19]. BDD is practically water-insoluble (3.6 μg/ml in water at 25°C) and its dissolution rate is extremely slow, resulting in very low bioavailability (20–30%) [20]. For enhancement of the solubility and bioavailability of BDD after oral administration, an Emulsion system composed of oil (Neobee M-5®), surfactant (Tween 80), and cosurfactant (Triacetin), was prepared, and its physicochemical properties and pharmacokinetic parameters were evaluated in comparison to commercial product G-cell and 0.5% calcium-carboxymethylcellulose (Ca-CMC) suspension of BDD.

3.1 Solubility of BDD in various ratios of surfactant to oil

The solubility of BDD sharply increased from 0.7 to 5.02 mg/g as the ratio of Tween 80 to Neobee M-5® increased from 1:4 to 2:1 and leveled off above the ratio of 2:1 (Fig. 7). At the ratio of 2:1, the solubility of BDD increased 7-fold compared with that at the ratio of 1:4. Hence, the ratio of 2:1 (Tween 80 to Neobee M-5®) was used for further studies.

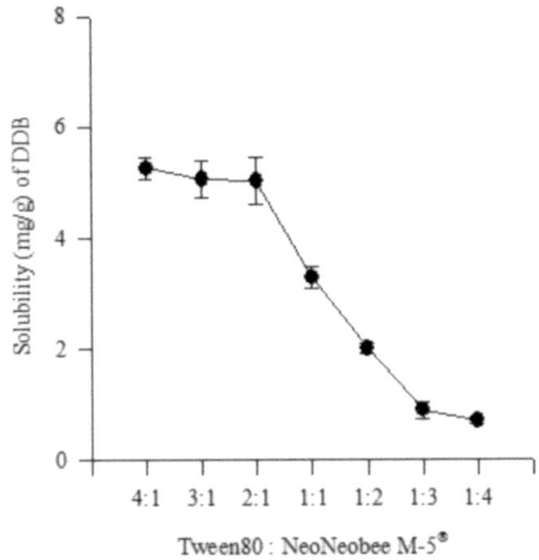

Fig. 7. Solubility of BDD in various ratios of Tween 80 to Neobee M-5®. Each point represents the mean ± SD of three experiments.

3.2 Solubility of BDD in an Emulsion system containing Triacetin

Triacetin has been used as oil to solubilize water-insoluble taxol [21] and as a food additive because of its good solubility in water and potential role as a parenteral nutrient [22]. In this study, the solubility of BDD in the absence of Tween 80 and Neobee M-5® increased proportionately with the added amount of Triacetin ranging from 0 to 50% (Fig. 8) Thus, to

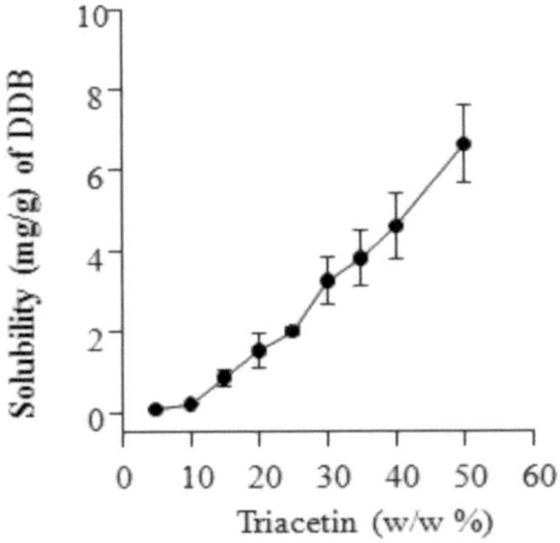

Fig. 8. Effect of Triacetin on the solubility of BDD. Each point represents the mean of four experiments.

further enhance the solubility of BDD, Triacetin was employed in a emulsion system composed of Tween 80 and Neobee M-5® at the ratio of 2:1. When the Triacetin was added to a emulsion system composed of Tween 80 and Neobee M-5® at the ratio of 2:1, the solubility of BDD gradually increased. With 55% of Triacetin, the solubility of BDD increased 40% compared to that without Triacetin, indicating the role of Triacetin as a cosurfactant in a emulsion system. Moreover, the presence of Triacetin in the emulsion system composed of Tween 80 and Neobee M-5® at the ratio of 2:1 led to the decrease in droplet size with higher physical stability and the increase in dissolution rate of BDD in aqueous media (data not shown). For further studies, 35% of Triacetin that led to 20% increase in solubility of BDD was chosen since the amount of Triacetin allowed for oral administration is less than 35%.

3.3 Droplet size of a microemulsion system

Droplet size of BDD emulsion composed of Tween 80 and Neobee M-5® at the ratio of 2:1, and 35% of Triacetin was kept constant both in distilled water and artificial gastric fluid without pepsin (pH 1.2) throughout 120 min incubation period The droplet sizes at 1 h after incubation in distilled water and artificial gastric fluid were 329.97 ± 24.31 and 284.07 ± 44.39 nm, respectively, without further changes at 2 h. This result suggests that the microemulsion system may be stable and well-dispersed in the GI tract.

3.4 Comparison of dissolution profiles among different formulations of BDD

The dissolution profiles of BDD powder, 0.5% Ca-CMC suspension of BDD, Commercial formulation G-cell and the emulsion formulation (SEDDS) of BDD composed of Tween 80 and Neobee M-5® at the ratio of 2:1, and 35% of Triacetin in distilled water were compared (Fig. 9). BDD in the emulsion formulation rapidly dissolved to a great extent whereas BDD in other formulations did hardly dissolve during 120 min incubation. About 50% of BDD in the emulsion formulation dissolved within 10 min.

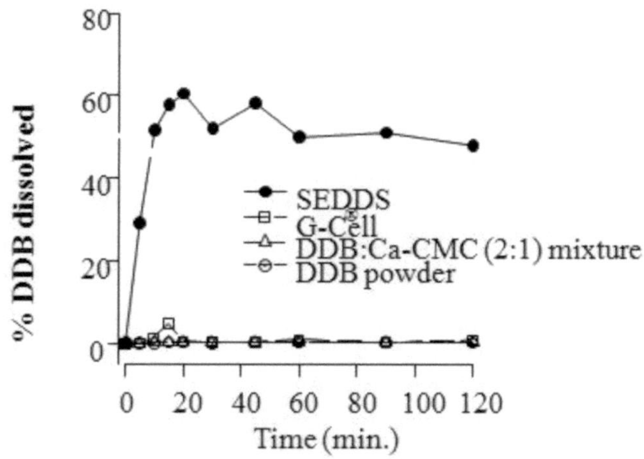

Fig. 9. Dissolution profiles of BDD from 0.5% Ca-CMC suspension of BDD, BDD powder, and the emulsion formulation of BDD (Tween 80 and Neobee M-5® at the ratio of 2:1, and 35% Triacetin; SEDDS).

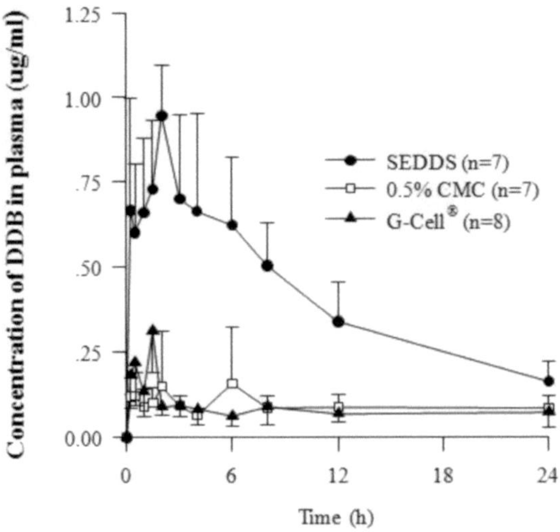

Fig. 10. Plasma concentration of BDD after oral administration to rats of 0.5% Ca-CMC suspension of BDD, BDD solution formulation and the emulsion formulation of BDD (Tween 80 and Neobee M-5® at the ratio of 2:1, and 35% Triacetin) at a dose equivalent to 12 mg of BDD/kg of body weight. Each point represents the mean ± SD of seven rats.

3.5 Pharmacokinetic analysis

An *in vivo* absorption study was undertaken to determine whether or not the enhanced solubilization and *in vitro* dissolution of BDD in a emulsion system could increase the GI absorption of drug after oral administration. Plasma concentration of BDD after oral administration of BDD in the emulsion formulation to rats increased rapidly and remarkably compared with 0.5% Ca-CMC suspension of BDD and BDD solution formulation (Fig. 10). $AUC_{0\rightarrow24h}$ of BDD in the emulsion formulation increased 5.0-fold and 1-fold compared with that of BDD in 0.5% Ca-CMC suspension and BDD solution formulation (9.829 vs. 1.955 μg·h/ml and 1.718 vs. 1.955 μg·h/ml) (Table 2). The emulsion formulation of BDD also enhanced C_{max} of BDD by 9.8-fold and 1.1-fold compared with

Parameters	Emulsion formulation	BDD in 0.5% Ca-CMC suspension	BDD solution formulation
C_{max} (μg/ml)	1.550 ±0.706**	0.158 ±0.165	0.1412 ±0.602
T_{max} (h)	1.833 ±1.125	1.254 ±1.025	1.788 ±2.777
AUC (μg·h/ml)	9.829 ±2.255**	1.955 ±0.712	1.718 ±0.536

** p < 0.01 when compared with 0.5% Ca-CMC suspension and BDD solution formulation by the ANOVA test.

Table 2. Noncompartmental pharmacokinetic parameters after oral administration of BDD products to rats.

0.5% Ca-CMC suspension of BDD and BDD solution formulation (1.550 vs. 0.158 μg/ml and 0.1412 vs. 0.158 μg/ml) (Table 2). However, T_{max} was not significantly different between two formulations (Table 2). These results indicate that the emulsion system of BDD considerably increases the bioavailability of BDD compared with 0.5% Ca-CMC suspension of BDD and BDD solution formulation. The enhanced bioavailability is probably due to the increase in solubility and dispersion of the drug in the GI tract. Nerurkar reported that surfactants, which are commonly added to pharmaceutical formulations, might enhance the intestinal absorption of some drugs by inhibiting this apically polarized efflux system [23]. In this study, the presence of a surfactant, Tween 80, in the emulsion system might have caused changes in membrane permeability, which could lead to an increase in bioavailability of the drug. It is thus expected that the increased bioavailability of BDD from a emulsion system may result in improved efficacy, allowing patients to be able to take less BDD.

3.6 Conclusions

Taken together, these results demonstrate that the microemulsion formulation of BDD composed of Tween 80 and Neobee M-5® at the ratio of 2:1, and 35% of Triacetin considerably improves the bioavailability of a poorly water-soluble BDD after oral administration, possibly due to the increase in solubility and dispersion of the drug in the GI tract. Therefore, the emulsion system may provide a useful dosage form for oral intake of water-insoluble drugs such as BDD.

4. Thermal revisable microemulsion for oral delivery of poorly water-soluble drugs

To avoid the precipitation of poorly water soluble drug in the pre-microemulsion, we prepared a solid or semi-solid dosage form and it would change into liquid state at body temperature. Poorly water soluble drug YH439 was loaded in a thermal reversible microemulsion system which can disperse rapidly in the aqueous contents of the stomach and form fine oil in water droplets, and thus leads to improve absorption of the poorly water soluble drugs as indication in Figure 11.

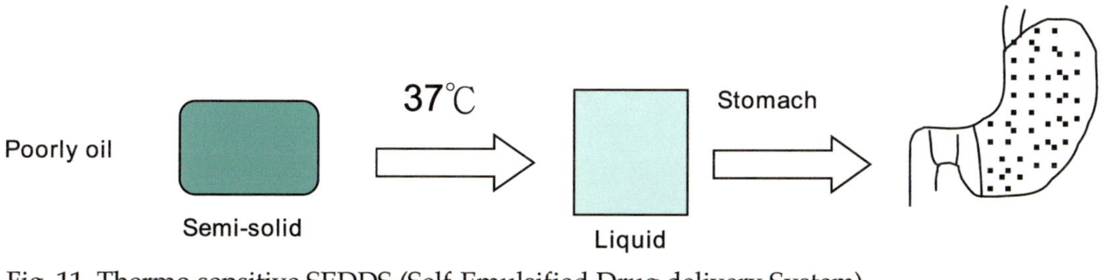

Fig. 11. Thermo sensitive SEDDS (Self-Emulsified Drug delivery System)

4.1 Effect of the composition ratio on melting point and particle size

The melting point of mixture of Pa-PEG (Palmitic-PEG400) and Ca-PEG(Capric-PEG400) was increased with increasing the composition ratio of Ca-PEG to Pa-PEG as shown in Figure 12. At the ratio of Ca-PEG to Pa-PEG was 1:3, the melting point was close to the body temperature as 36°C. Thus, firstly, the ratio of Ca-PEG to Pa-PEG was fixed at 1:3, then the effect of Cremorphor RH40® as a surfactant and Neobee M-5® as an oil on melting point of thermal reversible microemulsions was examined and the optimal composition ratio was determined for the thermal reversible microemulsion system. The melting point of thermal reversible microemulsions was raised up with increasing the amount of oil in the formulation as shown in Figure 13(a). Whereas, the melting point of microemulsions decreased with increasing the amount of surfactant. When the microemulsion was composed of lipid matrix, surfactant and oil with a ratio of 5:4:1, its melting point was close to the body temperature as 37°C. Thus this would be considered as the pre-concentration of microemulsion since the solid state could change to liquid state around 37°C and rapidly form fine droplets microemulsion in gastrointestinal tract.

Effect of each component of microemulsion systems on the resultant droplet size was investigated as shown in Figure 13(b). In the range of surfactant with more than 0.2 g and oil with less than 0.2 g, small droplet size (< 100nm) of microemulsion was obtained in this system. The droplet size of microemulsion was significantly reduced with increasing the amount of surfactant. At the composition ratio of lipid matrices:Cremorphor RH40®:Neobee M-5® with 5:4:1, the smallest droplet with 28 nm of mean size was obtained. The smaller sizes of droplets will have a larger interfacial surface areas per unit volume and correspondingly large free energy contribution from the liquid-liquid interfacial tension. As reducing the size of droplets, a larger interfacial surface areas per unit volume could be produced.

Thus from the results, the optimal physical properties including melting point and particle size of microemulsion were produced by the composition ratio of lipid matrices:Cremorphor RH40®:Neobee M-5® with 5:4:1. This formulation would be used as a formulation of the thermal reversible microemulsion system for further study in the work.

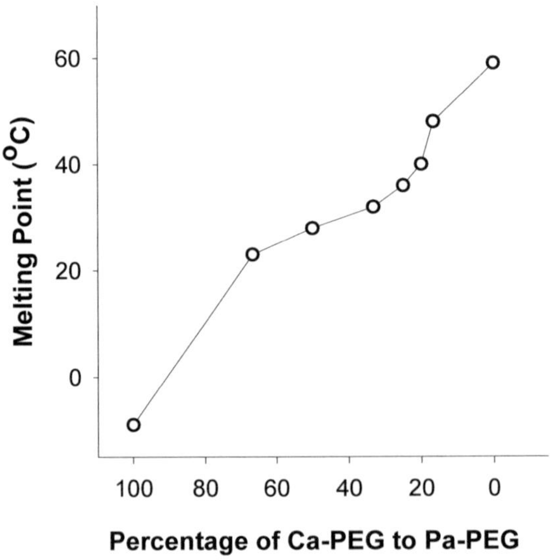

Fig. 12. The change of melting point was function of altering the composition ratio of Capric-PEG400 to Palmitic-PEG400.

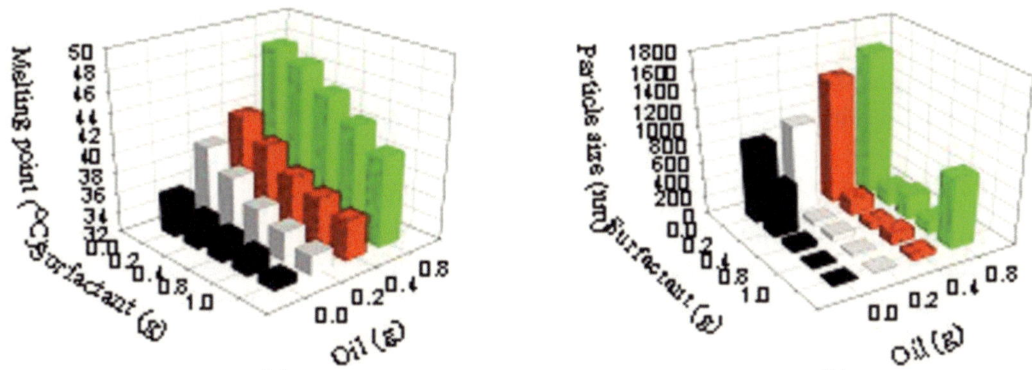

(a) (b)

Fig. 13. Effect of oil and surfactant on changing melting point (a) and particle size (b) of microemulsion system. The ratio of Capric-PEG400 to Palmitic-PEG400 was fixed at 1:3 and the lipid mixture of Capric-PEG400 and Palmitic-PEG400 was equivalent to 1 g.

4.2 Release profile of YH439 from thermal reversible microemulsion

The YH439 was released from various formulations including powder form, 5% of Ca-CMC suspension, Gelucire® and the thermal reversible microemulsion prepared in this work. The powder form of YH439 was released less than 5% of initial amount and after suspending in the 5% of Ca-CMC, the YH439 was released upto 20%. However, by additions of the Gelucire® formulation and the microemulsion to YHP439, 90% of YHP439 was released until 180 min as shown in Figure 14. The systems of Gelucire® and microemulsion remarkably improved the

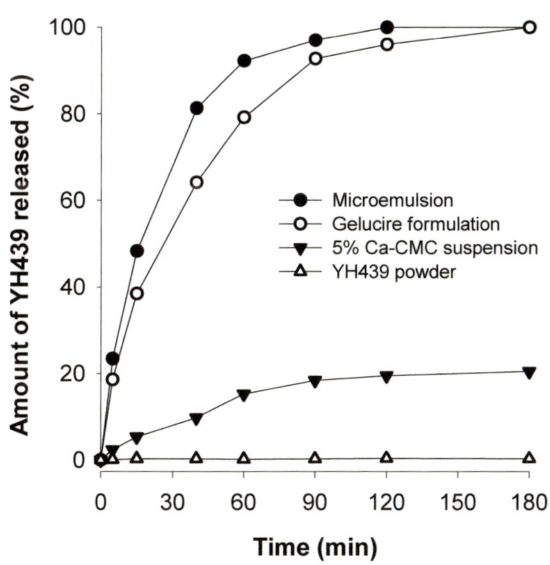

Fig. 14. Release profiles of YH439 from the thermal reversible microemulsion, Glucire® formulation, 5% Ca-CMC suspension and powder state of YH439. The thermal reversible microemulsion was composed of lipid matrices, Cremorphor RH40® and Neobee M-5® with a ratio of 5:4:1.

release property compared with powder form and 5% of Ca-CMC suspension. On the other hand, the fast release pattern was observed in the microsemulsion system compared with the Gelucire® formulation. This is desirable property since a slow dissolution rate of formulation might retard the absorption of drugs in the gastrointestinal tract.

4.3 The transport of YH439 across the caco-2 cell monolayer

The apparent permeability (Papp) of YH439 from the apical to the basolateral and basolateral to apical were calculated according the equation mentioned above. Papp of YH439 from the apical to the basolateral and basolateral to apparent direction were 2.08×10^{-5} cm/s, 4.71×10^{-5} cm/s for 0.63 uM, YH439. It was suggested YH439 was transported in the basolateral to apical direction is larger than from apical to basolateral direction in the present of a transport system as shown in Figure 4. It has been reported that the efflux of YH439 which is found in Caco-2 cells does not appear to influence the bioavailability of YH439 (Evidenced by the sufficiently high permeability in the absorption direction)[14]. YH439 was transported from microemulsion formulation larger than Gelucire® formulation and 5% Ca-CMC suspension as shown in Figure 15. The results were expected to response to *in vivo* animal experiments.

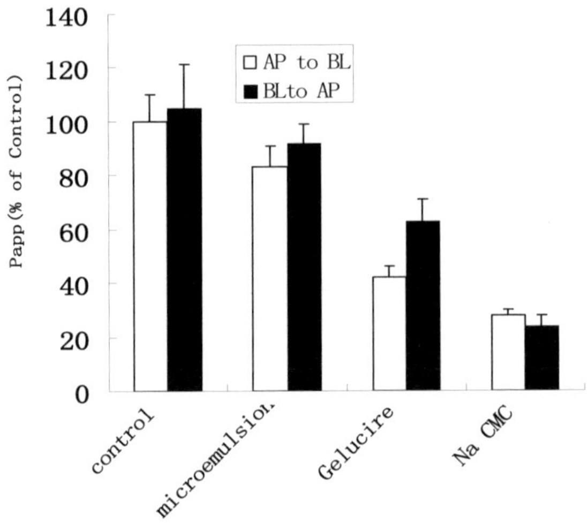

Fig. 15. Effect of YH439 transport from apical to basolateral and basolateral to apical in the microemulsion formulation, Glucire® formulation, and 5% Ca-CMC suspension (n=4).

4.4 Pharmacokinetic analysis

In vivo study was undertaken to examine the effect of a thermal reversible microemulsion system of YH439 on G.I. absorption after oral administration equivalent to 15 mg/kg of YH439 to rats. The plasma concentration of YH439 after oral administration of YH439 in the thermal reversible microemulsion formulation to rats increased compared with in the Gelucire® formulation and 5% Ca-CMC suspension as shown in Figure 5. AUC_{0-24} of YH439 in the thermal reversible microemulsion formulation increased 1.17-fold and 6.91-fold compared with that of Gelucire® formulation and 5% Ca-CMC suspension (28.14 vs 24.01 μg·h/ml and 28.14 vs 4.07 μg·h/ml) as listed in Table 2. The thermal reversible microemulsion formulation of YH439 also enhanced C_{max} of YH439 by 1.59-fold and 7.29-fold compared with Gelucire® formulation and 5% Ca-CMC suspension (2.26 vs 1.42 μg/ml and 2.26 vs 0.31) (Table 2). However, the significant difference was not observed in T_{max} value (Table 3). These results indicated that the thermal reversible microemulsion formulation of YH439 considerably increased the bioavailability of YH439 compared with Gelucire formulation and 5% Ca-CMC suspension. The enhanced bioavailability was very probably due to the increasing drug dispersion in the GI tract. Therefore, the increased solubility and enhanced bioavailability of YH439 from a thermal reversible microemulsion could result in improved drug efficacy.

Parameters	Intravenous	Oral		
		Microemulsion	Gelucire®	Suspension
C_{max} (mg/ml)	-	2.26±0.47[a,b]	1.42±0.10 [a]	0.31±0.85
T_{max} (min)	-	30	30	30
AUC (mg·h/ml)	6.11±0.28	28.14±2.89[a]	24.01±2.78 [a]	4.07±032
Absolute Bioavailability (F%)[c]	-	76.75 [a,b]	65.49 [a]	11.10

[a]P<0.001 by the AVOVA when compared to suspension
[b]P<0.05 by the AVOVA when compared to suspension
[c]$F=[AUC_{oral}/Dose_{oral}]/[AUC_{i.v}/Dose_{i.v}]$

Table 3. Plasma concentration of YH439 after oral administration of thermal reversible microemulsion, Glucire® formulation and 5% Ca-CMC suspension to rats at the dose of 15 mg/kg and after intravenous administration of thermal reversible microemulsion equivalent to 2.5 mg/kg as YH439 to rats (each group n=6).

4.5 Conclusion

The thermal reversible microemulsion system of YH439 was prepared from a lipid mixture of Ca-PEG and Pa-PEG, Cremophor RH40® and Neobee M-5® at the ratio of 5:4:1. The thermal reversible microemulsion promoted to increase the solubility of a poorly water-soluble YH439 and to enhance its bioavailability after orally administration to rat. The increased stability of poorly water-soluble YH439 by thermal reversible microemulsion system could lead to rapidly disperse as fine droplets inclusion drug in the gastrointestinal tract. Therefore, the thermal reversible microemulsion system could provide a useful dosage form for oral intake of water-insoluble YH439.

5. Acknowledgement

This work was supported in part by the National Nature Science foundation of China (No. 30873168).

6. References

[1] D. Höter and J.B. Dressman, Adv. Drug Deliver. Rev., 25 (1997) 3-14.
[2] P.P. Constantinides, Pharm. Res., 12 (1995) 1561-1572.

[3] A.C. Susan, N.C. Willam, C.R. Mark and D.W. Terry, Pharm. Res., 9 (1992) 87-93.
[4] W.A. Ritschel, Exp. Pharmacol., 13 (1991) 205-220.
[5] J.M. Sarciaux, L. Acar and P.A. Sado, Int. J. Pharm. 120 (1995) 127-136.
[6] Z.G. Gao, H.G. Choi, H.J. Shin, K.M. Park, S.J. Lim, K.J. Hwang, C.K. Kim, Int. J. Pharm. 161 (1998) 75-86.
[7] C.K. Kim, S.A. Ryuu, K.M. Park, S.J. Lim, S.J. Hwang, Int. J. Pharm. 147 (1997) 131-134.
[8] G.M. Eccleston, Microemulsions, in: J. Swarbrick, J.C. Boylan (Eds), encyclopedia of Pharmaceutical technology, Vol. 9, Marcel Dekker Publishers, New York, NY, 1992, pp. 375-421.
[9] C.W. Pouton, Adv. Drug Deliver. Rev., 25 (1997) 47-58.
[10] N.H. Shah, M.T. Carvajal, C.I. Patel, M.H. Infeld and A.W. Malick, Int. J. Pharm., 106 (1994) 15-23.
[11] P.P. Constantinides, Pharm. Res. 12 (1995) 1561–1572.
[12] C.K. Kim, Y.J. Chao, Z.G. Gao, J. of Controlled Release. 70 (2001) 149-155.
[13] D.H Han, Z.H. Jin, Y.Z Jin, X.Z. Yin, Y.Y. Shen,, Z.G. Gao, Chem.Pharm.Bull 8 (2010) 11-15.
[14] N.J. Kale and L.V. Allen, Int. J. Pharm. 57 (1989) 87-93.
[15] S.A. Charman, W.N. Charman, M.C. Rogge, T.D. Wilson, F.J. Dutko, and C.W. Pouton, Pharm. Res. 9 (1992) 87-93.
[16] B.D. Tarr and S.H. Yalkowsky, Pharm. Res. 6 (1989) 40-43.
[17] H.S. Lee, Y.T. Kim, H.C. Jung, Y.B. Yoon, I.S. Song and C.Y. Kim, Korean J. Intern. Med. 40 (1991) 173–178.
[18] J.H. Kim, Y.H. Ahn and M. Ohsawa, Biol. Pharm. Bull. 18 (1995) 24–27.
[19] S.G. Kim, S.Y. Nam, H.C. Jang, S.Y. Hong and K.H. Jung, J. Pharm. Pharmacol. 47 (1995) 678–682.
[20] S.J. Gu, X.L. Wang, W.W. Gao, P.X. Qiao, A.G. Wang, Z.Y. Qiang and Z.Y. Song, Yao Hsueh Hsueh Pao 25 (1990) 215–219.
[21] G.B. Park, C.K. Chung and K.P. Lee, J. Korean Pharm. Sci. 26 (1996) 1–11.
[22] B.D. Tarr, T.G. Sambandam and S.H. Yalkowsky, Pharm. Res. 4 (1987) 162–165.
[23] J.W. Bailey, R.L. Barker and M..D. Karlstad, J. Nutr. 122 (1992) 1823–1829.

7

The Mucosal Immune System: Modulation by Microemulsion

Eduardo Luzia França and Adenilda Cristina Honorio-França
Institute of Health and Biological Science, Federal University of Mato Grosso
Barra do Garças, Mato Grosso
Brazil

1. Introduction

Microemulsion techniques are capable of delivering organic and inorganic nano-sized particles with minimal agglomeration because the reaction occurs in nano-sized domains. However, microemulsion techniques present several disadvantages, including the use of a large amount of oil and surfactant and low yield. Nevertheless, microemulsion techniques are an alternative method for synthesising several types of inorganic and organic nano-sized particles[1, 2, 3].

The use of microemulsions as vehicles for pharmaceutical preparations and drug release systems has aroused significant interest due to their potential advantages[4], which can be attributed to their ability to improve the solubility of hydrophilic and lipophilic substances and their capacity to increase the stability and absorption of drugs, allowing greater bioavailability[5].

Furthermore, microemulsions can provide sustained drug release by compartmentalising active substances within structures organised by surfactants, water and oil, allowing the slow and continuous release of drugs over a long period of time[6].

Microemulsion is a suitable technique for obtaining nanometre-sized inorganic particles with minimal agglomeration[7]. Microemulsion is the thermodynamically stable, transparent, isotropic dispersion of two immiscible liquids such as water and oil, which are stabilised by surfactant molecules at the water/oil interface. In water-in-oil microemulsions, nano-sized water droplets are dispersed in the continuous hydrocarbon phase and are surrounded by a monolayer of surfactant molecules[7]. The diameter of aqueous droplets is usually 5–20 nm[1,8]. The aqueous droplets act as a microreactor or nanoreactor, and reactions occur when droplets containing suitable reactants collide with each other. First, precursor particles of hydroxide or oxalate are formed in a microemulsion system. After drying and calcination of the precursor powder at an appropriate temperature, the desired oxide system is obtained.

A number of biological barriers are present between the application site and the place where substances exert their pharmacological effect; thus, many pharmacologically active substances fail to achieve the appropriate concentration in the target tissue, and normal tissues of the body are exposed to the effects of potentially toxic substances. The mucosal immune system is a biological barrier that allows the passage of substances under certain conditions.

Microemulsions may interact with the stratum corneum and disorganise the lipid bilayer due to the presence of surfactants. Thus, these lipids pass as a fluid in a disorderly fashion, increasing the permeability of skin. As a result, the penetration of substances is facile. The control of drug release by microemulsions is an important advantage when the active compound seeks specific targets and therapeutic action must be extended.

The mucosal barrier control is important for immunomodulatory effects. The gastrointestinal and respiratory tract are the most important sites of infection and can act as open windows that microorganisms can use for the entrance, invasion and colonisation of tissue[9]. Mucosal surfaces represent an important physical-biological barrier between internal and external *millieu*[10]. During feeding and respiration, the body comes into contact with allergens, harmful substances, and opportunistic and pathogenic microorganisms. Notwithstanding, the body-environment equilibrium is almost perfect, and the state of illness is exceptional.

Intestinal mucosa is colonised by numerous bacteria derived from more than 500 different species[11]. Despite high bacterial colonisation and frequent allergen contact, acute inflammatory and allergic reactions are rarely observed in the mucosa[12].

Gastrointestinal associated lymphoid tissue (GALT), which is composed of discrete inductive and effector sites, is able to discriminate between harmful and harmless antigens while maintaining homeostasis.

Intestinal epithelial cells are the major producers of multiple peptides and proteins with antimicrobial activity in the intestine, which are key effector molecules of innate immunity[13]. The differential expression of diverse antimicrobial proteins in the gastrointestinal tract suggests that these proteins have distinct functional niches in mucosal innate defence, allowing the pharmacological exploitation of their antimicrobial properties[14].

2. Microemulsion

In 1943,[15] Hoar and Schulman described water-in-oil microemulsions, which they referred to as transparent water-in-oil dispersions. Subsequently, these systems have been investigated by others authors,[16] who calculated the size of droplets found in microemulsions.

Microemulsion is a suitable technique for obtaining nanometre-sized inorganic particles with minimal agglomeration[29]. Oxide and carbonate nanoparticles have been successfully synthesised by microemulsion techniques[17, 18, 19, 20]. A microemulsion is a thermodynamically stable, transparent, isotropic dispersion of two immiscible liquids (such as water and oil) that are stabilised by surfactant molecules at the water/oil interface. In water-in-oil microemulsions, nano-sized water droplets are dispersed in the continuous hydrocarbon phase and are surrounded by a monolayer of surfactant molecules[21].

Microemulsions improve therapeutic efficacy, reduce the volume of the drug delivery vehicle, mimimise toxic side effects[16,22] and improve immunological response[5,23,24]. In addition to these advantages, microemulsions are easy to administer and offer several benefits for oral administration. Namely, microemulsions are ideal for the oral and nasal delivery of drugs and vaccines[25,26,27]. Microemulsion techniques are capable of delivering inorganic and organic nano-sized particles with minimal agglomeration because the reaction occurs in nano-sized domains.

However, the major disadvantage of using microemulsion formulations as vehicles for drug delivery systems is the high concentration of surfactant and cosurfactants required to create these formulations. In microemulsion formulations used for technological applications in oil recovery, nanoparticles are composed of incompatible surfactants, cosurfactants, and oils that are compatible with human physiology and are non-irritating. Phospholipids, particularly lecithin, are nontoxic alternative emulsifiers that can be biocompatible[22]. Biocompatible surfactants and cosurfactants used in pharmaceutical microemulsions are important technological challenges for the development of adequate delivery systems that can be used in pharceutical formulations and vaccines for treating diseases and improving the immune system.

3. The mucosal immune system

In general, mucosal tissues are a specialised immune network composed of inductive and effector sites. The latter include lamina propria mucosae, the stroma of exocrine glands and surface epithelia, whereas inductive sites comprise mucosa-associated lymphoid tissue (MALT) and local and regional draining lymph nodes. The histological architecture of MALT is similar to the structure of the lymph nodes; however, MALT lacks afferent lymphatics. Antigens are captured and processed directly from the mucosal luminal side through specialised follicle-associated epithelium (FAE) containing microfold or membrane (M) cells. These cells deliver antigens to MALT antigen presenting cells (APCs), which are able to stimulate naive B and T cells[11].

Intestinal mucosa is one of the primary tissues that comprise the mucosal immune system and possess the highly elaborate architecture of the intestinal system. Intestinal mucosa is constantly challenged with the considerable task of allowing the exchange of nutrients, ions and liquids across the intestinal epithelium in the presence of an enormous density of potentially harmful luminal antigens and microbes[27].

Intestinal mucosa constitutes a barrier equipped with local defence mechanisms against invading pathogens; however, the mucosa must be selectively permeable to allow the uptake of nutrients. This dual function becomes even more challenging when we consider that the intestinal lamina propria contains a large number of immune cells with potent effector functions[26]. The unique architecture of the gastrointestinal tract facilitates both of these functions. In particular, the gastrointestinal tract possesses multiple levels and an immense overall surface area, allowing maximal nutrient absorption while housing the largest number of immune cells in the body[27].

The recognition of invading microorganisms is paramount to the survival of the host, and the innate immune system has evolved as the first line of defence in the immune response. At the mucosal surface, the host physically interacts with a nonsterile environment, and the ability to detect and contain invading pathogens is regularly tested[27,28].

Gastro-intestinal associated lymphoid tissue (GALT), which is composed of discrete inductive and effector sites, is able to discriminate between harmful and harmless antigens while maintaining homeostasis. Inductive sites are organised into specialised aggregations of lymphoid follicles called Peyer's Patches (PP), while effector sites are more diffusely dispersed[28,29].

The separation of these sites limits and controls immune responses. The human gastrointestinal tract is colonised by an abundance of bacteria, which are in constant interaction with the epithelial lining, leading to an intricate balance between tolerance and immunological response. Ample evidence suggests that the abundant presence of bacteria plays a role in the maintenance of human health and the induction of chronic inflammatory diseases of the gastrointestinal tract[30].

4. Defence of innate and adaptive immunity against infections

The mucosal barrier is formed by highly adapted epithelial cells, which are interconnected with tight junctions and are covered with mucus and bactericidal peptides. As a result, only a small number of bacteria are allowed to penetrate the intestinal epithelium. Primary defects in the barrier function can trigger intestinal inflammation[13, 31]. Mucosal surfaces have an efficient anti-infective protection system, which consists of many nonspecific mechanisms such as peristaltic movements and the transportion of mucus-fimbri, mucosal enzymes and antimicrobicidal proteins.

The mucosa surface presents intestinal epithelial cells, especially paneth cells. Paneth cells are the major producers of multiple peptides and proteins with antimicrobial activity in the intestine, which are key effector molecules of innate immunity. The most abundant and diverse antimicrobial proteins are the defensins, which are highly microbicidal in vitro and are likely important in vivo. However, the physiological functions of defensins remain incompletely understood. Paneth cells also produce cathelicidin, which contributes to the mucosal defence against epithelial-adherent bacterial pathogens, helps to set a threshold for productive infection, and is expressed constitutively by neutrophils, mast cells and differentiated epithelial cells in the colon and stomach but not the small intestine[32].

In contrast, adaptive mechanisms are represented mainly by the mucosal immune system (MALT, mucosal associated lymphoid tissue). This system consists of immuno-competent cells that infiltrate the mucosae, nodules that form lymphoid structures such as Peyer's plates and its equivalent in bronchial mucosae, and regional lymph nodes such as the mesenterics[32].

Under normal conditions, human gut mucosa is infiltrated by a large number of mononuclear cells due to the continuous stimulation of luminal antigens[33]. Approximately 70% of all lymphocytes in the human body are concentrated in the intestinal intra-epithelial and subepithelial layers, and the largest pool of tissue macrophages is located in the intestinal wall[30].

Mucosal tissue possesses the largest activated B-cell system of the body and contains 80–90% of all immunoglobulin (Ig)-producing cells. The major product of these lymphocytes is secretory (S) IgA (mainly dimers) with associated J chain[18]. Polimeric IgA contains a binding site for polymeric Ig receptor (pIgR) or secretory component (SC), which is required for their active external transport through secretory epithelia. When produced by local lymphocytes, the pIgR/SC binding site depends on the covalent incorporation of the J chain into the quaternary structure of the polymer. This important differentiation appears to be sufficient functional justification for the expression of the J chain by most B-cells terminating at IgD- or IgG-producing secretory effector sites. These cells likely represent a spin-off from sequential downstream CH switching for subsequent pIgA expression, apparently reflecting an effector B-cell clone maturational stage that is compatible with homing to these sites[11].

The differential expression of diverse antimicrobial proteins in the gastrointestinal tract suggests that they occupy distinct functional niches in mucosal innate defence, allowing the pharmacological exploitation of antimicrobial properties associated with active substances in microemulsions, which can modulate the mucosal immunological system.

5. Mucosal immune system and microemulsion

Over the past several decades, microemulsions have been used in various pharmaceutical technologies. For instance, the storage, stability, dosage, viability, side effects, controlled release, biological response and homogenous distribution of drugs in microemulsions have been explored for their potential use as drug or vaccine delivery systems [23,24].

Immunisation strategies have largely focused on the use of microemulsions. However, the efficiency and/or cost of vaccines must also be improved. In the literature, the theoretical basis of current research is to rationally design immunization methodologies for nanoparticle-based delivery systems. These methodologies may reduce the required dose and enhance the breadth and depth of protective immune responses.

In previous studies, experimental methods for the design and characterisation of nanoparticles directly from oil-in-water microemulsion precursors have been developed, and the breadth and depth of immune responses after immunisation with nanoparticles were enhanced[34, 35.36]. The intradermal administration of novel cationic nanoparticle-based DNA vaccine delivery systems using a jet injection device led to significantly enhanced Th1/Th2-balanced immune responses.

To elicit effective mucosal and systemic immune responses, a novel cationic DNA-coated nanoparticle engineered from a microemulsion precursor was modified, optimised and applied intranasally as a vaccine delivery system. DNA-coated nanoparticles significantly enhanced specific serum IgG and IgA. An enhanced splenocyte proliferative response was also observed after immunisation with pDNA-coated nanoparticles; thus, these particles may be used for immunisation via the nasal route[25] and inhalable nanoparticles may enhance immune responses[25,28].

6. Conclusion

The exact mechanism of the modification of mucosal immune response by microemulsions remains to be elucidated. However, the effect of microemulsions may be attributed to an enhancement in nanoparticle uptake by cells present in the mucosal immune system. The incorporation of antigens into nanoparticles may be a promising approach because colloidal formulations protect antigens from degrading milieu in the mucosa and facilitate their transport across barriers.

The use of nanoparticles for vaccine delivery provides beneficial effects, and good immune responses can be achieved. Although the mechanism of induction of mucosal immunity after vaccination has not been fully elucidated, antigens associated with microemulsion nanoparticles may enhance mucosal immune responses and improve vaccination methods.

7. References

[1] Paul BK, Moulik SP. Microemulsions: an overview, J Disper Sci Technol.18:301–67, 1997.

[2] Lim GK, Wang J, Ng SC, Gan LM. Formation of nanocrystalline hydroxyapatite in nonionic surfactant emulsions, Langmuir.15:7472–7,1999.

[3] Bose S, Saha SK Synthesis and characterization of hydroxyapatite nanopowders by emulsion technique. Chem Mater.;15: 4464–9, 2003.

[4] Júnior ASC, Fialho SL, Carneiro LB, Oréfice F. Microemulsões como veículo de drogas para administração ocular tópica. Arq Bras Oftalmol.66:385-91,2003.

[5] Honorio-França AC, Moreira CM, Boldrini F, França EL. Evaluation of hypoglicemic activity and healing of extract from amongst bark of "Quina do Cerrado" (Strychnos pseudoquina ST. HILL). Acta Cirúrgica Brasileira.23:504-10, 2008.

[6] Lawrence MJ, Rees GD. Microemulsion-based media as novel drug delivery sustems. Adv Drug Deliv Rev.45:89-121, 2000.

[7] Arriagada FJ. Synthesis of nanosize silica in a nonionic water-in-oil microemulsion, J Colloid Interf Sci. 211:210–20,1999.

[8] Karagiozov C, Momchilova D. Synthesis of nano-sized particles from metal carbonates by the method of reversed mycelles. Chem Eng Process.44:115–9,2005.

[9] Gomes TAT, Rassi V, MacDonald KL, Silva- Ramos SRT, Trabulsi LR, Vieira MAM, Guth BEC, Candeias JAN, Ivey C, Toledo MAM, Blake PA. Enteropathogens associated with acute diarrheal disease in urban infants in São Paulo, Brazil. J Infect Dis. 164:331-7, 1991.

[10] Giron JA, Ho Asy, Schoolnik GK. An inducible bundle-forming pilus of enterpathogenic Escherichia coli. Science. 254:710-13,1991.

[11] Brandtzaeg, P, Pabst R. Let's go mucosal: communication on slippery ground. Trends Immunol. 25: 570–7, 2004.

[12] Bjarnason I, Macpherson A, Hollander D. Intestinal permeability: an overview. Gastroenterology. 108: 1566–81,1995.

[13] Eckmann L, Nebelsiek T, Fingerle AA, Dann SM, Mages J, Lang R, Robine S, Kagnoff MF, Schmid RM, Karin M, Arkan MC, Greten FR. Opposing functions of IKKbeta during acute and chronic intestinal inflammation. Proc Natl Acad Sci U S A. 30:15058-63,2008.

[14] Lim GK, Wang J, Ng SC, Gan LM. Processing of fine hydroxyapatite powders via an inverse microemulsion route, Mater Lett. 28:431–6,1996.

[15] Hoar TP, Schulman JH. Transparent water-in-oil dispersions: the oleopathic hydro-micelle. Nature. 152:102, 1943.

[16] Schulman JH, Friend JA. Light scatttering investigation of the transparent oil-water disperse system II. J. Colloid Interface Sci. 4:497, 1949.

[17] M.P. Pileni, The role of soft colloidal templates in controlling the size and shape of inorganic nanocrystals. Nat Mater. 2:145–50, 2003.

[18] Sun Y, Guo G, Tao D, Wang Z. Reverse microemulsion-directed synthesis of hydroxyapatite nanoparticles under hydrothermal conditions. J Phys Chem Solids. 68: 373–7, 2007.

[19] S. Singh, P. Bhardwaj, V. Singh, S. Aggarwal and U.K. Mandal, Synthesis of nanocrystalline calcium phosphate in microemulsion—effect of nature of surfactants. J Colloid Interf Sci. 319: 322–9, 2008.

[20] Lim GK, Wang J, Ng SC, Gan LM. Processing of fine hydroxyapatite powders via an inverse microemulsion route. Mater Lett. 28:431–6, 1996.

[21] Karagiozov C, Momchilova D. Synthesis of nano-sized particles from metal carbonates by the method of reversed mycelles. Chem Eng Process. 44:115–9, 2005.

[22] Bagwe RP, Kanicky BJ, Patanjali PK, Shah DO. Improved drug delivery using microemulsions: rationale, recent progress and new horizons. Crit Rev Therap Drug Carr Systems. 18:77-140, 2001.

[23] Zhengrong Cui, Lawrence Baizer, Russell J Mumper. Intradermal immunization with novel plasmid DNA-coated nanoparticles via a nudle-free injection device. J Biotechnol. 102:105-15, 2003.

[24] Nastruzzi C, Gambari R. Antitumor activity of (trans) dermally delivered aromatic tetraamidines. J Control release. 29:53-7, 1994.

[25] Cui ZR, Mumper RJ. Intranasal administration of plasmid DNA-coated nanoparticles results in enhanced immune responses J Pharm Pharmacol. 54:1195-203 ,2002.

[26] Bhargava HN, Narurkar A, Leib LM. Using microemulsions for drug delivery. Pharm Technol. 11:46-9, 1987.

[27] Aguiar JC, Hedstrom RC, Rogers WO, Charoenvit Y, Sacci JB, Lanar DE, Majam VF, Stout RR, Hoffman SL, Enhancement of the immune response in rabbits to a malaria DNA vaccine by immunization with a needle-free jet device. Vaccine. 20: 275–80, 2002.

[28] Ali J, Ali M, Baboota S, Potential of Nanoparticulate Drug Delivery Systems by Intranasal Administration Sahni JK, Ramassamy C, Dao L,Bhavna . Cur Pharma Design. 16:1644-53, 2010

[29] Bjarnason I, Macpherson A & Hollander D. Intestinal permeability: an overview. Gastroenterology. 108: 1566–81, 1995.

[30] Macpherson AJ, Harris NL. Interactions between commensal intestinal bacteria and the immune system. Nat Rev Immunol. 4: 478–85, 2004.

[31] Carneiro-Sampaio MMS, Da Silva ML, Carbonare SB, Palmeira P, Delneri MT, Honório AC, Trabulsi LR. Breast-Feeding protection against Enteropathogenic *Escherichia coli*. Rev Microbiol. 27:120-5,1996.

[32] Law D. Adhesion and its role in the virulence of enteropathogenic Escherichia coli. Clin Micrbiol Rev. 7:152-73,1994.

[33] Macpherson AJ, Uhr T. Compartmentalization of the mucosal immune responses to commensal intestinal bacteria. Ann N Y Acad Sci. 1029: 36–43,2004.

[34] Arriagada FJ. Synthesis of nanosize silica in a nonionic water-in-oil microemulsion. J Colloid Interf Sci. 211:210–20, 1999.

[35] Baizer L, Hayes J, Lacey C, D'Antonio L. Needle-free injectors: advantages, current technologies, and future innovations. Pharm. Manu. Pack. Resou. Spring. 96–100, 2002.

[36] Zhengrong Cui, Russell J Mumper The effect of co-administration of adjuvants with a nanoparticle-based genetic vaccine delivery system on the resulting on the immune responses. Eur J of Pharm Biopharma. 55:11-18, 2003.

Part 4

Applications in Oil Industry and Preparation of Nanostructured Materials

8

The Use of Microemusion Systems in Oil Industry

Vanessa Cristina Santanna, Tereza Neuma de Castro Dantas and Afonso Avelino Dantas Neto
Federal University of Rio Grande do Norte
Brazil

1. Introduction

Microemulsions are thermodynamically stable, isotropic, and macroscopically homogeneous dispersions of two immiscible fluids, generally oil and water, stabilized with surfactant molecules, either alone or mixed with a cosurfactant, as shown in Figure 1 (Robb, 1981). Cosurfactant is a nonionic molecule (e.g. a short-chain of alcohols or amine) that has the function of stabilizing a microemulsified system by decreasing the repulsion forces between the hydrophilic parts of the surfactant.

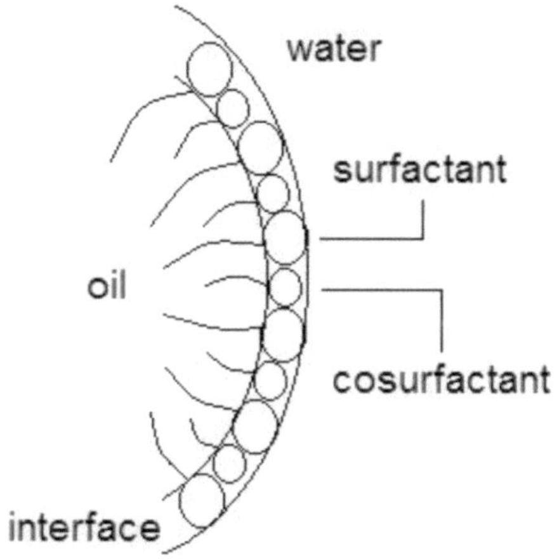

Fig. 1. Microemulsion droplet (Reversed micelle)

The potential of microemulsions in technological applications, however, has not yet fully been explored. Presently, there are many established industrial activities that rely on the use of emulsions as a mean to achieve effectiveness. It is important to consider the core differences between emulsions and microemulsions to improve the design of their applications, as shown in Table 1.

Microemulsion equilibriums, according to Winsor's classification (Friberg & Bothorel, 1987), vary according to the nature of the phases involved (Figure 2). This classification establishes four types of systems: WINSOR I (WI), where the microemulsion phase is in equilibrium with an organic phase in excess; WINSOR II (WII), where the microemulsion phase is in equilibrium with an aqueous phase in excess; WINSOR III (WIII), where the microemulsion phase is in equilibrium with both aqueous and organic phases (three-phase system); and WINSOR IV (WIV), which is a one-phase system, in a macroscopic scale.

Emulsions	Microemulsions
Unstable, with eventual phase separation	Thermodynamically stable
Relatively large-sized droplets (1-10 μm)	Small aggregates (around a few tens of nanometers)
Relatively static systems	Highly dynamic systems
Moderately large interfacial area	Very high interfacial area
Small amount of surfactant required for stabilization	Large amount of surfactant required for stabilization
Low curvature of the water-oil interface	Interfacial film may be highly curved

Table 1. Main differences between emulsions and microemulsions (Dantas Neto et al., 2009)

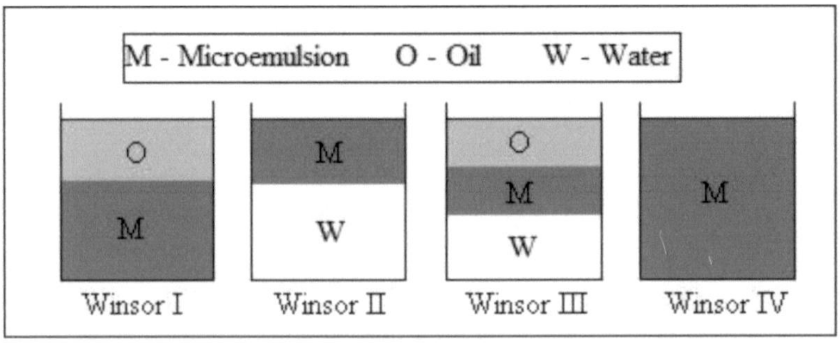

Fig. 2. Schematic showing Winsor's classification

The microemulsions, composed of four constituents (water, oil, surfactant, and/cosurfactant), can be represented by pseudoternary diagrams, built and displayed in a user-friendly manner. In order to simplify the microemulsion's representation, two constituents are kept in a fixed proportion, the ''pseudoconstituent''. The constant molar ratio (cosurfactant/surfactant) is usually applied during this phase as demonstrated in Figure 3. To determine the Winsor's regions in a pseudoternary diagram, the oil phase was mixed with the surfactant/cosurfactant phase, and the mixture was titrated with water to observe the changes in the Winsor's regions. The volume of water used was determined for each region change. The pseudoternary phase diagram was constructed by plotting the amounts of water, oil, and surfactant/cosurfactant phases.

Microemulsified media are of great interest due to its wide potentialities for industrial applications, such as in the oil industry. Therefore, this paper aims to show the applications of microemulsions in various areas of the oil industry, such as in wells' hydraulic fracturing (stimulation operation), as corrosion inhibitors in pipelines, in the breakdown of water/oil emulsions, in enhanced oil recovery, and as an alternative fuel.

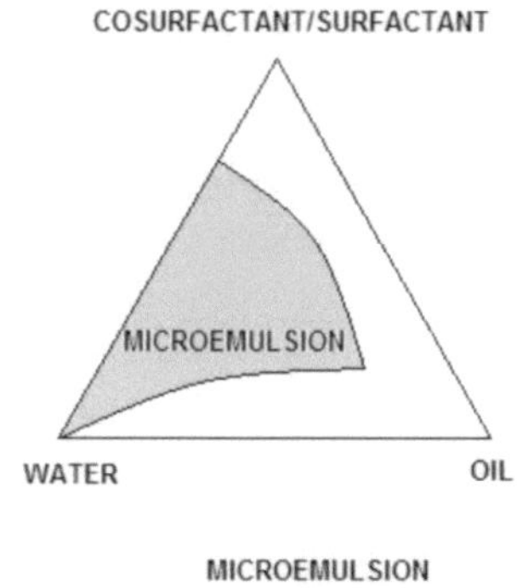

Fig. 3. Pseudoternary diagram with constant cosurfactant/surfactant ratio

2. Microemulsions in oil industry

2.1. Hydraulic fracturing

Hydraulic fracturing is a technique that aims to increase well productivity. The fracturing fluid is applied against the reservoir rock under high differential pressure to create fractures. A proppant (sand, bauxite, or ceramic) is pumped into the well with the fracturing fluid to keep the fracture open; creating a high conductivity way that eases the flow of fluids between the formation point and the well. In recent years, the insoluble residues of fracturing fluids left in the fracture have been the object of more specific studies. These fluids can cause damage (permeability reduction) in the proppant pack or on the surfaces of the fracture itself. Hence, fracturing fluid, microemulsion-based category, has been reported to reduce the formation of damage. Surfactant-based fracturing gels (microemulsions) are considered clean gels due to the absence of insoluble residues in its composition.

Samuel et al. (1999) developed a surfactant-based polymer-free fluid (ClearFRAC). The surfactant used was a quaternary ammonium salt, derived from a long-chain fatty acid. Aspects such as structure, rheology, fluid loss, and conductivity of this surfactant fluid were studied. The results showed that, in brine, the fluid's viscosity and viscoelasticity occurs due to the formation of highly entangled worm-like micelles. The micelles have a gross structure similar to a polymer chain. Since the viscosity of the fluid depends on the nature of micelles, the fluid can be broken by changing its micellar structure. The breaking occurs when the fluid is exposed to hydrocarbons or diluted with formation water. Therefore, conventional breakers are not required, and the produced oil or gas can act as breakers for this fluid system.

Castro Dantas et al. (2003a) developed a surfactant-based fracturing gel (SBG). The formulation of a viscous microemulsion system (gel) was obtained from the pseudoternary phase diagram as illustrated in Figure 4. The gel's composition was 18 wt. % soap, 9 wt. % isoamyl alcohol, 14 wt. % pine oil, and 59 wt. % water. Comparisons between the SBG and conventional gel (Hydroxypropylguar - HPG) were performed. The API (American Petroleum Institute) fluid loss test was performed and the fluid loss values for the SBG were 21 times larger than the ones for the conventional gel. This was because surfactant-based gels are free of solids and do not form a filter cake and the interactions between the microemulsion microdroplets are very weak, allowing the gel to flow through the filtration medium with little resistance, if compared to HPG. Two fluid-loss additives were used:

silica-flour (6kg/m³) and a urea solution (1–8%). The authors observed a 14% reduction of filtrate for SBG and a 20% reduction for HPG, when silica-flour was used in the experiment. The urea is a bi-functional and cationic salt that links an anionic micelle to another, increasing the viscosity of the gel and reducing the filtrate. With the addition of a urea solution, the leak-off coefficient had a reduction of up to 50% for the SBG. The capacity of gels to sustain the ceramic 16/20 mesh in function of time also was observed, and the SBG presented a great sustaining capacity for the proppant (Table 2). The authors concluded that the SBG presented compatible characteristics when compared with the HPG gel.

Gel type	v_s (cm/s)
HPG	1.63
SBG	1.17×10^{-3}

Table 2. Results of setting rate assays of surfactant-based fracturing gel (SBG) and Hydroxypropylguar gel (HPG) (Castro Dantas et al., 2003a)

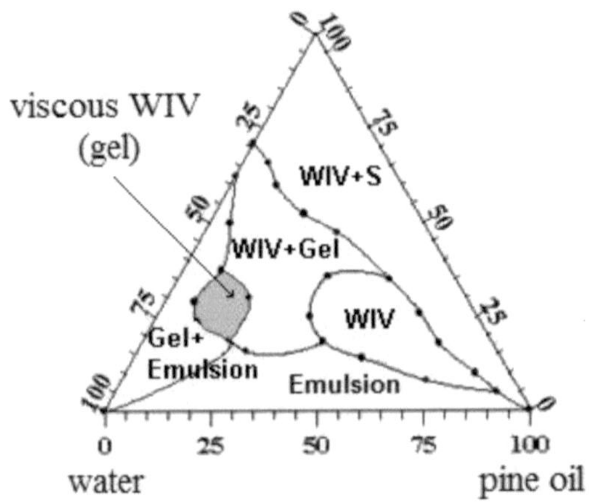

Fig. 4. Pseudoternary system comprising distilled water, commercial surfactant (S), pine oil, isoamyl alcohol (C), with C/S ratio = 0.5, constructed at 26 °C (Dantas et al., 2003a)

Depending on the surfactant concentration and temperature, micelles of different geometry are formed, such as spheres or rods. The determination of the inner structure of surfactant solutions through optical measurements is possible only when the surfactant concentration is very low and the solution is at rest. Rheological measurements offer an indirect way to determine the inner structure of surfactant solutions over a wide concentration range, and often can provide evidence of structural changes. On this subject, Castro Dantas et al. (2003b) performed rheological studies to evaluate the inner structure of anionic surfactant-based gels. Steady and oscillatory shear experiments were carried out. To prepare the surfactant-based gels, the gel region (viscous microemulsion) in the pseudoternary phase diagram was determined (Figure 5). In steady shear experiments, the temperature ranged from 26 to 86 °C. In oscillatory shear experiments the shear stress was maintained constant (1 Pa) and all of the experiments were done at 66 °C. In order to check the pseudoplastic (shear thinning) behavior of the gel, rheological tests were performed by varying the shear rate from 0.01 to 120 s^{-1}. Four compositions of gels were chosen as shown in Table 3.

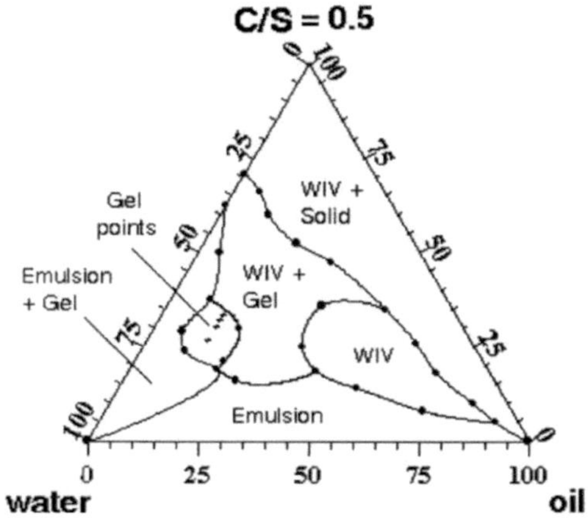

Fig. 5. Pseudoternary phase diagram showing the gel region with the four compositions chosen (26ºC) (Castro Dantas et al., 2003b)

Gel Cosurfactant/Surfactant (wt %)	Surfactant Soap (wt %)	Cosurfactant isoamyl alcohol (wt %)	Oil-pine oil (wt %)	Water (wt %)
27	18	9	14	59
30	20	10	14	56
32	21.3	10.7	14	54
34	22.7	11.3	14	52

Table 3. Compositions of gels for rheological studies

Based on oscillatory shear experiments, the authors concluded that the micellar structure exhibits a viscoelastic response typical of gel-like materials, characterized by loss and storage modulus slightly dependent on frequency. In steady shear experiments, they verified that the viscosity of the gels depends on C/S concentration and temperature as illustrated in Figure 6. The increase in viscosity with C/S concentration is attributable to the dispersed fraction in the solutions (micelles), to the nature of the micelle aggregates, and to the structure-forming interactions between the aggregates. From the gel-activation energies, the authors verified, that the activation energy is positive when there is a small variation of the volume of the micelles and negative when the volume variation of the micelles prevails in the system.

Castro Dantas et al. (2006) developed a laboratory break methodology for surfactant-based fracturing gel (SBG). According to the authors, fracturing gel should provide sufficient viscosity to suspend and transport the proppant into the fracture, and should break into a low-viscosity fluid after the fracturing is completed. This break facilitates the fracture cleaning by allowing a rapid counter flow of fluids to the surface. Surfactant-based fracturing gels can be broken when exposed to hydrocarbons or formation water. Therefore, conventional breakers, commonly used in polymer-based gels, are not required and the oil or produced gas can act as breakers for surfactant-based gels. The gel break test consisted basically of the injection of fluids into sandstone core samples as seen in Figure 7. To simulate reservoir conditions, brine and oil were injected into the core sample (Figure 8).

The surfactant-based gel (18 wt. % soap, 9 wt. % isoamyl alcohol, 14 wt. % pine oil, and 59 wt. % water) was then injected at a constant flow rate, in the opposite direction to oil production and samples were collected. Initially, the samples had some oil content, then, limpid samples were observed. These samples were collected at periods of time determined previously. The gel break was verified by assessing its viscosity in a Brookfield rheometer at 26 °C. The authors observed that the gel broke within the first hours of the tests due to the initial presence of oil and brine in the core. When oil and/or water were introduced into surfactant-based gels, the micelles moved apart and the viscosity decreased. They observed that the studied gel presented optimum break results, due to a viscosity decrease to 10 mPas in the first 6h of the test as seen in Figure 9.

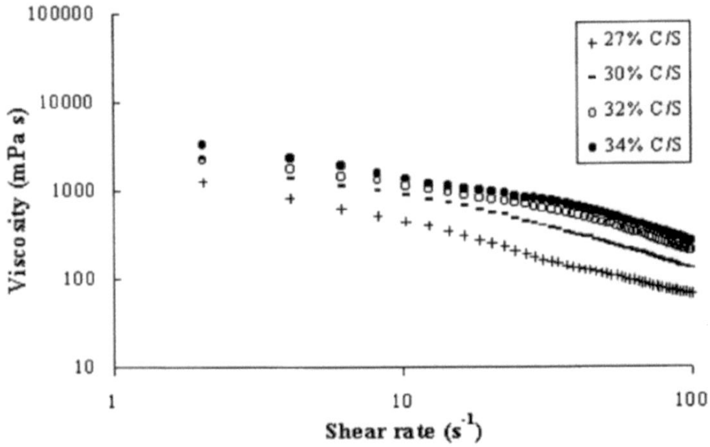

Fig. 6. Results of assays showing rheological behavior of surfactant-based gel (Castro Dantas et al., 2003b)

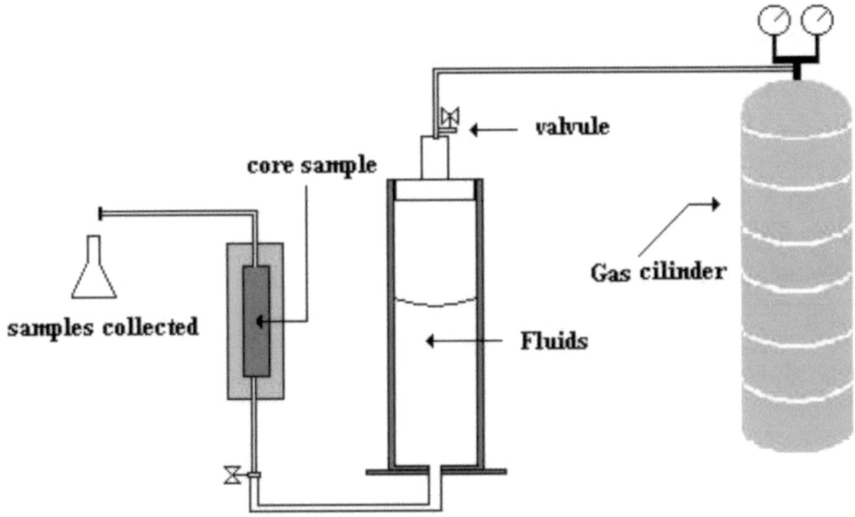

Fig. 7. System developed for the break tests with surfactant-based gels (Castro Dantas et al., 2006)

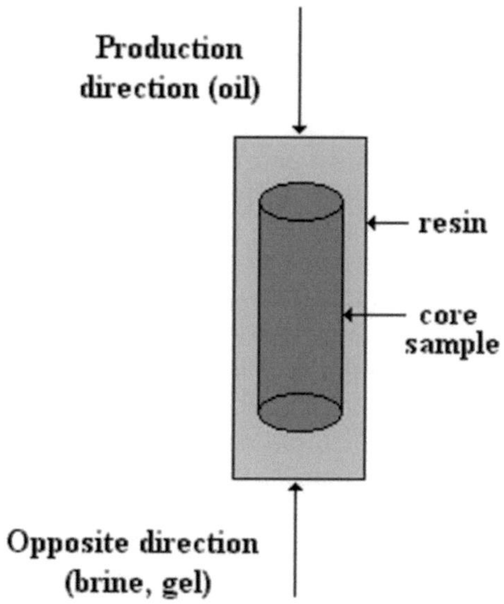

Fig. 8. Schematic diagram showing the flow directions used in the gel break tests (Castro Dantas et al., 2006)

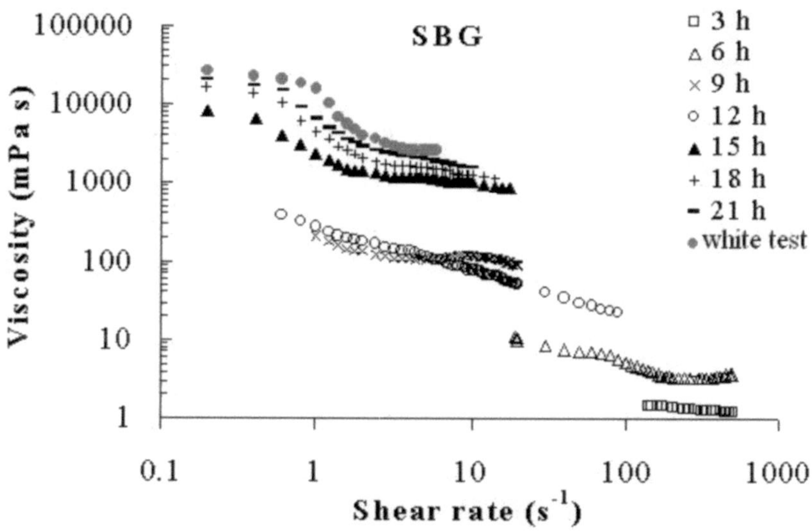

Fig. 9. Surfactant-based gel break results (Castro Dantas et al., 2006)

Recently, Liu et al (2010) observed the benefits of applying the microemulsion together with a polymer in hydraulic fracturing. The authors constructed a fracturing fluid which they named SPME-Gel with a combination of a single phase microemulsion (SPME) and a gelable polymer system. The formulation of a single phase microemulsion system was obtained from the pseudoternary phase diagram. The SPME composition was 22-27% w/w Tween-80 and Span-80 (2:1 ratio), 24-26% w/w kerosene, 21-23% w/w isopropanol and 27-29% w/w tap water. The SPME-Gel formulations were prepared by adding SPME into a gelable polymer system (Hydroxypropyl guar gum- HPGG) at various concentrations (1000–5000 mg/L). A series of comparative experiments were performed with the SPME-Gel and conventional gel (without SPME). API fluid loss was performed and the fluid-loss values of SPME-Gels were half of the conventional gel ones. This difference can be attributed to the assembly of micelles in a group forming microscopic spheres, rods, and plates that can create a deformable barrier, reducing leak off protection on the surface of the fractured rock. The effectiveness of breakage was studied. The results showed that in the broken SPME-Gel systems, the HPGG residues were stably suspended, while for the broken conventional gel, the HPGG residues were deposited. This indicates that the broken SPME-Gel system has remarkable suspension capability for HPGG residues. This property shows that in the SPME-Gel system less HPGG residues remained in formation after gel breakage. Core permeability regaining tests indicated that 83 to 91% of the original matrix permeability was regained after exposure to SPME-Gel, while only 51 to 62% was regained by the conventional gel. Proppant conductivity tests were performed and the results indicated that 90 to 96% of the proppant's conductivity was regained for SPME-Gels, against 68 to 77% for the conventional gels. From the measurements obtained, it was observed that properties of SPME-Gels were more predominant than those of conventional gels. Thus, SPME-Gel is a promising formula that offers advantages for reducing formation damage, lowering the initial cleanup pressure, and maintaining the original core regain for fracturing treatment.

2.2 Break down emulsions

Fluids injected during the drilling and completion process, with the aim of cleaning up the well, can cause formation damages due to rock pore clogging. Acid solutions, added later to restore the permeability of the rock, can promote the formation of highly viscous emulsions and dregs. This will cause a significant reduction in production and even compromise viable economic exploitation. The formation of emulsions and dregs is a serious problem during the well stimulation process and can result in large losses of oil (Castro Dantas et al., 2001). Many studies have proposed alternatives to break down emulsions, such as application of microemulsions.

Castro Dantas et al. (2001a) studied the performance of microemulsion systems as an alternative in the breakdown of petroleum emulsions. The crude oils used were characterized by the different Balance Sheet of Sediment and Water (BSW) values of 48%, 36%, and 32%. The microemulsion systems studied were composed of HCl 5.2% solution; toluene; and isopropyl alcohol (C)/surfactants (S), with a ratio C/S of 9.0. The commercial surfactants used were Dissolvan, NE-8, Dentrol, OC595, Fenoil, Polimus-43, NE-9, Polimus-22, Dehydet, and Ultrapan. The microemulsion efficiency to break down oil emulsions was evaluated by using the direct contact method between the microemulsions and crude (W/O) emulsions in graduated centrifuge tubes for 30 min, under stirring at 70°C. The results showed the achievement of a good percentage of emulsions breakdown for all systems. The results indicated that the breakdown efficiency is directly dependent on the microemulsion composition and the physicochemical properties of the oil. Dissolvan can efficiently be used in microemulsion systems, which their compositions are in the oil-rich phase and with C/S percentages from 60% to 82%, to break down 100% of the oil emulsion. The NE-8 showed to be efficient in aqueous phase rich microemulsion regions for all C/S percentages, being able to completely break down the oil emulsion. With the Polimus-43 in oil-rich and/or aqueous-rich phase microemulsion regions, 100% of emulsion breakdown was obtained. The authors concluded that some commercial surfactants used for the formation of microemulsion systems were able to completely break down oil emulsions formed during production operations of Brazilian petroleum fluids.

2.3 Corrosion inhibitor

In the petroleum industry, saline media is the main agent that leads to corrosion. It acts in the inner walls of oil pipelines, leading to a particular type of corrosion, which is commonly referred to as localized pitting corrosion. Surfactant molecules have been employed as corrosion inhibitors in order to minimize and control such phenomenon. Surfactant molecules form films on metallic surfaces that protect them from corrosion by impairing the action of electrolytes. The adsorption of surfactants on metal surfaces depends on the structure and concentration of surfactant molecules in the contacting medium determining the final adsorption layout with the formation of monolayers (Figure 10) or multilayers of surfactant molecules (Reyes et al., 2005).

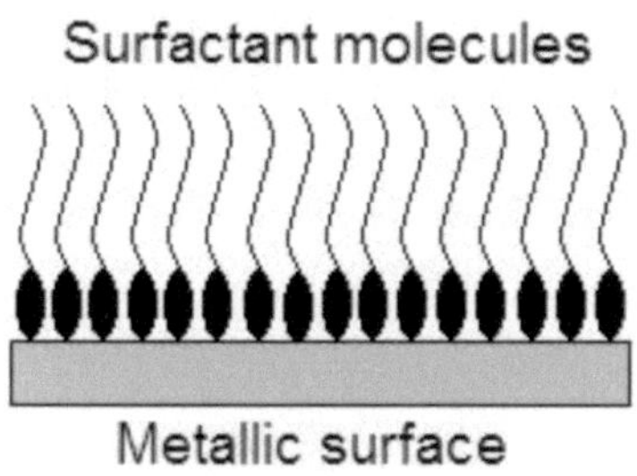

Fig. 10. Monolayers formed by adsorbed surfactant molecules onto a metallic surface

The microemulsions also have the characteristic of adsorbing onto interfaces. When microemulsions are used, an oil film is adsorbed onto the surface with the surfactants tails oriented towards it, in view of the usually positive character of the surface.

The anticorrosion capacity can be tested by electrochemical cells. Three electrodes normally are used in such assays, a reference electrode, a counter electrode, and a work electrode. These electrodes are immersed in the corrosive medium, which can be a saline or acidic solution, with or without an inhibitor. The reference electrode is involved directly in the corrosion potential measurements, from which anodic and cathodic polarizations are affected. The counter electrode is used as an auxiliary to complete the cell and balance charges. The efficiency of the system in inhibiting corrosion (E %) is estimated by means of Equation 1:

$$E\% = [100 \times (I_{corr} - I'_{corr})]/I_{corr} \quad (1)$$

where, I_{corr} and I'_{corr} denote the corrosion current densities in the absence and presence of inhibitor, respectively.

Using potentiometry measurements, Moura et al. (2009) studied the anticorrosion ability of three novel surfactant molecules in solution and microemulsion media. The surfactants were synthesized from ricinoleic acid, a castor oil derivative (sodium 12-*N*, *N*-diethylamino-9-

octadecenoate– AR1S, sodium 12-*N*, *N*-diethylamino-9, 10-dihydroxy-octadecanoate– AE2S and sodium 12-*N*, *N*-diethylamino-9-octadecanoate– AE1S) as shown in Figure 11. The ability to inhibit corrosion was investigated by preparing specific micellar solutions and microemulsion systems. The microemulsions components used were: the synthesized surfactants (AR1S, AE2S and AE1S); a saline aqueous solution (NaCl 0.1M); *n*-hexane and butan-1-ol, with a constant C/S concentration ratio equal to 2. Adsorption phenomenon was studied electrochemically and the Frumkin isotherm model indicated that the surfactant solutions tested can inhibit corrosion with levels as high as 95%. The authors verified that the chemical groups attached to the hydrocarbon chains of the AE2S and AE1S molecules enable adsorption phenomena with different levels of intensity at the surface. When in solution, surfactants behave as better inhibitors, if compared to microemulsion systems. This comparision provides clear evidence that the adsorption promoted by micellar solutions is stronger than microemulsion systems. Microemulsion systems also interact with the metallic surface, although via a less intense physical mechanism. On the other hand, the microemulsion systems, although featuring relatively lower performance, are advantageous considering that they are able to dissolve more active matter.

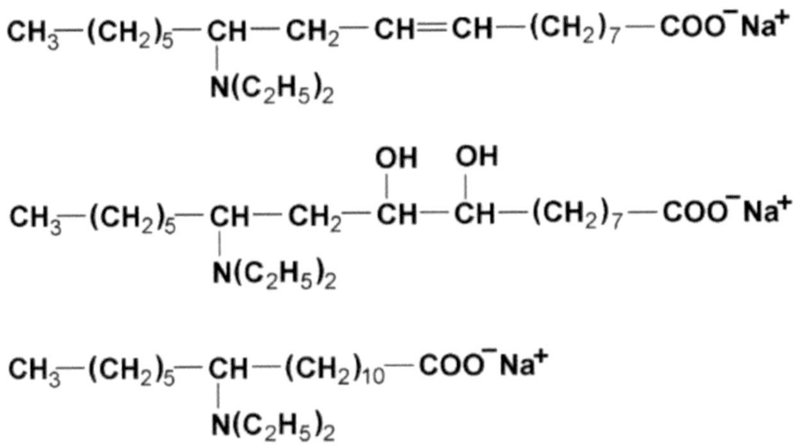

Fig. 11. Molecular structures of the synthesized surfactants AR1S, AE2S and AE1S, respectively (Moura et al., 2009)

2.4 Enhanced oil recovery

The application of enhanced oil recovery methods (tertiary oil displacement process) has been investigated with the purpose of responding to the demands of energy supply. (Santanna et al., 2009).

Petroleum recovery methods basically consist of the injection of fluids onto the rock with the objective of displacing oil out of the rock pores, where it is impregnated. The injected fluid may be known as displacing fluid and it pushes the oil out of the rock. At the same time, the injected fluid occupies the empty space left by the displaced fluid that was expelled. Enhanced oil recovery methods are classified in three distinct categories: thermal methods, miscible methods, and chemical methods. Microemulsion flooding is included into chemical methods classification.

In enhanced oil recovery, the microemulsion flooding has been suggested as an alternative method. It displays the unique properties of microemulsion systems, such as high viscosity and the ability to induce low interfacial tension, increasing oil extraction efficiency (Santanna et al, 2009). According to Austad and Strand (1996), very low interfacial tensions may be reached with microemulsion systems. According to Gurgel et al. (2008) microemulsions are potential candidates in enhanced oil recovery, especially because of its ultra-low interfacial tension values, attained between the contacting oil and water microphases that form them. Under such circumstances, microemulsions flow more easily through the porous medium, which enhance oil extraction performance rates.

Babadagli (2005) has written a review about development of mature oil fields. According to his review, the most common chemical injection technique, as a tertiary oil recovery method, is the surfactant solution injection due to its relatively lower cost when compared to micellar or microemulsion injection. The way by which the fluid is injected, when the chemical method is employed, is an important parameter for optimization of the technique. Continuous injection of a chemical solution may increase operation costs and/or reduce the amount of treated material. In view of this, injection of chemical solution considering the porous volume number (PV) is required in any efficient process. Thomas and coworkers (in: Babadagli, 2005) injected porous volumes of microemulsion in sandstone plugs containing 35% of residual oil, observing a linear relationship between the values of injected PV and the oil recovery. Results typically showed a 45% residual oil recovery when injecting 10 PV of microemulsion.

Santanna et al. (2009) studied the application of different types of microemulsion for enhanced oil recovery, one was prepared with a commercial surfactant (MCS), and another contained a surfactant synthesized in laboratory (MLS), (Table 4). The experiments consisted of the injection of fluids into cylindrical plug samples. During the microemulsion flooding, samples were collected as a function of time and the mass of oil recovered by the microemulsion was determined. The chemicals used to prepare the microemulsion systems were: commercial anionic surfactant (soap-sodium salt) derived from fatty acids; anionic surfactant (soap-sodium salt) synthesized in laboratory, derived from fatty acids (100wt.% of vegetable oil containing 12 carbon atoms); isoamyl alcohol; pine oil; and distilled water. From the results obtained, one could conclude that the use of microemulsion prepared with the commercial MCS allowed for recovery indexes as high as 87.5%, whilst the use of the MLS microemulsion permitted recovery indexes as high as 78.7% (Figures 12 and 13). This was due to the difference in microemulsion viscosities, corroborated by the fact that the MCS microemulsion (32 cP viscosity) could recover more oil than the MLS microemulsion (27 cP viscosity).

Microemulsion type	Surfactant Soap (wt. %)	Cosurfactant isoamyl alcohol (wt. %)	Oil pine oil (wt. %)	Water (wt. %)
MCS	13.3	6.7	50	30
MLS	16.7	8.3	45	30

Table 4. Compositions of microemulsions for enhanced oil recovery assays.

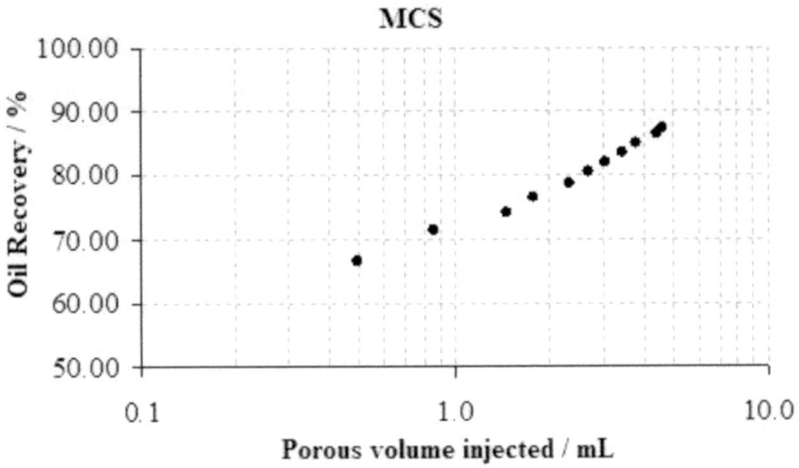

Fig. 12. Recovery factor for MCS microemulsion (Santanna et al., 2009)

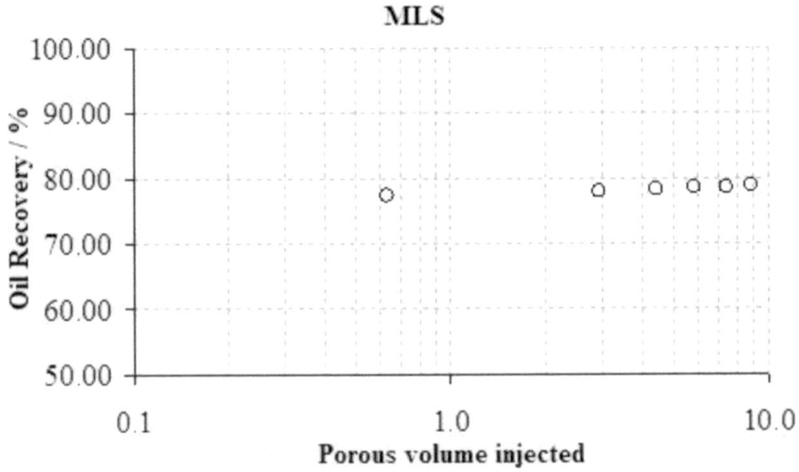

Fig. 13. Recovery factor for MLS microemulsion (Santanna et al., 2009)

2.5 Microemulsion as alternative fuel

The necessity of reducing gas emissions and improving the performance of engines has increased the interest in microemulsified systems as alternative fuels (Castro Dantas et al., 2001b; Ochoterena et al., 2010; Lif et al., 2010; Dantas Neto et al., 2011).

Castro Dantas et al. (2001b) verified the possibility of using microemulsified systems as alternative fuels. The authors studied new microemulsion systems containing diesel and different percentages of vegetable oils (soy, palm and castor). The main parameters that affect the formation of microemulsions are the nature of the surfactant and cosurfactant, the cosurfactant/surfactant mass ratio and composition of the oil phase. The results showed that it was possible to obtain new microemulsion systems with different oil phase composition using mixtures of diesel and vegetable oils. Among all studied systems, microemulsions containing diesel and soy oil could be formed over the widest composition range, indicating the possibility to apply them as alternative fuels.

Dantas Neto et al. (2011) used microemulsion systems as alternative fuels. After preparation and characterization of the systems, engine performance and emissions with the use of this new fuel were assessed. They verified that the density and viscosity of all studied systems were higher than those obtained for neat diesel. The authors also observed that these properties increased with increasing water content. According to their results for specific fuel consumption, the presence of water in microemulsions improves diesel fuel combustion. Carbon dioxide emissions were higher for the diesel/surfactant blend as well as all microemulsion fuels. Nevertheless, this difference decreased with increasing engine power when compared with neat diesel. The values of NOx emissions increased with increasing engine power, and decreased with increasing water content in the microemulsion fuels. They also observed a reduction in black smoke emissions for all microemulsion fuels tested, as compared with neat diesel. This was attributed to a better combustion reaction effected in the presence of water and surfactant, thereby reducing the formation of black smoke.

3. Conclusion

In this chapter, we have focused on how several microemulsion systems may be used in different areas of the oil industry, for example, in wells hydraulic fracturing, as corrosion inhibitors in pipelines, in the breakdown of water/oil emulsions, in enhanced oil recovery, and as an alternative fuel. When the microemulsion is used as a fracturing fluid, it has properties that are compatible with those of polymeric fluids, with the advantage of reducing the formation of damage. As corrosion inhibitor in pipelines, the microemulsion interacts with the metallic surface, although less intensively than if compared to surfactants in solution. Microemulsions showed good anticorrosive potential. In the breakdown of water/oil emulsions, the microemulsions presented great de-emulsifying characteristics. In enhanced oil recovery, microemulsion systems proved to be efficient in the oil extraction process. The microemulsified systems containing neat diesel or a mixture of diesel with different percentages of vegetable oils can be used as alternative fuels showing improved combustion properties in diesel engines. Overall, this chapter has attempted to demonstrate that optimization of different microemulsion systems provides important contribution to the petroleum industry, with direct implications on material processability and recovery yields.

4. Acknowledgment

The authors acknowledge all students, coworkers, and researchers that contributed to the publication of the material presented in this chapter.

5. References

Dantas Neto, A. A.; Castro Dantas, T. N.; Alencar Moura, M. C. P.; Barros Neto, E.L. & Gurgel, A. (2009). Microemulsions as potential interfacial chemical systems applied in the petroleum industry. In: *Microemulsions: Properties and Applications*, Monzer Fanum, pp. 411-449, CRC Press, ISBN 978-1-4200-8959-2, Boca Raton.

Babadagli, T. Mature field development—a review. *SPE*, No.93884, (June 2005), pp. 1-20.

Castro Dantas, T. N.; Dantas Neto, A. A. & Moura, E. F. Microemulsion systems applied to breakdown petroleum emulsions. *Journal of Petroleum Science and Engineering*, No.32, (2001a), pp. 145-149, ISSN 0920-4105.

Castro Dantas, T. N.; Silva, A. C. & Dantas Neto, A. A. New microemulsion systems using diesel and vegetable oils. *Fuel*, No.80, (2001b), pp. 75-81, ISSN 0016-2361.

Castro Dantas, T. N.; Santanna, V. C.; Dantas Neto, A. A. & Barros Neto, E.L. Application of surfactants for obtaining hydraulic fracturing gel. Petroleum Science and Technology, Vol.21, No.7 & 8, (2003a), pp. 1145-1157, ISSN 1091-6466.

Castro Dantas, T. N.; Santanna, V. C.; Dantas Neto, A. A.; Barros Neto, E.L. & Alencar Moura, M. C. P. Rheological properties of a new surfactant-based fracturing gel. *Colloids and Surfaces A: Physicochem. Eng. Aspects*, No.225, (June 2003b), pp. 129-135, ISSN 0927-7757.

Castro Dantas, T. N.; Santanna, V. C.; Dantas Neto, A. A.; Curbelo, F. D. S. & Garnica, A. I. C. Methodology to break test for surfactant-based fracturing gel. *Journal of Petroleum Science and Engineering*, No.50, (janeiro 2006), pp. 293-298, ISSN 0920-4105.

Dantas Neto, A. A.; Fernandes, M. R.; Barros Neto, E. L.; Castro Dantas, T. N. & Moura, M. C. P. A. Alternative fuels composed by blends of nonionic surfactant with diesel and water: engine performance and emissions. *Brazilian Journal of Petroleum and Gas*, Vo.28, No.3, (2011), pp. 521-531, ISSN 0104-6632.

Friberg, S.E. & Bothorel, P. (1988). *Microemulsions: Structure and Dynamics*, CRC Press, ISBN 0849365988, Boca Raton.

Gurgel, A.; Moura, M. C. P. A.; Dantas, T. N. C.; Barros Neto, E. L. & Dantas Neto, A. A. A Review on chemical flooding methods applied in enhanced oil recovery. *Brazilian Journal of Petroleum and Gas*, Vo.2, No.2, (2008), pp. 83-95, ISSN 1982-0593.

Lif, A.; Stark, M.; Nydén, M. & Holmberg, K. Fuel emulsions and microemulsions based on Fischer–Tropsch diesel. *Colloids and Surfaces A: Physicochemical and Engineering Aspects*, Vo.354, No.1-3, (February 2010), pp. 91-98, ISSN 0927-7757.

Liu, D.; Fan, M.; Yao, L.; Zhao, X. & Wang, Y. A new fracturing fluid with combination of single phase microemulsion and gelable polymer system. *Journal of Petroleum Science and Engineering*, No.73, (july 2010), pp. 267-271, ISSN 0920-4105.

Moura, E. F.; Wanderley Neto, A. O.; Castro Dantas, T. N.; Scatena Júnior, H. & Gurgel, A. Applications of micelle and microemulsion systems containing aminated surfactants synthesized from ricinoleic acid as carbon-steel corrosion inhibitors. *Colloids and Surfaces A: Physicochem. Eng. Aspects*, No.340, (March 2009), pp. 199-207, ISSN 0927-7757.

Ochoterena, R.; Lif, A.; Nydén, M.; Sven, A. & Denbratt, I. Optical studies of spray development and combustion of water-in-diesel emulsion and microemulsion fuels. *Fuel*, No.89, No.1, (January 2010), pp. 122-132, ISSN 0016-2361.

Reyes, Y.; Rodriguez, F. J.; del Rio, J. M.; Corea, M. & Vazquez, F. Characterisation of an anticorrosive phosphated surfactant and its use in water-borne coatings. *Progress in Organic Coatings*, No.52, (2005), pp. 366-371, ISSN 0300-9440.

Robb, I.D. (1981). *Microemulsions*, Plenum Press, ISBN 0306408341, New York.

Samuel, M. M.; Card, R. J.; Nelson, E. B.; Brown, J. E.; Vinod, P. S.; Temple, H. L.; Qi Qu & Fu, D.K. Polymer-free fluid for fracturing applications. *SPE Drilling and Completion*, Vo.14, No.4, (1999), pp. 240-246, ISSN 10646671.

Santanna, V. C.; Curbelo, F. D. S.; Castro Dantas, T. N.; Dantas Neto, A. A.; Albuquerque, H. S. & Garnica, A. I. C. Microemulsion flooding for enhanced oil recovery. *Journal of Petroleum Science and Engineering*, No.66, (January 2009), pp. 117-120, ISSN 0920-4105.

9

Mesostructured Polymer Materials Based on Bicontinuous Microemulsions

Masashi Kunitake[1,2*], Kouhei Sakata[1] and Taisei Nishimi[3]
[1]*Graduate School of Science and Technology, Kumamoto University, Kurokami, Kumamoto*
[2]*Core Research for Evolutional Science and Technology*
Japan Science and Technology Agency (JST-CREST), Honcho, Kawaguchi, Saitama
[3]*FUJIFILM Corporation, Kaisei-machi, Ashigarakami-gun*
Japan

1. Introduction

A bicontinuous microemulsion (BME, Winsor III), also called a middle-phase microemulsion, is a low-viscosity, isotropic, thermodynamically stable, spontaneously formed solution phase composed of water, organic solvent, and surfactants. The dynamic morphology of a microemulsion (ME) is generally determined by the hydrophilicity-lipophilicity balance (HLB) of the surfactants in the emulsion system, as shown in Figure 1. When the hydrophilicity and lipophilicity of a surfactant are well balanced in an ME system, the ME frequently possesses a bicontinuous structure, in which the water phase and the oil phase coexist on a microscopic scale. In this chapter, polymer materials prepared from BMEs are introduced. Unique polymer morphologies such as continuous porous monolithic, bicontinuous hybrid, and nanosheet structures were prepared by polymerization or gelation in BMEs.

Polymer emulsion or ME systems have a wide range of industrial applications, e.g., in adhesives, paints, paper coating and textile coatings. Emulsions are widely used in the production of polymer materials, especially polymer particles. Most polymer particles are formed by suspension or emulsion[1,2] polymerization. Recently, surfactant-free polymerization, so-called soap-free polymerization, using an ionic radical initiator in aqueous solvents, homogeneous polymerization, dispersion polymerization,[3-9] and precipitation polymerization[10,11] have become popular because the polymer products are surfactant-free.

Methods of forming monolithic bulk polymer products with continuous pores, based on spinodal decomposition processes, have been researched intensively; a major advantage of monolithic supports is that mass transfer can take place through their pores. Porous polymer monoliths first emerged as a new class of stationary phases for high-performance liquid chromatography in the early 1990s after the development of inorganic monoliths prepared from silica using a sol–gel reaction.[12] The spinodal decomposition method has also

* Corresponding Author

been used for control of pore structures in commercial filters; this is a very important industrial application. These monoliths are typically prepared from a mixture comprising monomers, an initiator, and a porogenic solvent, using a simple molding process carried out in a mold such as a tube or a capillary. [13] Methods using spinodal decomposition have been used to prepare polymer membranes for separation. Polyether sulfone porous membranes have an asymmetric structure, leading to good permeation performance. Membranes prepared by these methods are commercially available. [14] Kumar and coworkers produced monolithic columns consisting of porous polyacrylamide, and they reported that the polymer columns were effective as affinity-chromatography supports for cell separation. [15]

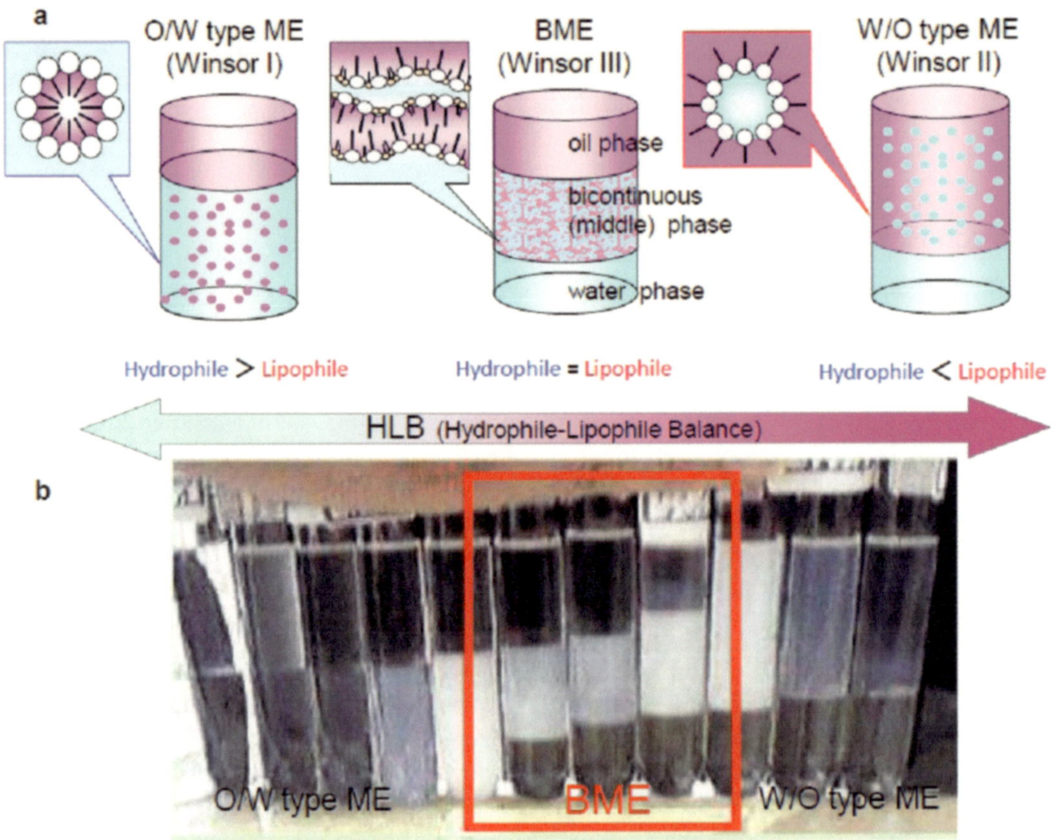

Fig. 1. Schematic representation and corresponding photographs of a series of MEs controlled by HLB (a, three types of ME model; b, functional ME (toluene/SDS + 1-butanol/saline). The phase structures of MEs were controlled by saline concentration.

Continuous porous polymer materials (molded monolithic porous polymers) can also be produced by decomposition of one component from a bicontinuous "gyroid" block-copolymer structure. Hashimoto and coworkers succeeded in obtaining continuous porous polystyrene (PS) with pores of diameter several tens of nanometers by dismantling polyisoprene by selective decomposition with ozone after having formed bicontinuous structures using PS/polyisoprene or polyisoprene/poly(2-vinylpyridine) block copolymers. [16,17] As described below, spinodal decomposition should be considered for BME polymerizations. It is worth noting that BMEs and spinodal-decomposition structures possess essentially the same "sponge-like" structure and that the only difference between them is size. Hashimoto and coworkers have summarized the associated theories on a physicochemical basis.

2. Construction of continuous porous polymer materials by BME polymerization

Thermal polymerization of BMEs consisting of a monomer liquid (oil) and water with an initiator is the simplest way to immobilize a BME structure. Simple BME polymerization leads to formation of continuous porous structures based on a bicontinuous-solution structure. In 1988, Haque and Qutubuddin reported the first preparation of porous PS based on a BME (styrene + 2-pentanol/SDS/water).[18]

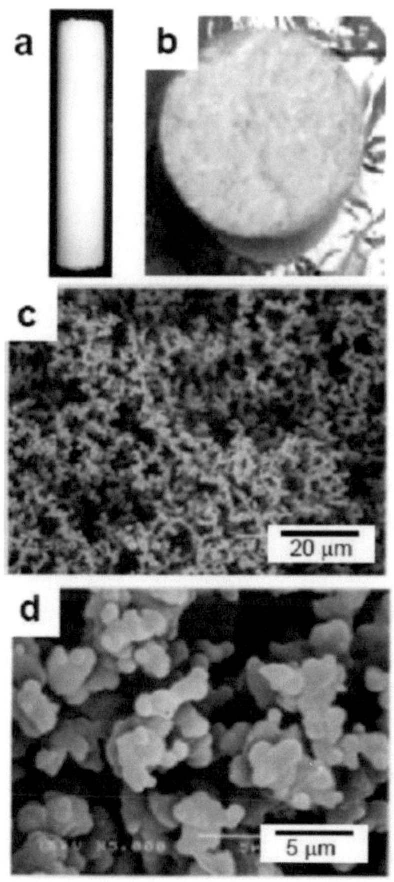

Fig. 2. Typical photographs (a, outer surface; b, torn surface) and SEM images (c and d) of the PS product prepared by BME thermal polymerization (styrene/SDS + 1-butanol/saline solution) in a glass tube.

We conducted a similar thermal polymerization of styrene/SDS + 1-butanol/saline solution; the HLB was controlled by the cosurfactant and salt concentrations. A hard white cylinder (Figure 2a) with a porous structure was formed. A porous structure was clearly observed in scanning electron microscopy (SEM) images, as shown in Figure 2c and d. The observed network structure consisted of connected particles, indicating percolation. When the BME polymerization was conducted using a BME solution in which two electrodes were separately placed, the electroconductivity of the polymer product was similar to that before polymerization, indicating the presence of a continuous micro saline-phase. However, the obtained pore size was 10 times larger than that expected, based on the solution structure. BME solution thicknesses have been determined using small-angle neutron scattering,[19] transmission electron microscopy with a freeze-fracture replica technique,[20,21] and simulations.[22] The reported thicknesses were from several tens of nanometers to 100 nm, and depended on the emulsion system and the measurement method. Transmission electron microscopy observation of a freeze-fracture replica sample has been used as a direct visual method. Abe and coworkers observed typical textures with thicknesses less than 100 nm for polymeric silicone BME systems.[20]

Continuous increases in pore sizes and polymer-wall thicknesses are generally observed as BME polymerization progresses. This phenomenon is the result of phase separation induced by a shift in the HLB equilibrium during polymerization. This phase separation is essentially the same as spinodal decomposition from the liquid phase. The driving force of this phase separation is a decreasing affinity of the surfactant for the oil phase consisting of monomer and newly formed polymer. This was proved by the appearance of an oil/water phase when PS was added to a styrene BME solution. The nano/meso-structures of chemically immobilized BME polymers are therefore predominantly determined by competitive reactions between immobilization and meso-phase separation during polymerization. In other words, the pore size should be controlled by the polymerization speed.

The production of porous structures with submicron continuous pores has been a goal since research in the field of BME polymerization began. Such products are expected to be transparent polymers with continuous pores. In order to construct such polymer products, polymerization-induced phase-separation should be avoided and/or suppressed. Generally, the meso-structures of polymer products prepared by BME polymerization are regulated kinetically by the competitive relation between immobilization by polymer growth and meso-phase separation. In order to achieve this, acceleration of immobilization (polymerization) and inhibition of phase separation are crucial. The immobilization speed should be increased. Increasing the initiator concentration and introduction of cross-linking agents are advantageous in terms of immobilization speed. It is worth noting that the polymerization temperature involves a trade-off because the speeds of both immobilization and phase separation increase with increasing temperature. Photo-induced polymerization at relatively low temperatures is therefore frequently used to suppress phase separation.

The most efficient approach will involve the use of polymerizable surfactants. The immobilization of polymerizable surfactants located at saline/oil interfaces is highly effective in suppressing the phase separation induced by polymerization. Transparent polymer products with continuous pores of diameter with several hundred nanometers or less are produced by polymerization of polymerizable surfactants and/or monomer (oil) phases.

Gan and coworkers[23] reported photopolymerization of the BME system (methyl methacrylate (MMA) and (2-hydroxyethyl) methacrylic acid (HEMA) mixed monomer solvent (1:1) /polymerizable cationic surfactant/water). They confirmed BME conditions of their system from the solution transparency and electroconductivity. Since the late 20th century, nanostructured proton-exchange membranes prepared by photopolymerizing BMEs have been a focus of attention because of their potential use in fuel cells. The continuous pores in membranes possess ionic passages. The inherent BME structure is essentially transferred to the matrix of the polymer membrane, without obvious phase separation, by polymerization. Gan[24] also reported photopolymerization of the BME solutions MMA/3-[(11-acryloyloxyundecyl)imidazoyl]propyl sulfonate (AIPS)/water, conducted between glass plates, to prepare transparent porous polymer membranes (Figure 3). The transparent membranes possessed proton-exchange properties as a result of introduction of a polymerizable zwitterionic surfactant. They reported continuous pores (channels) on the 20-50-nm scale. Liu et al.[25] reported the synthesis proton-exchange membranes with nanopores (diameter 1.5–2 nm) by thin-layer photopolymerization in a BME solution system (acrylonitrile/AIPS + cosurfactant/water). The pore scale is smaller than expected based on the size of the water channels in the BME solution before polymerization. Yang and coworkers[26] developed transparent, nanostructured, NIPAAm-based thermosensitive polymer membranes prepared from a BME system.

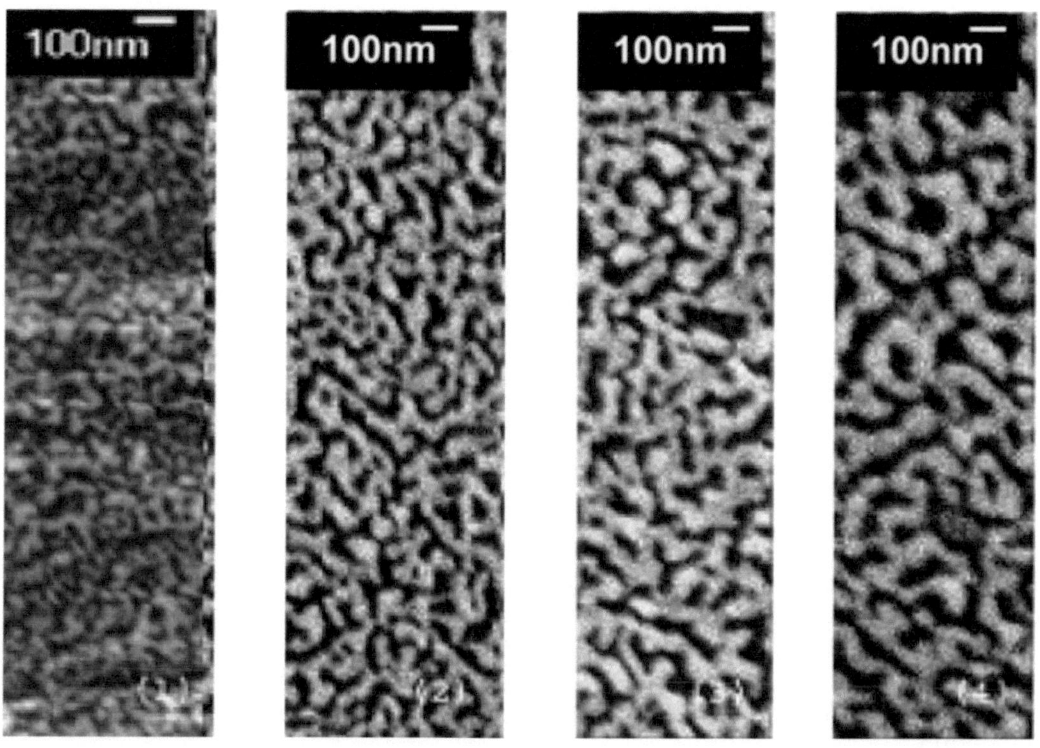

Fig. 3. SEM images of porous membranes prepared by polymerization with a polymerizable zwitterionic surfactant, AIPS (35 wt%), MMA (35 wt%), and water (30 wt%).[24]

3. Construction of hybrid soft materials by BME gelation

BMEs are used to produce not only porous condensed polymer products but also a new class of unique soft gels. Soft materials (soft matter), which are easily deformed, are important in a wide range of technological applications. Stimuli-responsive (intelligent) gels,[28] double-network (DN) gels,[29] nanocomposite gels,[30] and topological gels[31] have been extensively researched in the past decade. Such gels are expected to have potential as polymer materials for actuators[32] and drug-release systems;[33] these are key polymer technologies of the future.

Soft "wet" gels are categorized as organogels or hydrogels, according to the solvent used, which is a major component of the gels in terms of weight fraction. Generally, organogels and hydrogels are prepared by gelation to form three-dimensional networks based on physical aggregation (physical gelation) or chemical polymerization with cross-linking (chemical gelation). Gelation in emulsions or MEs has been investigated for composite gel formation since the 1980s.[34] Hydrogel particles prepared by hydrogelation in water/oil (W/O) emulsions have been investigated.[35] Bulk organogels bearing water microdroplets, prepared by organogelation in W/O emulsions, have also been investigated because of their medical applications such as dermal drug-delivery.[36] As expected, the converse combinations are possible but have rarely been reported.[37]

We have conducted gelation in BMEs (toluene/SDS and cosurfactants/saline) to produce hydro-organogels. Organogelation and hydrogelation were respectively achieved by physical gelation with an organogelator, namely 11-hydroxystearic acid, and polymerization (chemical gelation) of acrylamide with a cross-linker. Then, three composite gel systems, namely a BME organogel, a BME hydrogel, and an organo/hydro hybrid BME gel, were produced by hydrogelation and/or organogelation of each solution phase.

Figure 4 shows typical products obtained by gelation of macroscopic three-phase solutions including a BME phase. The BME phase (a middle-phase ME) is the middle phase of the macroscopic three-phase solutions. The upper and lower phases are a toluene phase and an aqueous phase, respectively. These gels, i.e., a BME organogel, BME hydrogel, and an organo/hydro hybrid BME gel, were self-supporting and very soft and elastic. Even in "one-side only" gelation products such as the BME organogel and the BME hydrogel, the solvent phases without a gelator were also macroscopically immobilized. No macroscopic pores were observed for any of the three BME gels. Moreover, all the BME gels had ionic conductivity, proving that the aqueous phase was continuous and that the bicontinuous structure was retained in the composite gels.

SEM images of freeze-dried one-side-gelated BME samples, i.e., the BME organogel and BME hydrogel, had uniform sponge-like porous structures, as shown in Figures 4h–k. The walls of the porous structures consisted of fiber networks, which were essentially similar to those in bulk organogels or hydrogels. In contrast, the morphology of the both-sides-gelated BME organo/hydro hybrid gel was entirely different, with a pore-free and smooth surface (Figure 4k).

In the case of soft swelling gels, a dried sample is necessary for SEM observations, and it is very difficult to eliminate the influence of the drying process on the structural analysis, even if the SEM sample is prepared by freeze-drying to minimize such influences.

In-situ imaging of "wet" BME gels was conducted using confocal laser scanning microscopy (CLSM). CLSM is a powerful nondestructive monitoring tool that provides three-dimensional mapping at the submicron scale by discrimination between the saline phase and oil phase under wet conditions.[38] By using hydrophilic and lipophilic fluorescent dyes, the aqueous region, oil region, and well-mixed region can be distinguished as different colors on the submicron scale. In Figures 4d–g, the red and green areas indicate phase-separated regions for the saline and toluene micro-phases, respectively. The yellowish color indicates that both oil and saline phases were uniformly mixed on a submicron scale. Because of the resolution limit, phase separation in BME gels cannot be estimated below 100 nm by CLSM.

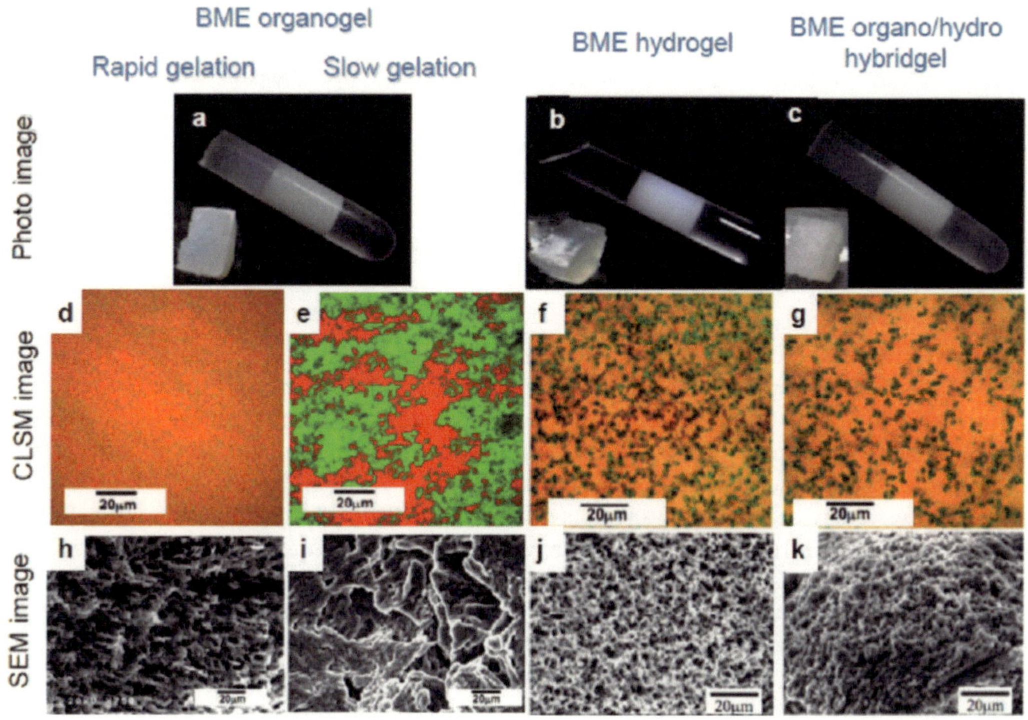

Fig. 4. Photographs (a–c), CLSM images (d–g), and SEM images (h - k) of a BME organogel (a, d, e, h, and i), BME hydrogel (b, f, and j), and BME organo/hydro hybrid gel (c, g, and k). The BME organogels were produced by "rapid" (d and h) and "slow" (e and i) cooling.[27]

The structures of BME gels are influenced by the kinetics of the gelation process, rather than by thermodynamic control of the original BME solution structure; this is similar to the case of condensed gels prepared by BME polymerization. Kinetic control of the physical gelation process to produce BME organogels was achieved relatively simply by control of the rate of cooling from the "sol state" to the "gel state". The BME organogel prepared by "rapid gelation" cooling (Figure 4d) gave a well-mixed submicron-scale BME gel (a texture-less yellowish color). This clearly proved that "rapid gelation" achieved immobilization of the solution structure in BME at less than several hundred nanometers, although even the microstructure of a gel prepared by "rapid" cooling might be influenced by mesoscopic phase-separation. In contrast, "slow" gelation cooling, typically at ca. 0.3 °C min^{-1} from 60 to 5 °C, provided a mosaic pattern consisting of two large domains, distinguished as green and red, indicating mesoscopic phase-separation (Figure 4e). The aqueous regions and oil regions were separated bicontinuously on a scale of several tens of microns, and no yellowish region was observed.

Generally, mesostructures are regulated by a competitive relationship between gelation speed and meso-phase separation speed. The driving force of mesoscopic phase-separation is the lowering of the surfactant activity by gelation. The propagation of gel fibers in oil or aqueous micro-phases kinetically induces lowering of the surfactant activity following mesoscopic phase-separation, in which the aqueous and oil phases progressively separate.[39] The mesoscopic phase-separation mechanism is similar to that of spinodal decomposition from the viewpoint of kinetic control.

Similar to the situation for BME organogels prepared by physical gelation, chemical hydrogelation in BME, namely polymerization of acrylamide with N,N′-methylene bisacrylamide, allowed us to produce bicontinuous composite gels. Jinnai and coworkers have reported porous NIPAAm polymer gels prepared by spinodal decomposition.[38,40] In the case of chemical gelation, the mesostructures are also kinetically regulated by gelation speed. It is well known that the use of a chemical accelerator, tetramethylethylenediamine (TEMED), increases gelation speed significantly.

However, the use of an accelerator represents a trade-off in terms of pore size because immobilization of the BME solution starts immediately on addition of TEMED to the BME solution, before the emulsion system can reach thermodynamic equilibrium. The ME system needs a standing time to reach a true thermodynamic equilibrium. Even when the solution seems to be macroscopically homogeneous, the thickness of the liquid layer of the bicontinuous phase changes continuously toward an equilibrium state after mixing. For instance, when the BME hydrogel was prepared by "rapid" gelation with TEMED without considering the stabilization time, the reaction started instantly and gelation was complete in less than 15 min. The CLSM image (Figure 4g) reveals the formation of a homogeneous bicontinuous structure consisting of isolated micron-scale green and red regions. There was almost no yellowish region, and the green and red regions did not overlap on the CLSM scale. In addition, it was possible to control the pore and wall sizes of the bicontinuous texture, which was on a scale between submicron and several tens of microns, by the reaction temperature.

By means of double gelation in oil and saline micro-phases, bicontinuous organo/hydro hybrid gels, in which the both the organic and aqueous micro-phases were immobilized, were produced as novel hybrid soft gels. Gong and coworkers reported DN hydrogel systems that were very tough and showed low friction, and discussed the similarities of these DN gels to biosystems.[39] The structure of the BME hybrid gel consisted of a homogeneous BME gel phase (yellowish) and separated islands (green), as shown in Figure 4f. The texture in the CLSM image was very similar to that of the BME hydrogel. As hydro chemical gelation was conducted prior to organo physical gelation, the mesostructure of the hybrid gels was predominantly determined by the hydrogelation step.

The hierarchical structures of polymer products (including soft gels) produced from BMEs were regulated by competition between three synergetic rate factors, namely immobilization of the BME structure by gelation, mesoscopic phase-separation by gelation, and time-dependent transformation of the solution/solution structure toward a thermodynamic structure in the equilibrium state.

4. Determination of BME solution structures at solid/liquid interfaces by electrochemical analysis

BME nanostructures are also attractive as media for electrochemical studies.[42,43] Rusling et al. conducted pioneering work on the electrochemistry of BME solutions using a glassy carbon (GC) electrode.[44] The electrochemical approach to BMEs shows the dynamic solution structure of BMEs at solid/liquid interfaces. The electrochemistry of redox molecules, namely $K_3Fe(CN)_6$ and ferrocene, in a BME (saline/SDS + butanol/toluene) was investigated in detail using various electrodes such as indium tin oxide (ITO), Au disk, GC disk, highly oriented pyrolytic graphite (HOPG), and alkanethiol-modified Au electrodes (Figure 5). The electrochemical contact with the micro aqueous and organic solution phases in a BME is alternately or simultaneously achieved by controlling the hydrophilicity and lipophilicity of the electrode surfaces.[45,46]

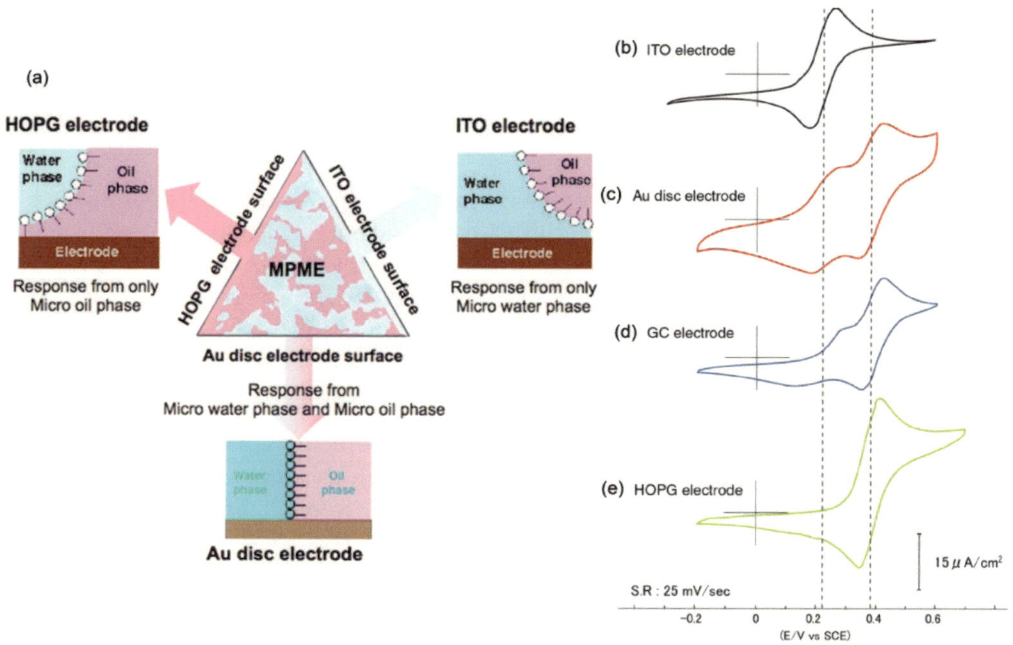

Fig. 5. Schematic representation of BME solution structures at solid/oil/saline interfaces (a) and typical cyclic voltammograms measured using ITO (b), polished Au disk (c), polished GC (d), and cleaved HOPG (e) electrodes in BME in the presence of $K_3Fe(CN)_6$ and ferrocene.[41]

The amphiphilic electrode (Au) revealed redox peak couples for both $K_3Fe(CN)_6$ and ferrocene in BME solutions simultaneously.[45] In contrast, strong hydrophilic or lipophilic electrodes (ITO and HOPG) produced only one of the peak couples of $K_3Fe(CN)_6$ or ferrocene, depending on the affinities for the detecting electrodes. For instance, an ITO electrode possesses a strongly hydrophilic, negatively charged surface. No redox peaks resulting from ferrocene in a micro oil-phase were observed, although the redox peaks of $K_3Fe(CN)_6$ in a micro saline-phase were clearly observed. Conversely, strongly lipophilic (hydrophobic) electrodes such as HOPG, with a neutral surface, showed the opposite electrochemical response in the BME.

The solution structure of a BME around the electrode changes thermodynamically in response to the HLB of the electrode surfaces. A well-balanced BME solution structure is easily converted to a biased or one-sided structure on an electrode surface to minimize surface energy. As mentioned already, the unique characteristics and structures of BMEs are based on the very delicate balance between the hydrophilicity and lipophilicity of the surfactant system; this balance is moderated by the concentrations of the cosurfactant and salt, or by the temperature.

It is worth emphasizing that an electrochemical response of the redox species in the micro toluene phase in a BME was observed even if there was no electrolyte in toluene phase . This indicates that the ionic conductivity is maintained via a continuous saline phase in the BME; this phase acts as an ion-conducting passage. This suggests that electrochemistry in nonpolar solvents without an electrolyte, becomes possible in BMEs.

These alternative electrochemical responses in BMEs indicate that the bulk bicontinuous structure is thermodynamically discontinuous near the surface, as a result of the influence of the hydrophilicity or lipophilicity of the electrode surface. Hydrophilic and lipophilic (hydrophobic) surfaces predominantly face the micro saline-phase and the micro organic-solvent phase, respectively, in BMEs.

The dynamic solution structures of three-phase interfaces (oil/water/solid) of BMEs have been discussed in terms of the apparent diffusion coefficients (D_{app}) of each redox species and electrode. Despite the complex solution structures, the scan-rate dependence of the peak currents for $K_3Fe(CN)_6$ and ferrocene in the BME indicated that the system is generally regulated by relatively simple diffusion control. The D_{app} for each redox species is calculated from the scan-rate dependence of the peak currents according to the Randles–Ševčik equation.

Lindman and coworkers reported that the self-diffusion coefficients of both solution species, water and toluene molecules, in the BME solutions, which were obtained by Fourier-transform pulsed-gradient spin-echo NMR measurements, were slightly smaller than those in homogeneous solutions.[47] In principle, the diffusion coefficients in a homogeneous solution should be independent of the type of electrode. However, the D_{app} values obtained using the hydrophilic ITO and lipophilic pyrolytic graphite (PG-b) electrodes were less than half of those obtained with the amphiphilic Au electrode. The ratios of the D_{app} values of ferrocene for Au against those for PG-b in the BME decreased with increasing temperature. Such deviations of D_{app} values from the ideal values provide information on the solution structure near the electrode surface, and the electrode area in contact with both phases.

These behaviors could not be explained by a simple sloped-structure model because the hydrophilic ITO and lipophilic PG-b electrodes consistently face either the micro saline-phase or the oil-phase, respectively. The results can be explained by an alternating W/O layered structure, as shown in the model (Figure 6).[49-51] The solution layers consist of micro aqueous and micro oil phases that form alternately, starting from the electrode surface. The first solution layer is adopted based on the affinity with the electrode surface. The second, counter-solution layer, either an oil layer or an aqueous layer, forms on the first aqueous or oil layer. The contribution of the layered structure will gradually decrease with increasing distance from the electrode surface. An alternating layered structure causes the lower D_{app} values of the hydrophilic and lipophilic electrodes rather than those of amphiphilic electrodes. Zhou et al. used small-angle neutron scattering to investigate the alternately layered solution structures of BMEs around solid surfaces.[49-51]

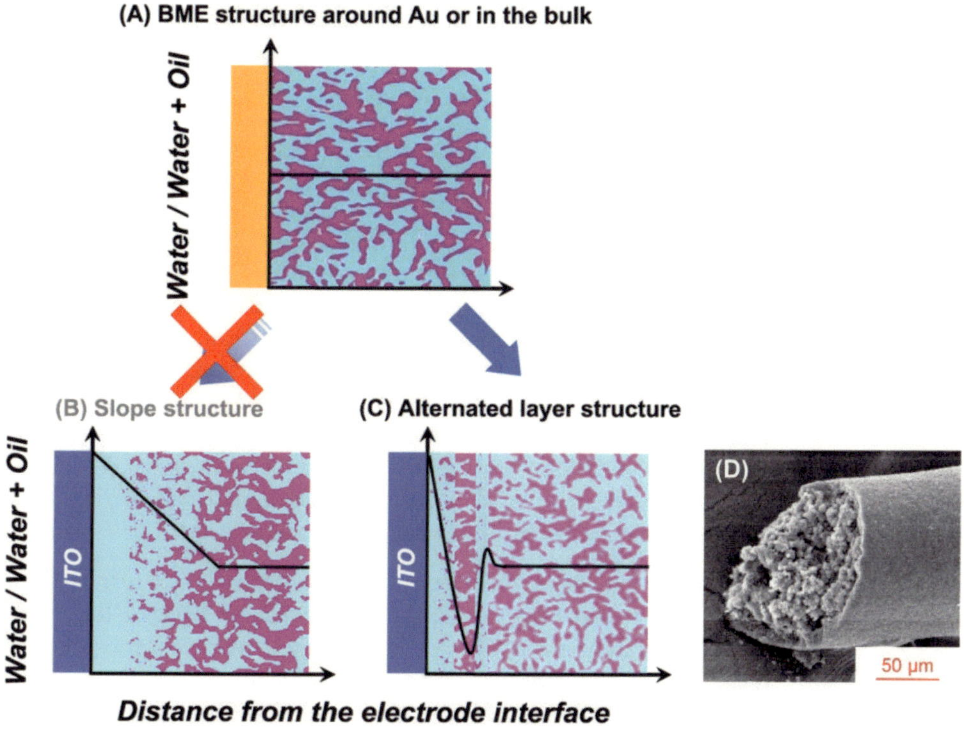

Fig. 6. Schematic illustrations with depth profiles of saline volume ratios for BME structures with bulk and amphiphilic Au electrodes (A), the slope structure (B), and the alternating layered structure around the hydrophilic ITO electrode (C) expected from the apparent diffusion coefficients of redox species and the SEM image (D) of a PS tube with a continuous porous structure and a pore-less surface prepared by thermal polymerization of a styrene BME in a capillary. The magenta and light cyan regions are the oil and saline phases, respectively.[48]

To visually evaluate the influence of the surface on the solution structure of the BME, thermal polymerization of a styrene-based BME was conducted in a capillary glass with an intact hydrophilic surface. The polymer formed in the capillary possessed a typical continuous porous structure based on a granular network, in accordance with that observed for bulk polymerization, and the porous structure was surrounded by a pore-less polymer sheet, as shown in Figure 6D. In addition, a gap between the polymer sheet and the inner wall of the capillary was frequently observed. The pore-less polymer sheets might prove the existence of an alternately layered structure in the BME solution before polymerization.

5. Construction of hierarchical structure based on BMEs and nanospaces

The formation of a polymer tube with continuous pores and seamless walls by simple polymerization in a capillary directed our attention to polymerization of BMEs in other microspaces, such as opal membranes, to construct hierarchical structures. At present, "structural transcription"[52] from inorganic materials to polymer materials, such as from a "silica opal membrane" to an "inverse opal polymer membrane", is very popular as a method for constructing unique polymer materials with regular, controlled, complex porous structures.[53-55] When thermal polymerization of styrene- or MMA-based BMEs was conducted in the submicron gaps of a silica opal membrane, surprisingly various fractal surface structures consisting of vertical polymer nanosheets were obtained instead of the expected continuous porous PS-filled silica opal membrane. Simple styrene polymerization in the silica opal membrane gave a PS-filled silica opal membrane.

Figure 7 shows typical SEM images of the "turf-like" fractal surface of the opal membrane after polymerization of the styrene BME. Unique morphologies such as fractal surfaces consisting of closely spaced nanosheets and vertically extended nanosheets were grown from the gaps of a silica opal membrane. The surface densities and lengths of the nanosheets were roughly controllable by the polymerization conditions, specifically the thickness of the opal membrane and the amount of BME applied to the membrane. The closely spaced nanosheets were typically ca. 200 nm thick, 0.5–2 μm wide, and 1–4 μm in length. Very long nanosheet ribbons were formed by polymerization in a relatively thick opal membrane (typical thickness > 2 mm). The lengths of the longest nanosheets were several ten of microns; however, the thicknesses and widths were 100 nm and 0.5–4 μm, respectively, almost the same as for nanosheets prepared under different conditions.

Although the concentrations of surfactant, cosurfactant, and salt required to form BMEs were quite different for styrene BMEs and MMA BMEs, the shapes and sizes of the "turf-like" nanosheets of polymethyl methacryrate (PMMA) were very similar to those of the PS nanosheets. These results indicate that the nanosheet structures originate from the solution/solution structure of the BME, regardless of the chemical structures of the monomers.

The nanosheet columns grow from the bottom of the column, not from the apex, analogous to the growth of fingernails. The emulsion species are drawn from the gaps in the opal to the roots of the polymer columns. No polymer species were observed in the silica gaps after polymerization. During the polymerization process, the emulsion, which is a mixture of monomers and polymers (oligomers), escapes through the submicron-scale silica gaps to the surface. The major driving force for escape of ME from the microspace, and the propagation of PS columns, is the lack of stability of the ME solution in the highly hydrophilic microspaces of the silica gaps. Furthermore, the nanosheet structure indicates the existence of a liquid-crystal (LC) phase as a template. Two factors are expected to influence formation of an LC phase from a BME during polymerization. A hydrophilic silica surface tends to trap water near the surface,[45,46] and the decrease in the water content of the ME causes enrichment of surfactants in the ME. The shift of the HLB of the surfactants resulting from the changes in the oil phase of the ME during polymerization, and hence changes in the monomer/polymer ratio and the degree of polymerization, might induce a phase change from a BME phase to an LC phase.

Finally, various fractal surfaces consisting of polymer nanosheets were prepared by simple thermal polymerization of BME in an opal membrane. The thicknesses of the nanosheets were roughly constant, but their sizes (lengths and widths) and surface densities were highly variable and were related to the thickness (depth) of the silica opal membrane and the amount of BME solution applied.

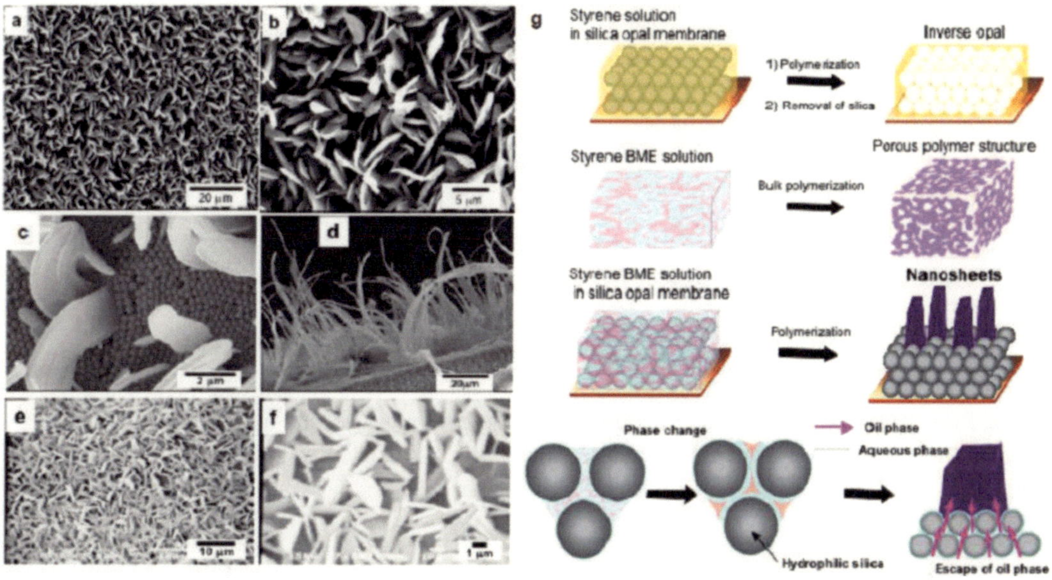

Fig. 7. Typical SEM images of PS (a–d) and PMMA (e and f) nanosheets prepared by polymerization of BME in silica gaps, and schematic representations (g) of the formation mechanism of nanosheets grown in the gaps of a silica opal membrane.[56]

These results prove that a combination of a regularly ordered "solid" microspace such as an opal membrane and the dynamic solution/solution structure of the ME, which can change in response to a variety of circumstances, has the potential for construction of hierarchical polymer structures by self-organization.

6. Conclusions

This chapter has introduced various unique polymer materials with controlled nanostructures based on polymerization or gelation of BMEs. Using bicontinuous solution structures in BMEs as a template, continuous porous materials and bicontinuous hybrid materials are prepared by polymerization or gelation in one side (water or oil phase) and in both sides, respectively. Similar structures can also be prepared by micro-phase separation of block copolymers and spinodal decomposition. Figure 8 summarizes the correlations among these three methods. The spinodal decomposition process is controlled kinetically, whereas micro-phase separation is completely thermodynamically controlled, with respect to the nature, shape, and length of each component. In the case of nanostructured polymer materials prepared from BMEs, the nanostructures are regulated both thermodynamically and kinetically. The solution structure of an ME is essentially ruled thermodynamically by the HLB. During the polymerization process, the structures of the polymers formed are controlled kinetically, as in spinodal decomposition, because the HLB is continuously changing as a result of polymer formation. The intentional regulation of competitive reactions between immobilization and phase separation allows us to control pore size or phase-separation size from several tens of nanometers to several microns.

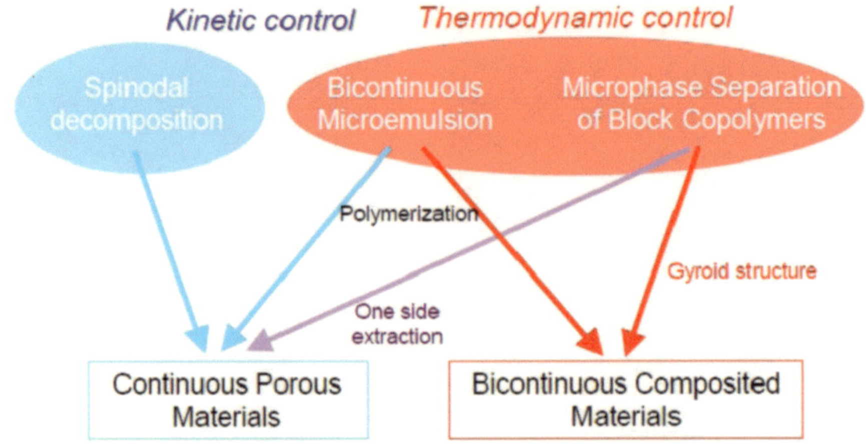

Fig. 8. Production of continuous porous materials and bicontinuous hybrid materials. The blue and red arrows (and areas) represent kinetically and thermodynamically controlled phenomena (or processes), respectively.

The structures of polymer products near a contacting surface change considerably. In particular, a BME is a precise well-balanced system, which can be easily deformed by changes in conditions. In other words, BMEs can produce a great diversity of hierarchical nanostructures by self-organization. Moreover, a lyotropic LC phase is used as a template instead of the BME. A lyotropic LC phase can be intentionally prepared from a BME with a relatively high surfactant-concentration. Frequently, changes in the HLB during polymerization induces a phase transition to a lyotropic LC phase, giving ribbon, tape, and plate structures. It will be possible to use these methods to produce highly advanced, polymer materials with hierarchical nanostructures, in a way similar to that found in living systems.

7. References

[1] Lovell P. A. & El-Aasser M. S. (1997). Emulsion Polymerisation and Emulsion Polymers, John Wiley and Sons (New York).
[2] Konno, M. Terunuma, Y. & Saito, S. (1991). J. Chem. Eng. Jpn. 24, 429-437.
[3] Tseng, C. M., Lu, Y. Y., El-Aasser, M. S. & Vanderhoff, J. W. (1986). J. Polym. Sci., Part A: Polym. Chem., 24, 2995-3007.
[4] Baines, F. L., Dionisio, S., Billingham, N. C. & Armes, S. P. (1996). Macromolecules, 29, 3096-3102.
[5] Minami, H., Yoshida, K. & Okubo, M. (2008). Macromol. Rapid Commun., 29, 567-572.
[6] Schmid, A., Fujii, S. & Armes, S. P. (2005). Langmuir, 21, 8103-8105.
[7] Dawkins, J. V., Neep, D. J. & Shaw, P. L. (1994) Polymer, 35, 5366-5368.
[8] Paine, A. J., Luymes, W. & McNulty, J. (1990). Macromolecules, 23, 3104-3109.
[9] Hong, J., Han, H., Hong, C. K. & Shim, S. E. (2008). J. Polym. Sci., Part A: Polym. Chem., 46, 2884-2890.
[10] Li, K. & Stover, H. D. H. (1993). J. Polym. Sci., Part A: Polym. Chem., 31, 3257-3263.
[11] Downey, J. S., Frank, R. S., Li, W.-H. & Stoever, H. D. H. (1999). Macromolecules, 32, 2838-2844.
[12] Nakanishi, K. (1997). J. Porous Mater., 4, 67-122.

[13] Hashimoto, T., Tsutsumi, K. & Funaki, Y. (1997). Langmuir, 13, 6869-6872.
[14] Ulbricht, M., Schuster, O., Ansorge, W., Ruetering, M. & Steiger, P. (2007). Separ. Purif. Tech., 57, 63-73.
[15] Kumar, A. & Srivastava, A. (2010). Nature Protocols, 5, 1737-1747.
[16] Jinnai, H., Hashimoto, T., Lee, D. & Chen, S.-H. (1997). Macromolecules, 30, 130-136.
[17] Hashimoto, T. (2005). Bull. Chem. Soc. Jpn., 78, 1-39.
[18] Haque, E. & Qutubuddin, S. (1988). J. Polym. Sci., Part C: Polym. Lett., 26, 429-432.
[19] Ryan, L. D. & Kaler, E. W. (1998). J. Phys. Chem. B, 102, 7549-7556.
[20] Sharma, S. C., Tsuchiya, K., Sakai, K., Sakai, H., Abe, M. & Miyahara, R. (2008). J. Oleo Sci., 57, 669-673.
[21] Ikeda, Y., Imae, T., Hao, J., Iida, M., Kitano, T. & Hisamatsu, N. (2000). Langmuir, 16, 7618-7623.
[22] Nagarajan, R. & Ruckenstein, E. (2000). Langmuir, 16, 6400-6415.
[23] Deen, G. R. & Gan, L. H., (2009) J. Polym. Sci., Part A: Polym. Chem., 47, 2059-2072.
[24] Gan, L. M., Chow, P. Y., Liu, Z., Han, M. & Quek, C. H. (2005). Chem. Commun., 4459–4461.
[25] Lim, T. H., Tham, M. P., Liu, Z., Hong, L. & Guo, B. (2007). J. Membr. Sci., 290, 146–152.
[26] Wang, L.-S., Chow, P.-Y., Phan, T.-T., Lim, I. J. & Yang, Y.-Y. (2006). Adv. Funct. Mater., 16, 1171–1178.
[27] Kawano, S., Kobayashi, D., Taguchi, S., Kunitake, M. & Nishimi, T. (2010). Macromolecules, 43, 473-479.
[28] (a) Osada, Y., Okuzaki, H. & Hori, H. (1992). Nature, 355, 242-244. (b) Holtz, J. H. & Asher, S. A. (1997). Nature, 389, 829-832. (c) Siegel, R. A. (1998). Nature, 394, 427-428. (d) Yoshida, R., Sakai, K., Okano, T. & Sakurai, Y. (1994). J. Biomater. Sci., Polym. Ed., 6, 585-598. (e) Hirose, M., Kwon, O. H., Yamato, M., Kikuchi, A. & Okano, T. (2000). Biomacromolecules, 1, 377-381.
[29] (a) Yang, W., Furukawa, H. & Gong, J. P. (2008). Adv. Mater., 20, 4499-4503. (b) Gong, J. P., Katsuyama, Y., Kurokawa, T. & Osada, Y. (2003). Adv. Mater., 15, 1155-1158.
[30] Haraguchi, K. & Takehisa, T. (2002). Adv. Mater., 14, 1120-1124.
[31] (a) Sakai, T., Murayama, H., Nagano, S., Takeoka, Y., Kidowaki, M., Ito, K. & Seki, T. (2007). Adv. Mater., 19, 2023-2025. (b) Okumura, Y. & Ito, K. (2001). Adv. Mater., 13, 485-487.
[32] Fukushima, T., Asaka, K., Kosaka, A. & Aida, T. (2005). Angew. Chem., Int. Ed., 44, 2410-2413.
[33] Sakata, S., Uchida, K., Kaetsu, I. & Kita, Y. (2007). Phys. Chem., 76, 733-737.
[34] Rees, G. D. & Robinson, B. H. (1993). Adv. Mater., 5, 608-619.
[35] (a) Hernandez-Barajas, J. & Hunkeler, D. J. (1995). Polym. Adv. Technol., 6, 509-517 (b) Dowding, P. J., Vincent, B. & Williams, E. (2000). J. Colloid Interface Sci., 221, 268-272.
[36] (a) Trickett, K. & Eastoe, J. (2008). Adv. Colloid Interface Sci., 144, 66-74. (b) Jadhav, K. R., Kadam, V. J. & Pisal, S. S. (2009). Current Drug Delivery, 6, 174-183. (c) Murdan, Sudaxshina (2005) Expert Opin. Drug Delivery, 2, 489-505.
[37] (a) Chen, H., Chang, X., Du, D., Li, J., Xu, H. & Yang, X. (2006). Int. J. Pharm., 315, 52-58. (b) Kreilgaard, M. (2001). Pharm. Res., 18, 367-373.
[38] Hirokawa, Y., Okamoto, T., Kimishima, K., Jinnai, H., Koizumi, S., Aizawa, K. & Hashimoto, T. (2008). Macromolecules, 41, 8210-8219.
[39] (2004). Advances in Polymer Science, Springer (Berlin and Heidelberg).

[40] Hirokawa, Y., Jinnai, H., Nishikawa, Y., Okamoto, T. & Hashimoto, T. (1999). Macromolecules, 32, 7093-7099.
[41] Kunitake, M., Murasaki, S., Yoshitake, S., Ohira, A., Taniguchi, I., Sakata, M. & Nishimi, T. (2005). Chem. Lett., 34, 1338-1339.
[42] Marckey, B. A., Texter, J. (1991). Electrochemistry in Colloid and Dispersions, Wiley-VCH (New York)
[43] Rusling, J. F. (2001). Pure Appl. Chem., 73, 1895-1905.
[44] Iwunze, M. O., Sucheta, A. & Rusling, J. F. (1990). Anal. Chem., 62, 644-649.
[45] Yoshitake, S., Ohira, A., Tominaga, M., Nishimi, T., Sakata, M., Hirayama, C. & Kunitake, M. (2002). Chem. Lett., 360-361.
[46] Kunitake, M., Murasaki, S., Yoshitake, S., Ohira, A., Taniguchi, I., Sakata, M. & Nishimi, T. (2005). Chem. Lett., 34, 1338-1339.
[47] Guéring, P. & Lindman, B. (1985). Langmuir, 1, 464-468.
[48] Makita, Y., Uemura, S., Miyanari, N., Kotegawa, T., Kawano, S., Nishimi, T., Tominaga, M., Nishiyama, K. & Kunitake, M. (2010). Chem. Lett., 39, 1152-1154.
[49] Zhou, X. L., Lee, L. T., Chen, S. H. & Strey, R. (1992). Phys. Rev. A, 46, 6479-6489.
[50] Zhou, X.-L. & Chen, S.-H. (1995). Phys. Rep., 257, 223-348.
[51] Olsson U. (2001). in Holmberg K., Handbook of Applied Surface and Colloid Chemistry, John Wiley & Sons, Chichester, 2, 333-356.
[52] Asakawa, K. & Hiraoka, T. (2002). Jpn. J. Appl. Phys., 41, 6112-6118.
[53] Schepelina, O. & Zharov, I. (2006). Langmuir, 22, 10523-10527.
[54] Rong, J. & Yang, Z. (2002). Macromol. Mater. Eng., 287, 11-15.
[55] Schroden, R. C., Al-Daous, M., Blanford, C. F. & Stein, A. (2002). Chem. Mater., 14, 3305-3315.
[56] Kawano, S., Nishi, S., Umeza, R. & Kunitake, M. (2009). Chem. Commun., 1688-1690.

10

Microemulsion Method for Synthesis of Magnetic Oxide Nanoparticles

A. Drmota[1], M. Drofenik[2,3], J. Koselj[1] and A. Žnidaršič[1,2]
[1]*Nanotesla Institute Ljubljana*
[2]*CO NAMASTE*
[3]*Faculty of Chemistry and Chemical Engineering*
Slovenia

1. Introduction

Among many of known nanomaterials, the special position belongs to magnetic nanoparticles, which are fundamentally differing from the classic magnetic materials with their domain structure. The magnetic properties of nanoparticles are determined by many factors, such as chemical composition, type and degree of defectiveness of the crystal lattice, particle size and shape, morphology (for structurally inhomogeneous particles), interaction of the particle with the surrounding matrix and the neighboring particles (Gubin, 2009). For this reason, it is important to develop techniques by which the size, shape and chemical homogeneity of the particles can be well controlled (Hessien et. al., 2008). Numerous nonconventional techniques, such as precipitation in aqueous solutions (Pankov et. al., 1993), glass crystallization method (Sato & Umeda, 1993), self-propagating high-temperature synthesis method (Elvin et. al., 1997), hydrothermal method (Ataie et. al., 1995), mechanical alloying (Ding et. al., 1998) and the sol-gel method (Surig et. al., 1994), were used to prepare magnetic nanoparticles.

Among them, a precipitation in microemulsion has been shown as a perspective method for the preparation of magnetic nanopartqicles of controlled size and morphology (Košak et. al., 2004).

A microemulsion can be defined as a thermodynamically stable isotropic dispersion of two immiscible liquids consisting of nanosized domains of one liquid in the other, stabilized by an interfacial film of surface-active molecules. The surfactant molecules provide a confinement effect that limits particle nucleation, growth and agglomeration. In this method, co-precipitation occurs in tiny droplets of water ("water pools") embedded with surfactant molecules and homogenously distributed in an oil phase. The size of these "water pools", so called reverse micelles, which act as micro reactors for the synthesis of the nanoparticles, is thermodynamically defined by the water-to-surfactant molar ratio (R) (Makovec & Košak, 2005).

Although there are many kinds of interesting magnetic nanoparticles, we have been focused our study on iron oxide nanoparticles.

First of all, we synthesized silica (SiO_2) coated maghemite (γ-Fe_2O_3) nanoparticles. Monodisperse magnetic nanoparticles with unique properties, which are dominated by superparamagnetism, have a great potential to use in biomedical applications such as cell labeling and magnetic cell separation, drug delivery, hyperthermia and MRI contrast enhancement due to of their non-toxic nature and lower susceptibility to physical and chemical changes. In this field the main challenge is to create the suitable surface of magnetic nanoparticles in order to functionalize and allow strong interactions with specific biological components. Therefore, chemical modification of the nanoparticle surface with biocompatible molecules, such as silica (SiO_2), dextran, polyvinyl alcohol (PVA) and phospholipids, is an important issue that provides bio functionality and resistance to physiological conditions such as pH and enzymes, binding sites between the particles and the target sites on the cell, cytotoxic drug attaching to a biocompatible magnetic nanoparticle carrier, etc. Superparamagnetic nature and a high degree of specific magnetization of these magnetic nanoparticles are demanded for their applications in drug delivery or magnetic separation to defeat hydrodynamic forces acting on the nanoparticles in the flowing solution. Small sizes are required to allow the transport through the vascular system or tisular diffusion of particles (Mornet et. al., 2002; Vidal-Vidal et. al., 2006).

On the other hand, we have been focusing our study on M-type strontium hexaferrite ($SrFe_{12}O_{19}$) nanoparticles because of their magnetoplumbite structure, which present a high magnetization and strong magnetic anisotropy along c axis. Such properties of material, which are directly related to crystallographic structure, are very important because of their use in the manufacture of microwave-absorbing materials. The development of novel electromagnetic (EM) wave absorption materials has great interest. Although, electric equipments make our life more convenient, the electromagnetic (EM) radiation restricted the continual development of our society because of their pollution to environment and harm to human beings.

The absorption of EM waves occurs in magnetic materials due to their magnetic losses (Lax & Button, 1962). Ferrites with a submicron grain size are some of the most promising materials in magnetic nanocomposites for the absorption of microwave radiation (Bregar, 2004). In nanocomposite form a ferrite can be used for the absorption of microwave radiation where the ferromagnetic nanopowders exhibit a higher absorption at low field strengths and a broader absorption range in the microwave region than multi-domain powders.

Ferrites exhibit substantial magnetic losses in the vicinity of their natural resonance (FMR). Because of this phenomenon they are one of the best materials for MW absorbers. Ferrites with a spinel crystal structure can be applied in the frequency range of several hundred MHz to several GHz (Pardavi Horvath, 2000). AM_2O_4 spinel ferrites are binary ferromagnetic oxides with compositions in the system Fe_2O_3-MO, where M is usually a transition element. Almost any divalent transition-metal ion can be replaced to form spinel ferrite. However, magnetite is an exceptional case, where the M is Fe^{2+}. The spinel ferrite's cubic unit cell comprises eight formula units with 16 M^{3+} cations on 16 octahedral sites and 8M^{2+} ions on the tetrahedral A sites. In the reversed spinel eight of 16 M^{3+} ions occupy all the tetrahedral sites. The magnetic properties of ferrites can be considered in terms of the Neel model of ferrimagnetism. One of the good properties of ferrites is the possibility to prepare different compositions and thereby modify the magnetic properties.

On the other hand, ferrites with a hexagonal structure are an extensive family of compounds with a complex magnetoplumbite structure within the ternary system AO–Fe_2O_3–MeO, where A = Ba, Sr, Ca(La) and Me is a bivalent transition metal. Due to their high natural frequency, these ferrites can be used across the whole GHz region (Pardavi Horvath, 2000). Furthermore, these ferrites do not require external bias fields in RF operation and are "self-biased" due to their very high intrinsic anisotropy and can be applied for the miniaturization of microwave devices (Ovtar et. al., 2011). Their operating range is around the remanence of the rectangular hysteresis loop. Due to the solely Fe^{3+} state in the hexaferrite lattice, they are good insulators with very low dielectric losses. The best-known members are the uniaxial permanent magnets of the M-type, i.e., Ba and Sr-hexaferrites, which can be used across the whole GHz region (Rodrigue, 1963).

By varying the chemical compositions, it is possible to control the electromagnetic properties such as the saturation magnetization, magnetocrystalline anisotropy, permeability and the permittivity of a ferrite composite. In addition, the microstructure of ferrites has an additional impact on their properties, and consequently on the EM absorption. Compared to the ferrites with a spinel structure, the hexagonal ferrites exhibit a larger intrinsic magnetocrystalline anisotropy field and it can be applied at higher frequencies. In general, MW absorbers can be prepared in the form of ceramics or as composites, where the ceramic phases are embedded in a polymeric matrix. Here, the EM properties of the composites can be very effectively tuned, simply by varying the volume fractions of the filler constituent phases.

In addition, a synergetic effect of the constituent phase properties may also be observed in some composites. For this reason magnetic composites are interesting for microwave applications (Kim et. al., 1994; Nakamura & Hankui, 2003; Pereira et. al., 2009; Verma et. al., 2002; Ghasemi et. al., 2008; Groningen, 2003). Furthermore, during the development of suitable absorbers, their composition and processing are equally important.

2. Experimental procedure

In the present investigation, silica (SiO_2) coated maghemite (γ-Fe_2O_3) nanoparticles were performed *in-situ* via the precipitation in two different microemulsion systems. First microemulsion system consisted from water phase (aqueous solution of Fe^{2+}/Fe^{3+} ions or aqueous solution of NH_4OH), sodium n-dodecyl sulfate (SDS) served as anionic surfactant, 1-butanol served as co-surfactant and cyclohexane used as an oil phase. The second microemulsion system consisted from water phase (aqueous solution of Fe^{2+}/Fe^{3+} ions or aqueous solution of NH_4OH), n-Hexadecyltrimethylammonium bromide (CTAB) served as cationic surfactant, 1-butanol served as co-surfactant and 1-heksanol used as an oil phase. Whatever of the chosen system, the synthesis of nanoparticles was performed at the same water-to-surfactant + co-surfactant molar ratio (R).

The influence of the concentration of reactants in aqueous phase, the temperature of reaction, the pH value after the precipitation of hydroxides and the type of surfactant and co-surfactant, on the maghemite (γ-Fe_2O_3) nanoparticles were investigated. The thickness of silica shell was carefully controlled by the amount of tetraethoxysilane (TEOS) added to the microemulsion after the precipitation step.

The precursors of strontium hexaferrite ($SrFe_{12}O_{19}$) were prepared with the classical co-precipitation method and with the microemulsion method in microemulsion system water/SDS, 1-butanol/cyclohexane. The precursors were calcined at different temperatures ranging from 400 °C to 1000 °C in air. The influence of the Sr^{2+}/ Fe^{3+} molar ratio and the calcination temperature on the product formation and magnetic properties were studied.

Furthermore, the composites based on the magnetic filler, composed of phases within the SrO - Fe_2O_3 system, embedded in the polyphenylene sulfide (PPS) matrix with a concentration range of 80:20 by weight destined for microwave absorption were investigated. The electromagnetic parameters of composites were measured with a vector network analyzer at 400 MHz - 32 GHz.

2.1 Synthesis of silica (SiO_2) coated maghemite (γ -Fe_2O_3) nanoparticles

The chemical reagents used in this synthesis process were iron (II) sulfate heptahydrate ($FeSO_4$ $7H_2O$, ACS, 99+%), iron (III) sulfate hydrate ($Fe_2(SO_4)_3$ xH_2O, Reagent Grade), ammonium hydroxide solution (25%, puriss p.a.), sodium n-dodecyl sulfate (SDS) ($CH_3(CH_2)_{11}OSO_3Na$, Fluka, 99%), n-Hexadecyltrimethylammonium bromide (CTAB) ($CH_3(CH_2)_{15}N(Br)(CH_3)_3$, Alfa Aesar, 99+%), 1-Butanol ($CH_3(CH_2)_3OH$, Alfa Aesar, 99%), cyclohexane (C_6H_{12}, ACS, Alfa Aesar, 99+%), 1-hexanol ($CH_3(CH_2)_5OH$, Alfa Aesar, 99%), tetraethoxysilane (TEOS) ($Si(C_2H_5O)_4$, Alfa Aesar, 98%) and ethanol (CH_3CH_2OH, Riedel-de Haën, 96%).

Maghemite (γ-Fe_2O_3) nanoparticles were prepared according to Schikorr´s reaction, eq 1 (Košak et. al., 2004):

$$Fe^{2+} + Fe^{3+} + 2OH^- + O_2 \rightarrow \gamma\text{-}Fe_2O_3 + H_2O \quad (1)$$

In this synthesis, the Fe (II) and Fe (III) hydroxides were precipitated during the reaction between two different microemulsions containing an aqueous solution of corresponding ions (MEI) or precipitating reagent (MEII). In the second step of the synthesis, the Fe (II) hydroxide is oxidized, resulted to the formation of the spinel maghemite (γ-Fe_2O_3) phase. The surface of γ-Fe_2O_3 nanoparticles prepared in microemulsion systems was functionalized with tetraethoxysilane (TEOS).

Synthesized SiO_2 coated γ-Fe_2O_3 nanoparticles were characterized using transmission electron microscopy (TEM), X-ray diffractometry (XRD) and specific surface area measurements (BET). The specific magnetization (DSM-10, magneto-susceptometer) of the prepared samples was also measured.

2.1.1 Experimental procedure

Two aqueous solutions were prepared (aqueous solution of Fe^{2+}/Fe^{3+} ions (0.4 or 0.15 M) and 5 % aqueous solution of NH_4OH). Aqueous solution of Fe^{2+}/Fe^{3+} ions with a molar ratio 1.85:1 was prepared by dissolving an appropriate amounts of iron (II) sulphate heptahydrate, $FeSO_4 \cdot 7H_2O$, and iron (III) sulphate hydrate, $Fe_2(SO_4)_3$ xH_2O.

To chose a suitable composition that formed stable water-in-oil (w/o) microemulsion in the systems:

- water/SDS, 1-butanol/cyclohexane (System A) and
- water/CTAB, 1-butanol/1-hexanol (System B).

The region of microemulsion stability, within which the microemulsion is optically transparent was determined by the titration method in connection with conductivity measurements. In Fig. 1 the microemulsion stability region for both systems is represented by solid line for 0.4 M aqueous solution of Fe^{2+}/Fe^{3+} ions and by dashed line for 5 % aqueous solution of NH_4OH, within which the compositions of microemulsion are optically transparent. These are pseudo-ternary phase diagrams, in which every point represents a quaternary system whose overall composition is entirely defined by any pair among the three weight fractions: f_w (aqueous phase), f_s (surfactant + co-surfactant) and f_o (oil phase).

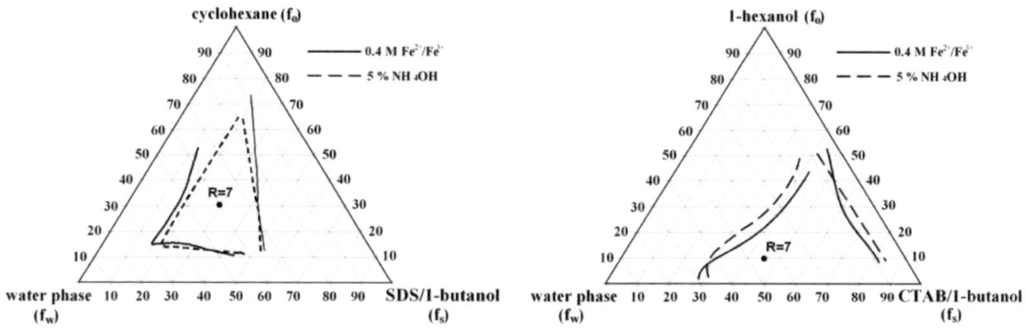

Fig. 1. Phase diagram of the water / SDS / 1-butanol / cyclohexane system (left) and water/CTAB, 1-butanol/1-hexanol (right) at 20 °C.

Prior to the synthesis, two microemulsions of the same composition were prepared. The microemulsion I (MEI) contained 0.4 M or 0.15 M aqueous solution of Fe^{2+}/Fe^{3+} ions, whereas the microemulsion II (MEII) contained 5 % aqueous solution of the ammonium hydroxide (NH_4OH) served as precipitation agent. The SDS to 1-butanol (System A) and CTAB to 1-butanol (System B) weight ratios were kept constant at 1:1.3 and 1.5:1, respectively. In the second step, MEI and MEII were mixed at different temperatures and different pH values for 1 h. A black precipitate was formed immediately. In the last step, the surface of γ-Fe_2O_3 nanoparticles prepared in microemulsion systems with the same molar ratio (R=7) and the following composition for system A: 39 wt. % of water phase, 31 wt. % of SDS/1-butanol, 31 wt. % of cyclohexane and for the system B: 45 wt. % of water phase, 10 wt. % of CTAB/1-butanol, 45 wt. % of 1-hexanol was functionalized using TEOS. The thickness of silica shell was carefully controlled by the amount of TEOS added to the microemulsion after the precipitation step. Finally, the product was centrifuged and washed several times with mixture of ethanol and water and dried at 100 °C.

2.1.2 Results and discussion

Fig. 2 shows the X-ray diffraction patterns of the prepared samples. Except in the case of sample S2 (pH = 6), the diffractograms confirmed the formation of the spinel structure, characteristic of magnetite (Fe_3O_4) and maghemite (γ-Fe_2O_3) with different Fe^{2+} content. The chemical analysis of the synthesized nanoparticles showed that they contain only around 0.5 wt. % of Fe^{2+}. From this we confirmed the formation of γ-Fe_2O_3 nanoparticles.

To estimate the average particle size (D_{XRD}) a Debye-Scherrer formula was used. The values are shown in Table 1.

The difference in the XRD patterns is in the broadness of the peaks. Broader peak indicates smaller crystallite size. From the graph, it is observed that the crystallite size for sample S1 is the smallest one and it is followed by samples S6, S4, C7, C8, C9 and S3, respectively. With increasing the concentration of aqueous solution of Fe^{2+}/Fe^{3+} ions from 0.15 M to 0.4 M the average size of the nanoparticles increased from 5.6 nm to 7.5 nm (samples S6 and S4), and with increasing the reaction temperature from 20 °C to 50 °C, the average size of the nanoparticles increased from 3.3 nm to 7.3 nm (samples S1 and S4). The pH value had a strong influence to average particle size and phase composition. At pH 6 a non-magnetic phase was formed (what about the magnetic properties of the products obtained in other pH conditions?). With increasing of the pH value from 8.6 to 10.2, the average size of the nanoparticles decreased from 12.8 nm to 7.3 nm (samples S3 and S4).

System	Sample	MEI (M)	$T_{(syn.)}$ (°C)	$pH_{(syn.)}$	BET (m^2/g)	D_{BET} (nm)	D_{XRD} (nm)	M (Am^2/kg)
A	S1	0.4	20	10.2	226	5.4	3.3	19
	S2		50	6.0	132	9.3	-	1.4
	S3			8.6	100	12.2	12.8	49
	S4			10.2	189	6.5	7.3	48
	S5			10.4	213	5.8	-	31
	S6	0.15		10.2	274	4.5	5.6	14
B	C7	0.4	20	9.5	102	12.0	7.6	51
	C8		50	9.5	92	13.3	8.8	60
	C9	0.15		9.5	101	12.2	8.8	52

Table 1. Influence of preparation conditions on the properties of prepared samples.

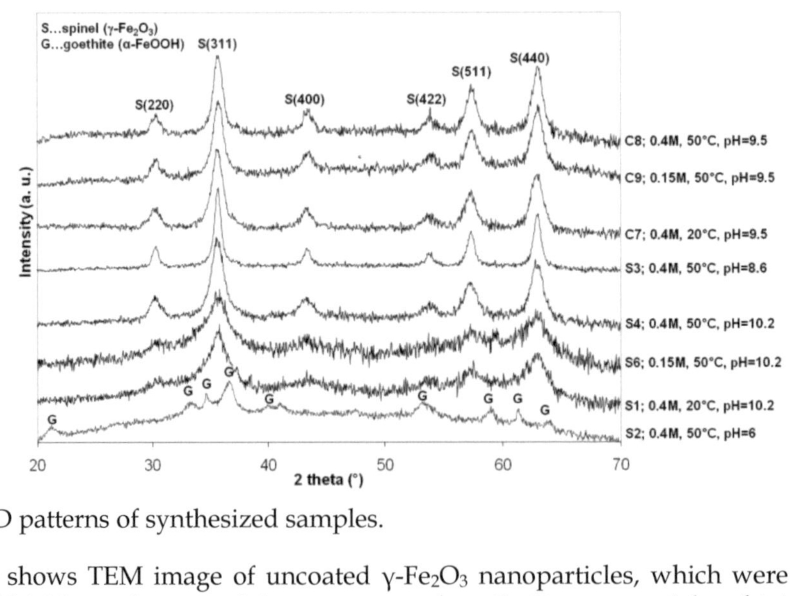

Fig. 2. XRD patterns of synthesized samples.

Fig. 3 left shows TEM image of uncoated γ-Fe_2O_3 nanoparticles, which were spherical in shape and highly agglomerated, in compare to the γ-Fe_2O_3 nanoparticles obtained after the surface coating (Fig. 3 right) with a thin layer of SiO_2.

The results show that the concentration of Fe^{2+}/Fe^{3+} ions in aqueous solution , the reaction temperature and the pH value had a strong influence on the average particle size distribution and as a consequence on the specific magnetization.

The results in Table 1 show that with decreasing of concentration of Fe^{2+}/Fe^{3+} ions in aqueous solution from 0.4 M to 0.15 M, the average particle size determined from BET measurements decreases from 6.5 nm to 4.5 nm in system A (samples S4 and S6) and from 13.3 nm to 12.2 nm in system B (samples C8 and C9). The smallest average particle size was obtained at 20 °C. With increasing of the reaction temperature from 20 °C to 50 °C, the average particle size increased from 5.4 nm to 6.5 nm in system A (samples S1 and S4) and from 12 nm to 13.3 nm in system B (samples C7 and C8). At pH 6 a nonmagnetic phase was formed with the average particle size of 9.3 nm and a specific magnetization of 1.4 Am^2/kg (system A, sample 2). With increasing the pH value from 8.6 to 10.3 in system A, the average particle size decreased from 12.2 nm to 5.8 nm (samples S3, S4 and S5).

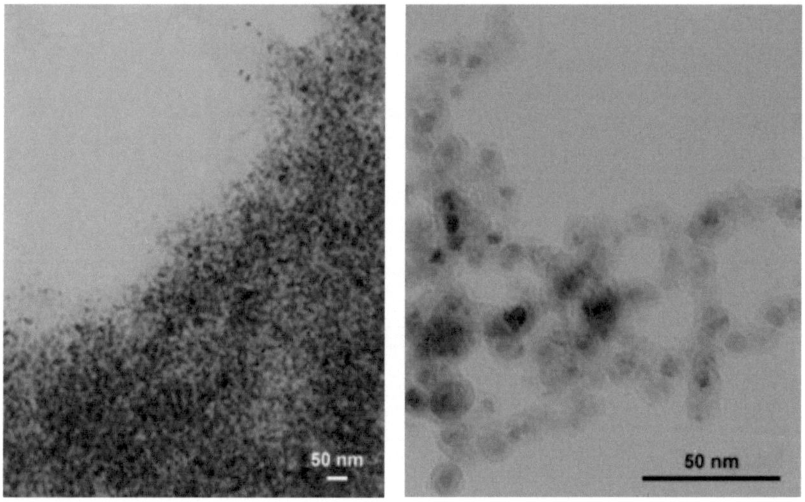

Fig. 3. TEM micrograph of (left) uncoated γ-Fe_2O_3 nanoparticles and (right) γ-Fe_2O_3 nanoparticles obtained with a thin layer of SiO_2.

Fig. 4 shows that specific magnetization (M) of the prepared γ-Fe_2O_3 nanoparticles increased from 19 Am^2/kg for the sample with average particle size of 5.4 nm (system A, sample S1) to 49 Am^2/kg for the sample with average particle size 12.2 nm (system A, sample S3) and from 51 Am^2/kg for the sample with average particle size 12 nm (system B, sample C7) to 60 Am^2/kg for the sample with average particle size 13.3 nm (system B, sample C8).

The thickness of silica shell was carefully controlled by the amount of TEOS added to the microemulsion mixture during the precipitation step. Specific magnetization of the uncoated γ-Fe_2O_3 nanoparticles prepared in microemulsion system A, was 48 Am^2/kg (sample S4) and decreased to 25 Am^2/kg for the nanoparticles with 3 nm thickness of silica shell (Fig. 5). Furthermore, the specific magnetization of the uncoated γ-Fe_2O_3 nanoparticles prepared in microemulsion system B was 60 Am^2/kg (sample C8) and decreased to 21 Am^2/kg, when the thickness of silica shell was 4 nm (Fig. 5).

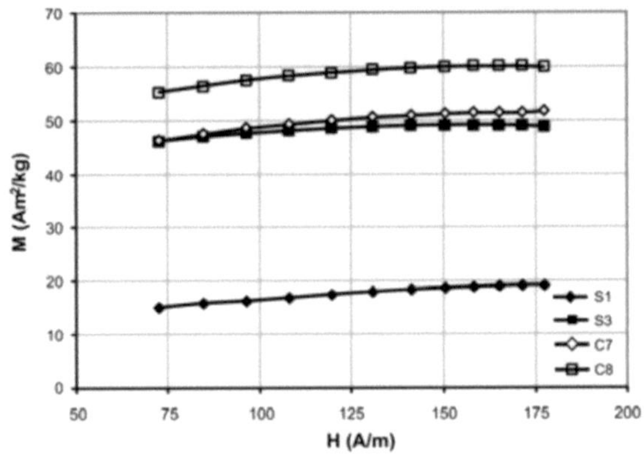

Fig. 4. Specific magnetization of uncoated γ-Fe_2O_3 nanoparticles prepared in microemulsion systems water / SDS / 1-butanol / cyclohexane (samples S1 and S2) and water/CTAB, 1-butanol/1-hexanol (samples C7 and C8).

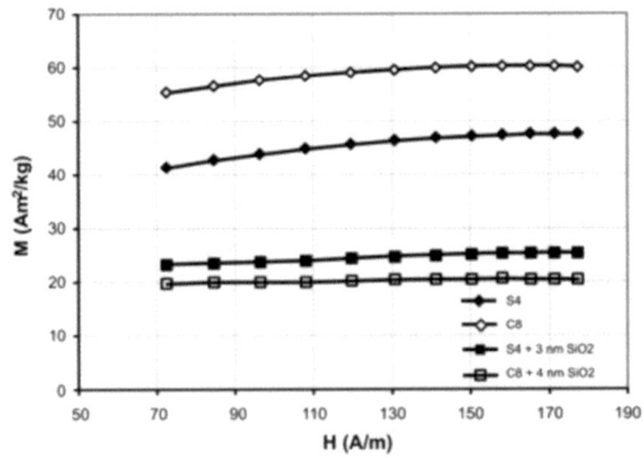

Fig. 5. Specific magnetization of uncoated and coated γ-Fe_2O_3 nanoparticles prepared in microemulsion systems water / SDS / 1-butanol / cyclohexane (samples S4 and S4 + 3 nm SiO_2) and water/CTAB, 1-butanol/1-hexanol (samples C8 and C8 + 4 nm SiO_2).

2.2 Synthesis of strontium hexaferrite $SrFe_{12}O_{19}$ nanoparticles

The chemicals used for the syntheses of the different samples were strontium nitrate anhydrous ($Sr(NO_3)_2$), 98%, Alfa Aesar, iron (III) nitrate nonahydrate ($Fe(NO_3)_3\ 9H_2O$), ACS, 98,0-101,0%, Alfa Aesar; tetramethylammonium hydroxid solution 25% (TMAH) ($C_4H_{13}NO$), Applichem; cyclohexane (C_6H_{12}), ACS, 99+%, Alfa Aesar; sodium n-dodecyl sulfate (SDS) ($CH_3(CH_2)_{11}OSO_3Na$), 99%, Sigma Aldrich; 1-butanol ($CH_3(CH_2)_3OH$), 99%, Alfa Aesar and ethanol (CH_3CH_2OH), 96%, Riedel-de Haën.

The precursors of strontium hexaferrite ($SrFe_{12}O_{19}$) were prepared according to the reaction, eq 2:

$$Sr(NO_3)_2 + 12Fe(NO_3)_3 + 38(CH_3)_4NOH \rightarrow Sr(OH)_2 + 12\,Fe(OH)_3 + 38((CH_3)_4N)NO_3 \quad (2)$$

In this synthesis, the Sr (II) and Fe (III) hydroxides were precipitated during the reaction between two different aqueous solutions or microemulsions containing an aqueous solution of corresponding ions (MEI) and precipitating reagent (MEII). In the second step of the synthesis, the hydroxides are oxidized during the calcinations process, resulting in the formation of the nano-crystalline hexagonal ($SrFe_{12}O_{19}$) phase according to the reaction, eq 3:

$$Sr(OH)_2 + 12Fe(OH)_3 \rightarrow 6Fe_2O_3 + SrO + 19H_2O \quad (3)$$

The dried precursors were characterized using thermogravimetric analyses (TGA) and differential thermal analyses (DTA) in air with a heating/cooling rate of 5 °C/min. The morphology of the powders after the calcination was investigated with transmission electron microscopy (TEM). The size of the nanoparticles was measured with dynamic light-scattering particle size analyses (DLS) and from the powders specific surface area A_s using the BET method and estimated using the relation $D_{BET} = 6\rho/A_s$ where $\rho = 5.1\ g/cm^3$ for spherical particles. The magnetic measurements of the calcined samples were carried out using a magneto-susceptometer (DSM-10).

2.2.1 Experimental procedure

The co-precipitation and the microemulsion methods were applied for the preparation of the $SrFe_{12}O_{19}$ precursors. Prior to the synthesis, aqueous solutions of strontium and iron nitrate with various Sr^{2+}/Fe^{3+} molar ratios (1:6.4, 1:8, 1:10 and 1:12) were prepared. Although a molar ratio of 1:12 should be sufficient, according to the stoichiometry, an excess of strontium nitrate was necessary, because the strontium hydroxide is partially soluble in water (Ataie, 1995).

In the co-precipitation method the Sr(II) and Fe(III) hydroxide precursors were precipitated during the reaction between the aqueous solution of metal nitrates and the 0.5 M aqueous solution of tetramethylammonium hydroxide (TMAH), which served as a precipitating agent (Table 2). The precipitation was performed at room temperature and a pH value of 12.7. The brownish precipitates were washed several times with a mixture of distilled water and ethanol (volume ratio 1:1) and dried at 100 °C.

	Aqueous solution	wt. %
Metal nitrates	0.01 M $Sr(NO_3)_2$ + 0.08 M $Fe(NO_3)_3$	50
Precipitating agent	0.5 M TMAH	50

Table 2. The composition of a typical co-precipitation system used for the synthesis of the precursor with a Sr^{2+}/Fe^{3+} molar ratio of 1:8.

The microemulsion system used in this study consisted of sodium n-dodecyl sulfate (SDS) as the surfactant, 1-butanol as the cosurfactant, cyclohexane as the continuous oil phase and an aqueous solution of reactants as the dispersed phase. Two microemulsions (I and II) with identical compositions and different reagents in the aqueous phase were prepared (Table 3). The aqueous phase in microemulsion I comprised of a mixture of strontium and iron nitrate aqueous solutions. The aqueous phase in microemulsion II comprised of a 0.5 M solution of tetramethylammonium hydroxide (TMAH), which served as a precipitation agent. The brownish precipitates of metal hydroxides appeared within the nanosized aqueous droplets after the two microemulsions were mixed at room temperature and pH 13.5. The precipitates were washed several times with a mixture of distilled water and ethanol (volume ratio 1:1) and dried at 100 °C.

Finally, these dried precursor powders prepared with different Sr^{2+}/Fe^{3+} molar ratios were calcined at different temperatures (400 to 1000 °C) for 1h with a heating/cooling rate of 5 °C/min. The calcinations were performed in a sample holder in the TG apparatus in a static atmosphere with a precise control of the weight change and the calcination temperature.

	Microemulsion I	Microemulsion II	wt. %
Aqueous phase	0.01 M $Sr(NO_3)_2$ + 0.08 M $Fe(NO_3)_3$	0.5 M TMAH	35
Surfactant	SDS	SDS	13
Cosurfactant	1-butanol	1-butanol	17
Oil phase	cyclohexane	cyclohexane	35

Table 3. The composition of the microemulsion system used for the synthesis of the precursor with a Sr^{2+}/Fe^{3+} molar ratio of 1:8.

2.2.2 Results and discussion

Typical TGA and DTA curves of the dried precursors prepared with the co-precipitation method and the microemulsion method at a Sr^{2+}/Fe^{3+} molar ratio of 1:8 are shown in Fig. 6.

The DTA thermograms of the samples prepared by the co-precipitation method show an endothermic peak at 116 °C and two exothermic peaks at 262 °C and 719 °C. The endothermic peak was attributed to the vaporization of the water from the precursor. On the other hand, the first exothermic peak was ascribed to the disintegration-oxidation of the residual organic precipitation agent in air at elevated temperatures. The second exothermic pick was attributed to the formation enthalpy and/or subsequent crystallization of the hexaferrite ($SrFe_{12}O_{19}$).

In the case that the precursor was prepared by co-precipitation in a microemulsion, the DTA thermograms indicated an endothermic peak at 116 °C and three exothermic peaks at 255 °C, 288 °C and 735 °C. Here, the endothermic peak is a consequence of the vaporization of the residual solvent and the water from the precursor. While the appearance of two close but separate exothermic peaks between 255 °C and 288 °C is consistent with the chosen method of synthesis, where we used two different organic agents, i.e., in addition to an organic precipitation agent (TMAH), an organic surfactant (SDS) was also employed. However, both of them exhibited different disintegration temperatures. The third exothermic peak at around 735 °C could, like in the former case, be attributed to the formation of crystalline $SrFe_{12}O_{19}$.

The TGA showed a continuous weight loss from room temperature to about 730 °C. Theoretically, the transformation of the dried precursor hydroxides $Sr(OH)_2$ and $Fe(OH)_3$ into their oxides SrO and Fe_2O_3 and further into the hexagonal ferrite structure, at a Sr^{2+}/Fe^{3+} molar ratio of 1:8, during the calcination in air led to a weight loss of around 24%, which is in good agreement with the experimental results (Fig. 6).

In order to investigate the effect of the calcination temperature on the formation of $SrFe_{12}O_{19}$ powders, co-precipitated samples and those prepared in microemulsions was calcined at various temperatures, ranging from 600 to 1000 °C, for a precursor Sr^{2+}/Fe^{3+} molar ratio of 1:6.4. Figs. 7 and 8 show the XRD patterns of these samples.

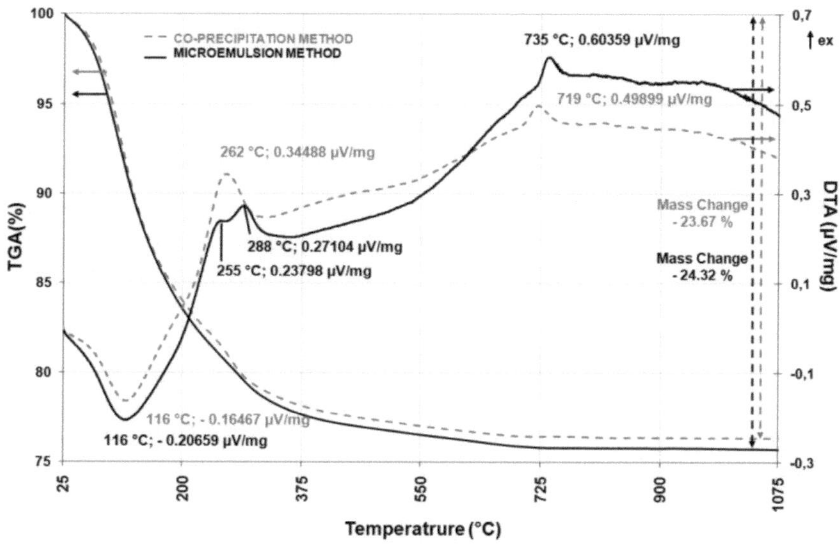

Fig. 6. DTA and TGA thermograms of precursor with the Sr^{2+}/Fe^{3+} molar ratio 1:8 prepared by the co-precipitation and microemulsion methods.

The XRD patterns of the produced powders showed the presence of strontium hexaferrite $SrFe_{12}O_{19}$, strontium peroxide SrO_2, hematite Fe_2O_3, and magnetite Fe_3O_4 phases depending on the calcination temperature and the method of synthesis.

At a low temperature of 600 °C, powder prepared by co-precipitation contained a nonmagnetic phase hematite Fe_2O_3, while the powder prepared with the microemulsion method contained the magnetic phase magnetite. Here, the temperature to form the Sr-hexaferrite is too low, while on the other hand, the second constitutive phase SrO is, in spite of its large excess at low calcinations temperatures, not observed due to its low crystallinity.

The formation of two different forms of iron oxides during the calcinations at 600 °C, i.e., hematite after the co-precipitation was used and magnetite when the microemulsion method was applied can be explained by the difference in the preparation method. The precipitation reaction should be slower in a reverse micelle medium, when considering the necessary coupling of the rate constant of the chemical reaction and the rate constant of the fusion of the reverse micelles, which must happen prior to the particular reaction (Natarajan et. al., 1996).

The slower rate of the precipitation, when the synthesis is performed within the reverse micelles, which leads to a compartmentalization of the bulk compared to that in the bulk conditions, could lead to the formation of Fe_2O_3 H_2O, which on further heating might yield a magnetic phase maghemite/magnetite. However, the main condition that is crucial for the formation of magnetite is the presence of organic aids, which during heating at elevated temperatures prior to the complete oxidation and the formation of carbon oxide, ensures the reduction conditions that yield the magnetite phase instead of the hematite. Furthermore, the reaction occurs in a closed system of the TGA apparatus, where any draught of air was excluded. On the other hand, in co-precipitated and calcined samples the $Fe(OH)_3$ yields hematite.

At a temperature of 800 °C it is mainly the formation of $SrFe_{12}O_{19}$ that can be observed for both types of samples, prepared by co-precipitation and/or by microemulsion-assisted synthesis. This is consistent with the DTA, where the maximum assigned to ferrite formation is observed at 735 °C. Besides, some XRD peaks of hematite Fe_2O_3 in the samples prepared via microemulsion the hematite phase can be detected.

The XRD analysis of the samples calcined at 900 °C showed the formation of the $SrFe_{12}O_{19}$ phase, with no other phase being detected, Figs. 7 and 8. This sample still has a large excess of SrO and the temperature is sufficiently high that all of the hematite and/or magnetite in the precursor reacts to form the hexaferrite phase, with the exception of the strontium oxide, which was designed to be in excess due to the partial solubility of the strontium hydroxide in the water/alcohol during the processing of the precursor. The SrO, which is otherwise prone to react with atmospheric moisture and carbon dioxide, forming hydrocarbonates with a low crystallinity, can be easily extracted with dilute hydrochloric acid and chemically analyzed, while the Sr-hexaferrite is poorly soluble in acids. From these results one can conclude that the Sr^{2+}/Fe^{3+} molar ratio of 1:6.4 is a little bit too high, since the excess of free SrO and/or SrO_2 can be chemically analyzed in the calcined samples.

On the other hand, the XRD patterns for the samples calcined at 1000 °C show identical diffraction patterns to those of the samples calcined at 900 °C. Therefore, one can conclude that the calcination temperature at 900 °C is sufficient for a complete reaction between stoichiometric ratios of the precursor components forming the target phase, i.e., $SrFe_{12}O_{19}$, while the excess of SrO can be easily removed from the product.

In order to complete the range of trial-and-error experiments the samples with Sr:Fe molar ratios of 1:8, 1:10 and 1:12 were calcined at 900 °C. Here, the influence of a relatively large excess of SrO in the precursor was gradually reduced and the phase formation depending on the calcinations temperature was followed by XRD (Fig. 9) and by measuring the magnetization of the products (Figs. 12 and 13).

Fig. 9 shows the XRD patterns for the samples prepared with the co-precipitation method and Sr^{2+}/Fe^{3+} molar ratios of 1:6.4, 1:8, 1:10 and 1:12. The pure crystalline single-phase $SrFe_{12}O_{19}$ was formed at a Sr:Fe molar ratio of 1:6.4. On the other hand, the XRD patterns of the produced samples at Sr^{2+}/Fe^{3+} molar ratios of 1:8, 1:10 and 1:12 showed, besides strontium peroxide SrO_2, the presence of the magnetite Fe_3O_4 phase.

With a decrease of the relatively large excess of Sr^{2+} the magnetite phase starts to appear, as observed from the XRD spectra and the magnetization measurements, Fig. 12. It is surprising that in spite of an overall large excess of SrO in the precursor the iron oxide phases can be detected in calcined samples with a molar ratio > 1:6.4. The magnetite Fe_3O_4 phase in the samples to a great extent determines the magnetic properties, as is clear from the hysteresis curves, Fig. 13. The phase compositions for the samples prepared with the microemulsion method were almost the same. The proposed reason for such a diverse phase formation, depending on the Sr^{2+}/Fe^{3+} molar ratio, is discussed latter.

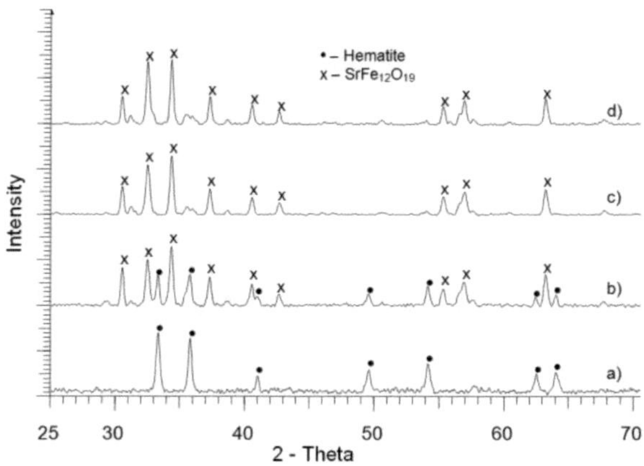

Fig. 7. XRD patterns of co-precipitated samples with a Sr^{2+}/Fe^{3+} molar ratio of 1:6.4 calcined at temperatures: a) 600 °C, b) 800 °C, c) 900 °C and d) 1000 °C for 1h.

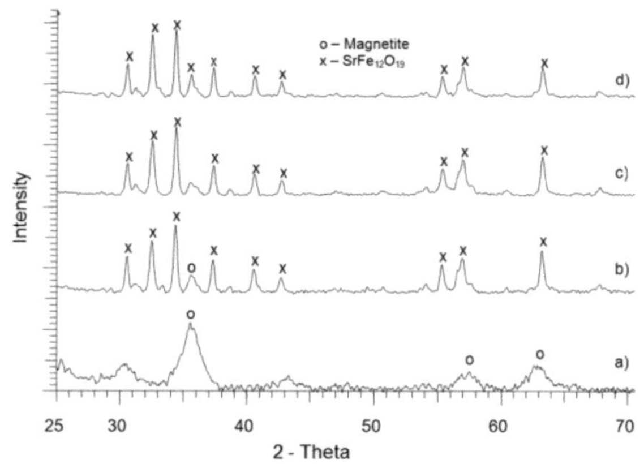

Fig. 8. XRD patterns of samples prepared by the microemulsion method with a Sr^{2+}/Fe^{3+} molar ratio of 1:6.4 calcined at temperatures: a) 600 °C, b) 800 °C, c) 900 °C and d) 1000 °C for 1h.

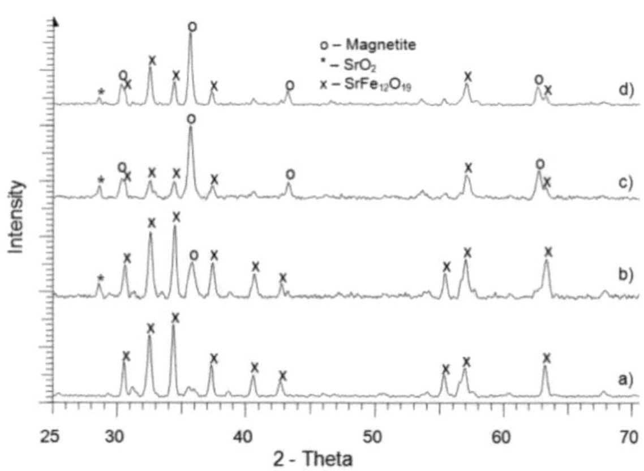

Fig. 9. XRD patterns for samples with various Sr^{2+}/Fe^{3+} molar ratios: a) 1:6.4, b) 1:8, c) 1:10 and d) 1:12) prepared with the co-precipitation method calcined at 900 °C.

Figs. 10 and 11 show the TEM images of $SrFe_{12}O_{19}$ particles prepared by the co-precipitation and microemulsion methods, respectively, with a Sr^{2+}/Fe^{3+} molar ratio of 1:8, calcined at 900 °C for 1h. There were no dramatic differences in both of the particle size and morphology when the samples were prepared using different synthesis methods.

The average size of the $SrFe_{12}O_{19}$ particles in the samples with a limited degree of agglomeration is in the range of 40–80 nm, from an inspection of the TEM images. However, when we compare this value with that of the average particle size obtained from the BET measurements, i.e., the d_{BET} = 61 nm (19 m^2/g) with the same Sr^{2+}/Fe^{3+} molar ratio, one can conclude that the agglomeration of particles during the processing is not remarkable. Namely, the agglomeration of the particles might strongly decrease the measured total surface area and consequently the average particle size estimated from it is apparently larger. The average size of the nanoparticles stabilized in a suspension, i.e., in butylene glycol using a sonificator, was also measured using a dynamic light-scattering particle size analysis (DLS). The results of the DLS analysis showed that the size of the nanoparticles was around 40 nm. Here we must stress that the high energy of the ultrasound radiation breaks the agglomerates and decreases their number. Taking into account the data from the TEM images, the DLS analyses and the BET measurements one can describe the morphology of the synthesized powders as being composed of mostly not agglomerated particles of a size around 40 to 80 nm. The critical diameter of a spherical strontium hexaferrite with a single domain is calculated using the eqs 4 and 5 (Rezlescu et. al., 1999):

$$D_{m(crit)} = 9\sigma_w/2\pi M_s^2, \tag{4}$$

where

$$\sigma_w = \sqrt{\left(\frac{2k_B T_c |K_1|}{a}\right)} \tag{5}$$

is the wall density energy, $|K_1|$ the magnetocrystalline anisotropy constant, T_c the Curie temperature, M_s the saturation magnetization, k_B the Boltzmann constant and a the lattice constant. For $D_m < D_{m(crit)}$ the particles are multi domains. For Sr-hexaferrite, using T_c = 749K, a = $5.88 \cdot 10^{-8}$ cm, $|K_1|$ = $3.6 \cdot 10^6$ erg/cm³ and M_s = 450 Gauss, the value of $D_{m(crit)}$ estimated was around 250 nm. Thus, all the strontium ferrite particles in this investigation are of single domain.

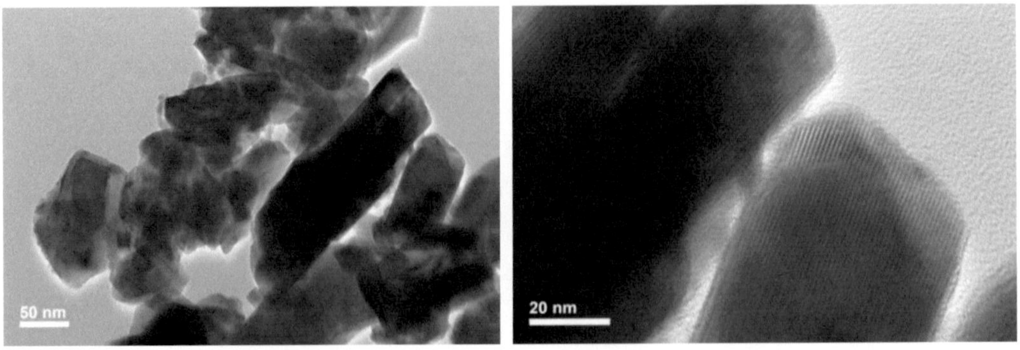

Fig. 10. TEM micrographs of $SrFe_{12}O_{19}$ prepared by the co-precipitation method and calcinated at 900 °C for 1 h.

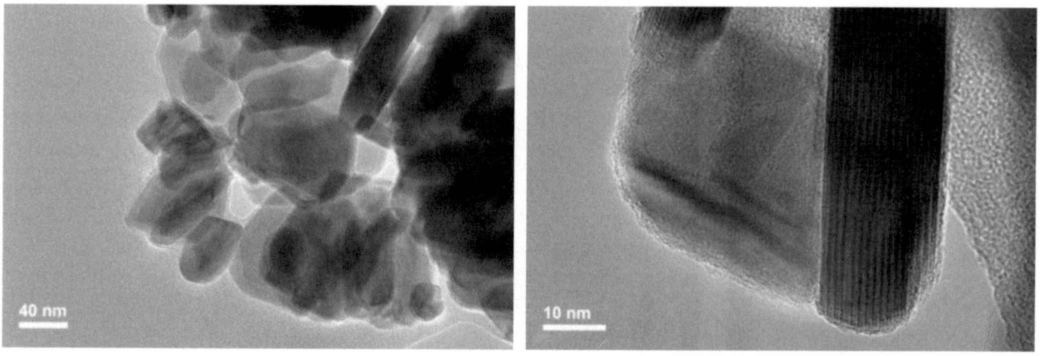

Fig. 11. TEM micrographs of $SrFe_{12}O_{19}$ prepared by the microemulsion method and calcinated at 900 °C for 1 h.

Therefore, we correlate the results of the magnetization measurements with the phase composition of the studied samples for a constant precursor Sr^{2+}/Fe^{3+} molar ratio heated at temperatures from 600 °C to 1000 °C. The saturation magnetization vs. calcination temperature for samples prepared by the co-precipitation and microemulsion methods, with a Sr^{2+}/Fe^{3+} molar ratio of 1:6.4 calcined at different temperatures for 1h is shown in Fig. 12. Following the magnetization of the product versus the temperature of the calcinations, we see a gradual increase of the magnetization, which is to be expected, according to the phase composition recorded with XRD spectra, Figs. 7 and 8. At 400 °C the low magnetization mostly reflects the composition of the precursor, where the ferrimagnetic phases were just starting to form, but could not be detected in the XRD spectra. They are, however, reflected in a slight magnetization. With an increase of the calcination temperature to about 600 °C the saturation magnetization increased from 4 Am^2/kg at 400 °C to 6 Am^2/kg in the case of the co-precipitation method, and from 10 Am^2/kg to 29 Am^2/kg at 600 °C for the samples prepared from microemulsions. With a further increase in the calcination temperature the magnetization steadily increases to 45 Am^2/kg for the co-precipitated samples and 55 Am^2/kg for the samples from the microemulsion. When the temperature increased to 900 °C and 1000 °C the magnetization of the magnetic particles stays in the range between 50 and 60 Am^2/kg.

The relatively large difference in the saturation magnetization for the co-precipitated and/or the microemulsion synthesized samples calcined at 600 °C are related to the phase composition of the powders examined by XRD analysis. The samples prepared with the co-precipitation method contained hematite, whereas those prepared with the microemulsion method contained the magnetic phase magnetite that increases the magnetization of the otherwise non-magnetic constitutive part (SrO) in the powder calcined at 600 °C. With a further temperature increase the target hexaferrite is being formed Figs. 7 and 8, which increases the magnetization of the calcined products.

Fig. 13 shows the plots of the saturation magnetization (M_s) as a function of the applied field (H_c) for samples prepared by the co-precipitation method and the microemulsion method with different Sr^{2+}/Fe^{3+} molar ratios and calcined at 900 °C for 1h. Here, the shape of the hysteresis loops indicates a drastic change in the phase composition depending on the precursor composition and the temperature.

The samples prepared with Sr^{2+}/Fe^{3+} molar ratios of 1:6.4 and 1:8 exhibited a relatively high saturation magnetization of 64 and 62 Am^2/kg, respectively, and wide hysteresis loops with coercivities of 5.4 and 5.0 kOe and remanent magnetizations of 39 and 36 Am^2/kg, respectively. With decreasing the Sr^{2+}/Fe^{3+} molar ratio from 1:8 to 1:10 and/or 1:12 the shape of the hysteresis loop changes drastically, indicating a crucial change in the phase composition, i.e., the magnetic properties of the samples change from high coercivity phases, indicating a hard magnetic character, to a soft magnetic phase. The XRD analysis indicates that the hard magnetic phase is Sr-hexaferrite and the soft magnetic phase is magnetite.

The formation of a submicron-sized strontium hexaferrite, responsible for a nearly square-shaped hysteresis loop in samples with a molar ratio Sr^{2+}/Fe^{3+} of 1:6.4 and 1:8, and a submicron-sized magnetite, exhibiting a hysteresis loop with a very small coercivity, indicating a soft magnetic character in samples with lower Sr^{2+}/Fe^{3+} molar ratios, 1:10 and 1:12, is consistent with the coarsening ability of fine, reactive, iron oxide particles. The essential element of this phenomenon is the fact that the coarsening of the iron particles occurs prior to the onset of the key solid-state reaction, which leads to the target compound. This process is associated with the passivity of the iron oxide particles due to a drastic decrease in the specific surface area needed for an effective solid-state reaction.

At molar ratios of 1:6.4 and 1:8 the solid-state reactions forming Sr-hexaferrites are straightforward, as is usually observed during the preparation of Sr-hexaferrites via a solid-state reaction. The large excess of SrO leads to a direct formation of the hexaferrite phase and can be easily extracted with dilute hydrochloric acid after the termination of the solid-state reaction.

However, at low molar ratios the course of the phase formation is widely different and is grounded on the well-known self-sintering phenomenon (Natarajan et. al., 1996). The soft magnetic particles of magnetite powder formed in that case represent the major part of the reaction product, exhibiting a hysteresis loop with a very low coercivity, Fig. 13. In addition, in these samples, with a low Sr^{2+}/Fe^{3+} molar ratio, the formation of $SrFe_{12}O_{19}$ at 750 °C, Fig. 6, is strongly hindered. The XRD diffraction spectra and the magnetic measurements show that magnetite is the major phase.

The observed phenomenon, i.e., with a large excess of Sr^{2+} the Sr-hexaferrite forms, while with a lower excess of Sr^{2+} mostly the iron oxides can be detected, can be linked to the high reactivity of the chemically synthesized iron oxide particles in the starting mixture. Such an otherwise homogeneous precursor mixture might favor, particularly at a low Sr^{2+}/Fe^{3+} molar ratio, the auto sintering of iron oxide powder (Urek & Drofenik, 1996), i.e., the coarsening of the reactive nanoparticles of iron oxide at a temperature lower than that needed to form Sr-hexaferrite at 750 °C, Fig. 6. This process is much more pronounced when the Sr^{2+}/Fe^{3+} molar ratio is low, where the coordination of the iron oxide particles with the same kind of particles is increased. The coarsening of the iron oxide particles leads to a drastic decrease in the iron oxide reactivity, inducing its passivity and strongly delaying the onset of the target solid-state reaction, the formation of Sr-hexaferrite.

Here, we recognize that in spite of the chemical method used in our studies the procedure by itself exhibits some peculiarities, which lead to the formation of unexpected phases.

During the preparation of fine-grained Sr-hexaferrite via chemical methods a large excess of SrO is required for the suppression of the auto sintering of iron oxide and represents a general demand during the synthesis of iron-oxide-based compounds. Thus, in order to avoid the need for a large excess of SrO during the syntheses of fine-grained particles of Sr-hexaferrite the *"in-situ"* methods (Drofenik et. al., 2010) have an advantage over other chemical methods.

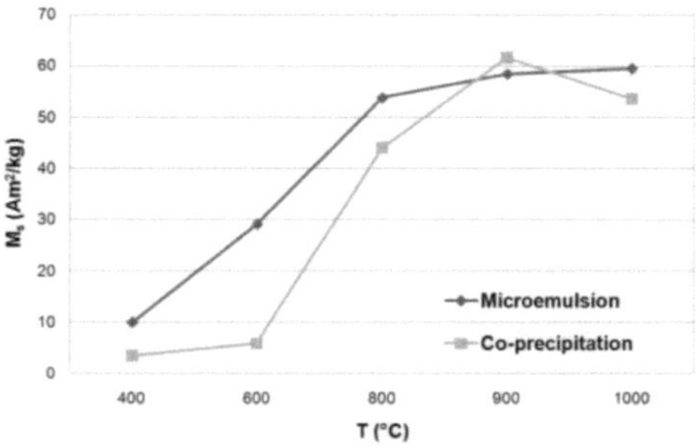

Fig. 12. The saturation magnetization (M_s) of samples prepared by the co-precipitation and microemulsion methods with Sr^{2+}/Fe^{3+} molar ratio 1:6.4 and calcined at different temperatures for 1h.

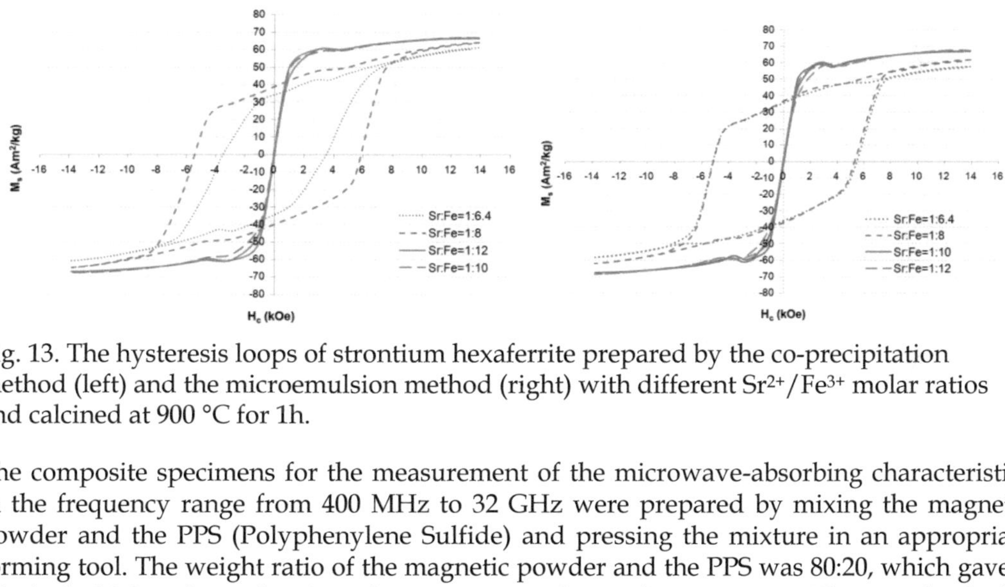

Fig. 13. The hysteresis loops of strontium hexaferrite prepared by the co-precipitation method (left) and the microemulsion method (right) with different Sr^{2+}/Fe^{3+} molar ratios and calcined at 900 °C for 1h.

The composite specimens for the measurement of the microwave-absorbing characteristics in the frequency range from 400 MHz to 32 GHz were prepared by mixing the magnetic powder and the PPS (Polyphenylene Sulfide) and pressing the mixture in an appropriate forming tool. The weight ratio of the magnetic powder and the PPS was 80:20, which gave a relatively high volume fraction of magnetic powder in the composite. The samples had a toroidal form with an inner dimension of 1.3 mm, an outer dimension 2.9 mm and a height of 1.3 mm, in order to fit well into the SMA[1] female or male adapter.

Both the relative complex permeability ($\varepsilon_r = \varepsilon'-j\varepsilon''$) and permittivity ($\mu_r = \mu'-j\mu''$) of the samples were measured using an Anritsu 37269D vector network analyzer in the frequency range from 400MHz to 32 GHz. The permeability and permittivity were calculated from the measured scattering parameters (Bregar et. al., 2007).

---

[1] SubMiniature version A.

The magnetic particles obtained via calcination, which serve as the magnetic/filler phase in the composites for the MW absorption, were magnetically characterized, as listed in Table 4.

The magnetization of sample S1 is 6 Am^2/kg, which is much higher than expected for hematite (α-Fe_2O_3), which at room temperature shows a weak magnetism of about 0.15 emu/g (Schieber, 1967). The observed magnetization of the sample is the result of the magnetic ingredients present in the product, which must consist of small amounts of maghemite and/or magnetite. These tend to form in small amounts as byproducts. Both of them contribute to the final magnetization of the sample, but they cannot be detected in the XRD pattern. However, when the powders are calcined at 800 °C and 900 °C, the particles changed from weakly magnetic to ferrimagnetic.

The saturation magnetization, remanent magnetization and coercivity increased with the calcination temperature from 38 Am^2/kg at 800 °C to 60 Am^2/kg at 900 °C. In sample S2 the magnetization increases due to the formation of Sr-hexaferrite phase. On the other hand, in sample S3, which consists of Sr-hexaferrite nanoparticles, the magnetization measured was 60 Am^2/kg, i.e., less than the bulk value (74.3 Am^2/kg). In part, the decrease in the magnetization of ultra-small nanoparticles can be attributed to their large surface-to-volume ratio and therefore the increased proportion of surface atoms that have an incomplete coordination leading to a non-collinear spin configuration (Kadama et. al., 1996).

The sample S4 consists of Sr-hexaferrite and magnetite. When considering the intensity of the diffraction peaks with the highest intensity it is possible to roughly estimate a mass ratio of 1:2 for the magnetite particles. The larger amount of magnetite in the composite contributes to a smaller coercivity and a lower remanent magnetization of the S4 sample. The results of the magnetic measurements are in agreement with the general principle that both phases with different magnetizations contribute to the final magnetization of the mixed product roughly in their mass proportions. Here, we have a mixture of a cubic magnetite phase, with soft magnetic properties, and a hexagonal ferrite, with hard magnetic properties, which in proportion to their mass ratios contribute to the average magnetization of the magnetic part of the composite in a magnetic field of 1T.

Sample code	Phase	Crystal structure	M_s (Am^2/kg)	M_r (Am^2/kg)	H_c (kA/m)
S1	α-Fe_2O_3	Cubic	6	/	/
S2	α-Fe_2O_3, $SrFe_{12}O_{19}$	Cubic, Hexagonal	38	25	399.4
S3	$SrFe_{12}O_{19}$	Hexagonal	60	36	464.2
S4	Fe_3O_4, $SrFe_{12}O_{19}$	Cubic, Hexagonal	63	7	15.4

Table 4. Magnetic assessment of the products investigated.

The complex permittivity ($\varepsilon_r' - j\varepsilon_r''$) and complex permeability ($\mu_r' - j\mu_r''$) are known to determine the absorption characteristics of the material. The real and the imaginary parts of the permeability of the composite specimens prepared for measurements of the microwave-absorbing characteristics in the frequency range from 400 MHz to 32 GHz with a weight ratio of magnetic powder to PPS = 80:20 are presented in Fig. 14. The real and the imaginary parts of the permeability for samples S1, S2 and S3 are almost constant over the whole frequency range. Just the opposite is the case for sample S4, where the real part of the permeability rapidly decreased from 400 MHz to around 6 GHz, and after that it is lower than for the samples S2 and S3. The imaginary part of the permeability for sample S4 increases from 400 MHz to around 4.5 GHz, and after that it slightly decreases from 5 GHz to 25 GHz. These results are in an agreement with the theory (Giannakopoulou et. al., 2002; Feng et. al., 2007) that materials with a spinel structure (sample S4) have a higher permeability at lower frequencies than anisotropic materials with a hexagonal structure. Based on the fact that microwave absorbers must have a large imaginary part of the complex permeability, we can conclude that the specimens with a hexagonal structure, exhibiting just a notable imaginary part of permeability up to 32 GHz (samples S2 and S3), might be used as microwave-absorbing materials in the frequency range higher than 32 GHz.

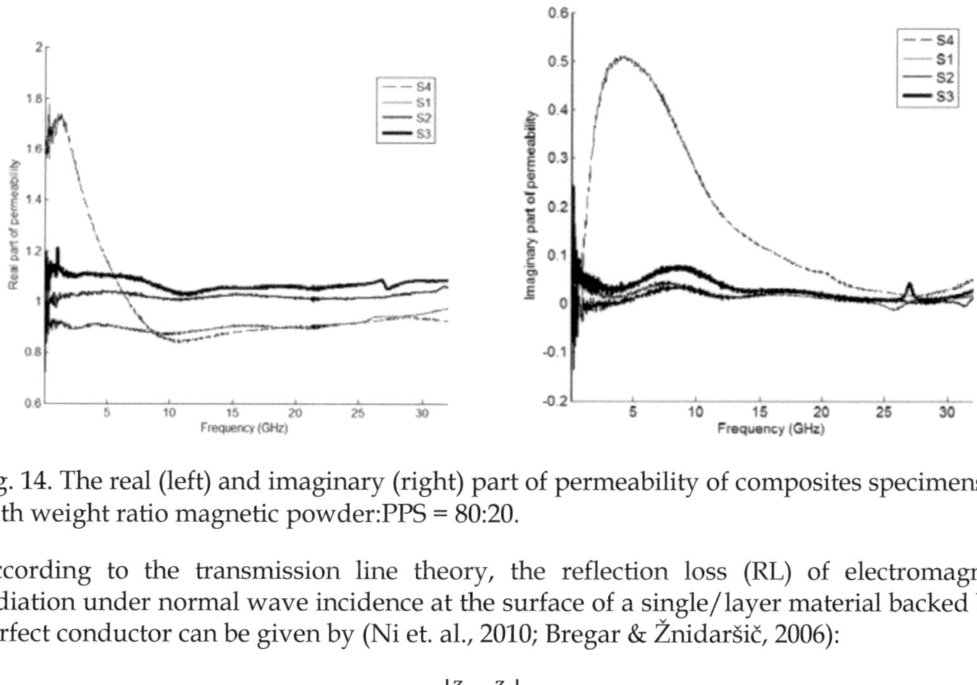

Fig. 14. The real (left) and imaginary (right) part of permeability of composites specimens with weight ratio magnetic powder:PPS = 80:20.

According to the transmission line theory, the reflection loss (RL) of electromagnetic radiation under normal wave incidence at the surface of a single/layer material backed by a perfect conductor can be given by (Ni et. al., 2010; Bregar & Žnidaršič, 2006):

$$R.L. = 20log_{10}\left|\frac{Z_{in}-Z_0}{Z_{in}+Z_0}\right| \text{ unit: decibel (dB)} \quad (6)$$

where Z_0 is the characteristic impedance of free space,

$$Z_0 = \sqrt{\frac{\mu_0}{\varepsilon_0}} \quad (7)$$

and Z_{in} is the input impedance at free space and material surface:

$$Z_{in} = \sqrt{\frac{\mu_r}{\varepsilon_r}} tanh\left[j\left(\frac{2\pi f t}{c}\right)\sqrt{\mu_r \varepsilon_r}\right] \quad (8)$$

where μ_r and ε_r are the relative complex permeability and permittivity of the composite medium, respectively, which can be calculated from the complex scatter parameter, where c is the velocity of light, f is the frequency of the incidence EW wave and t is the thickness of the composites. The impedance-matching condition is given by $Z_{in} = Z_0$, representing the ideal absorbing conditions.

Fig. 15 shows the dependence of the thickness of the absorber layer (d) on the reflection loss of the composite specimens with a weight ratio of the magnetic filler, i.e., PPS = 80:20. The absorber layer (d) was 2 mm (Fig. 4 left) and 4 mm (Fig. 4 right). It is clear that the reflection-loss peak in the case of samples with the spinel structure (samples S1 and S4) shifts to a lower frequency along with the increased thickness and the peak value becomes bigger and narrower. This shows that by changing the thickness of the material with the spinel structure the position and the attenuation-peak frequency can be easily manipulated in terms of the frequency range. In contrast, the reflection loss for samples with the hexagonal structure (samples S2 and S3) is found to depend sensitively on the absorber thickness in the frequency range from 400 MHz to 32 GHz. These results are in a good agreement with the phenomenon that a decrease in the coercivity may be responsible for the magnetic resonance reduction and as a consequence the thinning of the peaks of the reflection loss (Ghasemi et. al., 2008). We can conclude that composite materials with a larger fraction of the spinel phase/structure can be used as electromagnetic wave absorbers in the lower GHz range, while composite materials with a larger fraction of hexagonal phase might be used in the GHz range above 32 GHz, due to their dielectric and magnetic losses. However, there is a range of compositions that can cover a broad range of frequency absorption, which was the aim of this contribution.

Fig. 15 shows the variation of the reflection loss versus the frequency determined from the composite samples with filler phases (samples S_1–S_4) and two sample thicknesses. The electromagnetic wave absorption data based on Fig. 15 are summarized in Table 5. Here, the bandwidth is defined as the frequency width in which the reflection loss is less than -10 dB, which indicates that 90 % of the EM waves are absorbed by the material (Huo et. al., 2009). RL values of less than -10 dB were obtained with an absorber layer d = 4 in the case of the sample S1 and d = 2 and 4 for the sample S4. It is clear that a wider absorption width of 2.1 GHz and a minimum reflection loss peak of -16.5 dB were observed at 24.8 GHz with an absorber layer of 2 mm for sample S4.

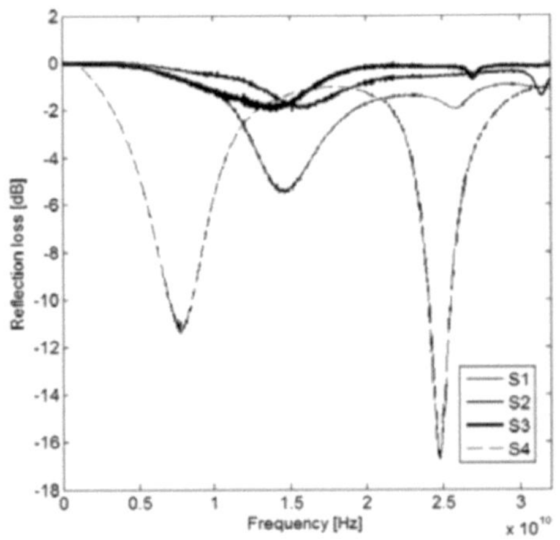

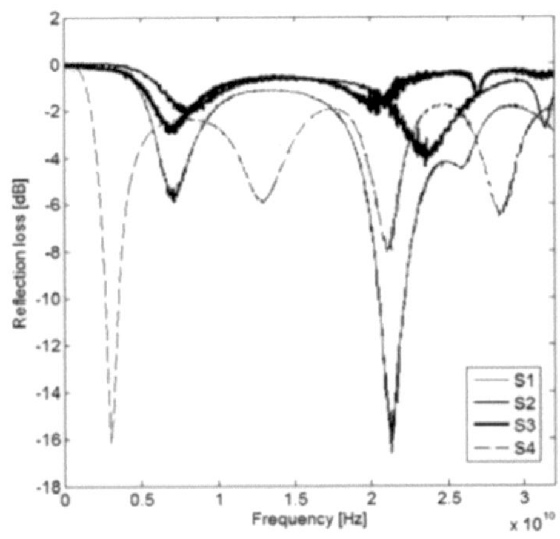

Fig. 15. The dependence of the thickness of absorber layer d (2 mm and 4 mm) on the reflection loss of the composites specimens with weight ratio magnetic-powder: PPS = 80:20.

Samples code	Thickness (mm)	Minimum RL frequency (GHz)	Minimum RL value (dB)
S1	2	14.3	-5.2
	4	7.0 21.3	-5.8 -16.5
S2	2	15.7	-1.8
	4	7.7 23.5	-2.0 -4.0
S3	2	14.0	-2.0
	4	7.0 20.5	-5.7 -2.0
S4	2	7.7 24.8	-11.2 -16.7
	4	3.0 13.0 21.5 28.5	-16.0 -5.8 -7.9 -6.3

Table 3. Electromagnetic wave absorption properties of composite samples.

3. Conclusions

The microemulsion method is a very adaptable technique which allows the preparation of a different kind of nanomaterials just alone or in combination with other techniques.

In the present investigation γ-Fe_2O_3-SiO_2 core-shell nanoparticles with a narrow particle size distribution were prepared via the precipitation in two different microemulsion systems: water/SDS, 1-butanol/cyclohexane and water/CTAB, 1-butanol/1-hexanol. Results obtained in this research, show that the type of surfactant, concentration of aqueous solution of Fe^{2+}/Fe^{3+} ions, the reaction temperature and the pH value, had a strong influence on average particle size distribution and as a consequence on specific magnetization. Specific magnetization of the γ-Fe_2O_3 nanoparticles is ranging from 14 Am^2/kg for a particle size of 5.6 nm to 60 Am^2/kg for a particle size of 8.8 nm. The specific magnetization of functionalized silica (SiO_2) coated maghemite (γ-Fe_2O_3) nanoparticles sharply decreased due to the non-magnetic nature of SiO_2 layer.

Furthermore, the precursors of the nano-crystalline strontium hexaferrite ($SrFe_{12}O_{19}$) powder were successfully synthesized using the co-precipitation and microemulsion methods. The DTA thermograms indicated an exothermic peak at 719.3 °C for the sample prepared by the co-precipitation method and at 735.1 °C for the sample prepared by the microemulsion method, which could be attributed to an exothermic reaction and the crystallization of the $SrFe_{12}O_{19}$ hexaferrite particles. When using both synthesis methods single-phase $SrFe_{12}O_{19}$ was obtained when the hydroxide precursor prepared at a Sr^{2+}/Fe^{3+} molar ratio of 1:8 was calcined at 900 °C.

The $SrFe_{12}O_{19}$ prepared with Sr^{2+}/Fe^{3+} molar ratios of 1:8 and 1:6.4 had a high saturation magnetization of around 62-64 Am^2/kg and wide hysteresis loops with a coercivity of 5.0-5.4 kOe and remanent magnetization of 36-39 Am^2/kg. Decreasing the Sr^{2+}/Fe^{3+} molar ratio below 1:8 caused a drastic change in the phase composition and consequently in the shape of the hysteresis loop. Here, the magnetite represents the major phase in the product as a consequence of the self-sintering phenomenon in the reaction mixture, where the iron oxide represents a large part of the constitutive phases.

In the last step we investigated composites designed for microwave absorption based on magnetic filler, composed of phases within the $SrO–Fe_2O_3$ system, embedded in a polyphenylene sulfide matrix with a concentration ratio of 80:20 by weight. With a manipulation of the molar ratio of Sr^{2+}/Fe^{3+} and a calcination of the co-precipitated precursor at an elevated temperature, one can realize filler within the composite with a multipart crystal structures and multifarious magnetic properties. From this reason it is possible to prepare materials with natural resonance in different frequency ranges. The ferrites with a spinel structure are useful for absorbing materials at lower GHz range while the hexagonal ferrites are special kind of absorbing materials in the range above 32 GHz due to their magnetic anisotropy.

4. References

Ataie A., Harris I.R. & Ponton C.B. (1995). Magnetic properties of hydrothermally synthesized strontium hexaferrite as a function of synthesis conditions, *Journal of Materials Science*, Vol.30, No.6, (June 1995), pp. 1429-1433, ISSN 1573-4803

Bregar V.B. (2004). Potential application of composite with ferromagnetic nanoparticles in microwave absorber, *IEEE Transactions on Magnetics*, Vol.40, No.3, (May 2004), pp. 1679-1684, ISSN 0018-9464

Bregar V.B. & Žnidaršič A. (2006). Analysis of electromagnetic noise suppression in microstrip lines with absorber sheets, *Proceedings of Asia-Pacific Microwave Conference*, pp. 540-543, ISSN: 0018-9464 Yokohama, Japan, December 2006

Bregar V.B. (2004). Potential application of composite with ferromagnetic nanoparticles in microwave absorber, *IEEE Transactions on Magnetics*, Vol.40, No.3, (May 2004), pp. 1679-1684, ISSN 0018-9464

Bregar V.B., Lisjak D., Žnidaršič A. & Drofenik M. (2007). The application of effective-medium theory for the nondestructive characterization of ceramic composites, *Journal of the European Ceramic Society*, Vol.27, No.2-3, (2007), pp. 1071-1076, ISSN 0955-2219

Ding J., Miao W.F., McCormick P.G. & Street R. (1998). High-coercivity ferrite magnets prepared by mechanical alloying, *Journal of Alloys and Compounds*, Vol.281, No.1, (November 1998), pp. 32-36, ISSN 0925-8388

Drofenik M., Ban I., Ferk G., Makovec D., Žnidaršič A., Jagličić Z. & Lisjak D. (2010). The concept of a low-temperature synthesis for superparamagnetic $BaFe_{12}O_{19}$ particles, *Journal of the American Ceramic Society*, Vol.93, No.6, (February 2010), pp. 1602-1607, ISSN 0002-7820

Elvin G., Parkin I.P.P., Bui Q.T., Barquin L.F., Pankhurst Q.A., Komarov A.V. & Morozov Y.G. (1997). Self-propagating high-temperature synthesis of $SrFe_{12}O_{19}$ from reactions of strontium superoxide, iron metal and iron oxide powders, *Journal of Materials Science Letters*, Vol.16, No.15, (January 1997), pp. 1237-1239, ISSN 0261-8028

Feng Y.B., Qiu T. & Shen C.Y. (2007). Absorbing properties and structural design of microwave absorbers based on carbonyl iron and barium ferrite. *Journal of Magnetism and Magnetic Materials,* Vol.318, No.1-2, (2007), pp. 8-13, ISSN 0304-8853

Ghasemi A., Hossienpour A., Morisako A., Liu X. & Ashrafizadeh A. (2008). Investigation of the microwave absorptive behavior of doped barium ferrites, *Materials & design,* Vol.29, No.1, (2008), pp. 112-117, ISSN 0264-1275

Giannakopoulou T., Kompotiatis L., Kontogeorgakos A. & Kordas G. (2002). Microwave behavior of ferrites prepared via sol-gel method, *Journal of Magnetism and Magnetic Materials,* Vol.246, No.3, (May 2002) pp. 360-365(6), ISSN 0304-8853

Groningen R. (November 2003). Magnetic properties of nanocrystaline materials for high frequency applications, *Doctoral dissertation,* ISBN 90 367 1953 4, Netherlands: University of Groningen

Gubin S.P. (1993). *Magnetic nanoparticles,* Edited by: Gubin S.P. © 2009 John Wiley & Sons, ISNB 3527407901, Weinheim

Hessien M.M., Rashad M.M. & El-Barawy K. (2008). Controlling the composition and magnetic properties of strontium hexaferrite synthesized by co-precipitation method, *Journal of Magnetism and Magnetic Materials,* Vol.320, No.3-4, (February 2008), pp. 336–343, ISSN 0304-8853

Huo J., Wang L. & Yu H. (2009). Polymeric nanocomposites for electromagnetic wave absorption, *Journal of Materials Science,* Vol.44, No.15, (2009), pp. 3917-3927, ISSN 0022-2461

Kadama R.H., Berkowitz A.E., McNiff E.J. & Foner S. (1996). Surface spin disorder in $NiFe_2O_4$, *Physical Review Letter,* Vol.77, No.2, (July 1996), pp. 394-397, ISSN 0031-9007

Kim S., Han D. & Cho S. (1994). Microwave absorbing properties of sintered Ni-Zn ferrite, *IEEE Transactions on Magnetics,* Vol.30, No.6, (November 1994), pp. 4554-4556, ISSN 0018-9464

Košak A., Makovec D. & Drofenik M. (2004). The preparation of MnZn-ferrite nanoparticles in a water/CTAB, 1-butanol/1-hexanol reverse microemulsion, *Physica status solidi (c),* Vol.1, No.12, (December 2004), pp. 3521-3524, ISSN 1610-1634

Košak A., Makovec D., Drofenik D. & Žnidaršič A. (2004). In situ synthesis of magnetic MnZn-ferrite nanoparticles using reverse micro emulsion, *Journal of Magnetism and Magnetic Materials,* Vol. 272-276 (2004), pp. 1542-1544, ISSN 0304-8853

Lax B. & Button K.J. (1962). *Microwave Ferrites and Ferrimagnetics,* Edited by : McGraww-Hill, ISBN 62011947, New York

Makovec D. & Košak A. (2005). The synthesis of spinel-ferrite nanoparticles using precipitation in microemulsions for ferrofluid applications, *Journal of Magnetism and Magnetic Materials,* Vol.289, (March 2005), pp. 32-35, ISSN: 0304-8853

Mornet S., Grasset F., Portier J. & Duguet E. (2002). Maghemite@silica nanoparticles for biological applications, *European Cells and Materials,* Vol.3, No.2, (2002), pp. 110-113, ISSN 1473-2262

Nakamura T. & Hankui E. (2003). Control of high-frequency permeability in polycrystalline (Ba,Co)-Z-type hexagonal ferrite. *Journal of Magnetism and Magnetic Materials,* Vol.257, No.2-3, (February 2003), pp. 158-164, ISSN 0304-8853

Natarajan U., Handique K., Mehra A., Bellare J.R. & Khilar K.C. (1996). Ultrafine metal particle formation in reverse micellar systems: effects of intermicellar exchange on the formation of particles, Langmuir, Vol.12, No.11, (May 1996), pp. 2670–2678, ISSN 0743-7463

Ni S., Sun X., Wang X., Zhou G., Yang F., Wang J. & He D. (2010). Low temperature synthesis of Fe_3O_4 micro-spheres and its microwave absorption properties, Materials Chemistry and Physics, Vol.124, No.1, (November 2010), pp. 353-358, ISSN 0254-0584

Ovtar S., Lisjak D. & Drofenik M. (2011) Preparation of oriented barium hexaferrite films by electrophoretic deposition. *Journal of the American Ceramic Society,* [in press, accepted manuscript].

Pankov V.V., Pernet M., Germi P. & Mollard P. (1993). Fine hexaferrite particles for perpendicular recording prepared by the co-precipitation method in presence of an inert component, *Journal of Magnetism and Magnetic Materials,* Vol.120, No.1-3, (1993) 69-72, ISSN 0304-8853

Pardavi Horvath M. (2000). Microwave application of soft ferrites. *Journal of Magnetism and Magnetic Materials,* Vol.215-216, No.1, (December 2000), pp. 171-183, ISSN 0304- 8853

Pereira F.M.M., Santos M.R.P., Sohn R.S.T.M., Almeida J.S., Medeiros A.M.L., Costa M.M. & Sombra A.S.B. (2009). Magnetic and dielectric properties of the M-type barium strontium hexaferrite ($Ba_xSr_{1-x}Fe_{12}O_{19}$) in the RF and microwave (MW) frequency range, *Journal of Materials Science: Materials and Electronics,* Vol.20, No.5, (2009), pp. 408-417, ISSN 0957-4522

Rezlescu L., Rezlescu E., Popa P.D. & Rezlescu N. (1999). Fine barium hexaferrite powder prepared by the crystallisation of glass. *Journal of Magnetism and Magnetic Materials,* Vol. 193, No. 1-3, (March 1999), pp. 288-290, ISSN 0304-8853

Rodrigue G.P. (1963). Magnetic materials for millimeter applications. IEEE Transactions on Microwave Theory and Techniques, Vol.MMT-11 pp. 351-356

Sato H. & Umeda T. (1993). Grain Growth of Strontium Ferrite Crystallized from Amorphous Phases, *Materials Transactions, JIM,* Vol.34, No.1, (January 1993), pp. 76-8, ISSN 0916-1821

Schieber M.M. (1976). *Experimental Magnetochemistry, in: Selected Topics in Solid State Physics,* vol. VIII, ed. E.P. Wohlfarth, North-Holland, Amsterdam

Surig C., Hempel K.A. & Bonnenberg D. (1994). Hexaferrite particles prepared by sol-gel technique, *IEEE Transactions on Magnetics,* Vol.30, No.6, (November 1994), pp. 4092-4094, ISSN 0018-9464

Urek S. & Drofenik M. (1996). Influence of iron oxide reactivity on microstructure development in MnZn ferrites, *Journal of Materials Science,* Vol.31, No.18, (1996) 4801-4805, ISSN 0022-2461

Verma A., Mendiratta R.G., Goel T.C. & Dube D.C. (2002). Microwave studies on strontium ferrite based absorbers. *Journal of Electroceramics,* Vol. 8, No. 3, (September 2002) , pp. 203-208, ISSN 1385-3449

Vidal-Vidal J., Rivas J. & López-Quintela M.A. (2006). Synthesis of monodisperse maghemite nanoparticles by the microemulsion method, *Colloids and Surfaces A: Physicochemical and Engineering Aspects,* Vol.288, No.1-5, (October 2006), pp. 44-51, ISSN 0927-7757

11

Synthesis of Nanocatalysts via Reverse Microemulsion Route for Fischer-Tropsch Reactions

N.A. Mohd Zabidi
Universiti Teknologi PETRONAS
Malaysia

1. Introduction

The current hike on fuel prices, the discovery of more natural gas fields and dwindling of oil reserves have renewed interests in the production of synthetic crude via the gas-to-liquid (GTL) conversion process. The chemical conversion of natural gas into fuels is also an attractive option for monetizing stranded natural gas. The GTL process consists of the following steps (Bao et al., 2010):

i. Production of synthesis gas (a mixture of carbon monoxide and hydrogen) from natural gas
ii. Synthesis of hydrocarbon via the Fischer-Tropsch synthesis (FT) reaction
iii. Product upgrading of the synthesized hydrocarbons to yield various products such as gasoline and diesel.

Fischer-Tropsch (FT) reaction involves conversion of syn-gas (a mixture of CO and H_2) into hydrocarbons such as olefins, alcohols, acids, oxygenates and paraffins. The Fischer-Tropsch (FT) reaction, shown in Equation (1), produces clean gasoline and diesel fuels. The fuels derived from GTL technology are environmental-friendly because they contain low particulates, sulfur and aromatics.

$$nCO + (2n+1)H_2 \rightarrow C_nH_{2n+2} + nH_2O \quad (1)$$

Catalysts used for the Fischer-Tropsch reaction are generally based on iron and cobalt (Khodakov et al., 2007). Ruthenium is an active catalyst for FT but is not economically feasible due to its high cost and insufficient reserves worldwide. Iron has been the traditional catalyst of choice for FT reaction. It is reactive and the most economical catalyst for synthesis of clean fuel from the synthesis gas mixture. Compared to cobalt, iron tends to produce more olefins and also catalyzes the water-gas shift reaction (Equation 2). Iron catalyst is usually employed in the high-temperature (573-623 K) FT operating mode (Steynberg, et al., 1999).

$$CO + H_2O \rightarrow CO_2 + H_2 \quad (2)$$

Cobalt has higher activity for Fischer-Tropsch reaction but more expensive compared to iron. The low-temperature (473-513 K) FT process usually employs cobalt-based catalysts due to their stability and high hydrocarbon productivity. Catalyst supports that have been utilized include silica, alumina, titania, zirconia, magnesium, carbon and molecular sieves. The cost of catalyst support, metal and catalyst preparation contributes to the cost of FT catalyst, which represents a significant part of the cost for the FT technology.

Various types of reactors have been installed in the FT industry such as fixed-bed reactor, multi-tubular reactor, adiabatic fixed-bed reactor, slurry reactor, fluidized bed reactor and circulating fluid-bed reactor systems (Bao et al., 2004; Steynberg et al.,1999). Since FT reaction is highly exothermic, temperature control and heat removal constitute the two most important design factors for the FT reactors.

FT research ranged from preparation of catalysts to design of reactors. Fundamental understanding on the relationship between the catalyst performance and its physical properties, such as particle size, surface area and porosity is vital. The deactivation behavior of cobalt has been linked to its crystallite size, therefore, control of crystallite size is of importance (Saib et al.,2006). A very stable and active catalyst is required to ensure the catalytic system is economically attractive. A model catalyst consists of well-defined catalytically active metal deposited on non-porous support (Moodley, 2008). Spherical model catalysts can be used to bridge the gap between the poorly-defined porous industrial catalysts and the well-defined single crystal surfaces. Knowledge on the relation between the rate of the reaction to the composition and morphology of the catalyst is still lacking (Sarkar et al., 2007). Thus, characterization of model catalysts can relate the physical properties, such as size and shape of particles, to the catalytic behavior of the catalytic materials.

The term nanoparticle is used for particles having diameters ranging from 2 to 50 nm with variable crystallinity, whereas well-defined crystalline nanoparticles are classified as nanocrystals (Hyeon, 2003). The commercially applied iron catalyst is in the fused form comprising large iron particles and therefore difficult to investigate using microscopy. The loss of catalyst activity is associated with changes of iron into a mixture of iron oxide and iron carbide during the Fischer-Tropsch synthesis reaction. The relation between deactivation and changes in composition and morphology are not fully understood for iron and cobalt catalysts (Sarkar et al., 2007). The application of electron microscopy techniques on the supported nanoparticles are well suited to investigate the morphology of supported catalysts and morphological changes that occur during Fischer –Tropsch synthesis. The knowledge on the effect of particle size on the product selectivity and yields for Fischer-Tropsch reaction is still lacking. It has been reported (Trepanier et al., 2010) that particle size of the catalyst material has an effect on the pressure drop in the reactor, and can influence the product distribution. Iron and cobalt nanoparticles of sizes less than 10 nm are expected to improve the kinetics of the Fischer-Tropsch reaction, selectivity for gasoline and stability of the catalyst.

This chapter provides an overview on the synthesis of FT catalysts using the microemulsion techniques as well as other common preparation techniques such as impregnation, precipitation and strong electrostatic adsorption method. It will focus on the results of catalyst characterization in terms of morphology, particle size distribution, and reducibility. FT reaction is considered as a surface-sensitive reaction in which the size of the active metal particles can influence the catalytic activity and product selectivity. The physicochemical properties of the FT catalyst prepared via microemulsion techniques shall be compared with those synthesized via other routes such as impregnation, precipitation and strong electrostatic adsorption methods. The performance of the cobalt-based and iron-based catalysts prepared via various formulations in the Fischer-Tropsch synthesis reaction will be discussed.

2. Methods of catalyst preparation

An extensive review on the development of cobalt Fischer-Tropsch catalysts was presented by Khodakov and co-workers (Khodakov et al., 2007). The performance of the cobalt catalyst in Fisher-Tropsch reaction is greatly influenced by the catalyst preparation method. The variables include suitable support, deposition method of the metal precursor, catalyst promoter, and the subsequent thermal treatments (Moodley, 2008). Cobalt has shown better resistant to deactivation and attrition, but it is also much more expensive compared to iron. Therefore, well-dispersed cobalt on the catalyst support is highly desired to gain economic attractiveness. The reactivity in FTS is correlated to the number of cobalt metallic particles exposed to the syn-gas molecules (Ling et al., 2011). This factor in turns depends on the cobalt loading, dispersion of cobalt species and its reducibility. Hence, an ideal supported catalyst would have uniformly distributed cobalt species that undergoes complete reduction, forming cobalt metallic nanoparticles at optimum size of 6~8nm, where high dispersion guarantees optimum use of cobalt without jeopardizing the FT performance. FT catalysts have been prepared by various methods such as impregnation, precipitation, colloidal method, strong electrostatic adsorption, and microemulsion method.

2.1 Impregnation

The incipient wetness impregnation method is a commonly used method for preparing cobalt and iron catalysts. Typically required amount of the precursor salt i.e. $Co(NO_3)_2.6H_2O$ is dissolved in deionized water and added dropwise to the support under constant stirring, followed by drying in an oven at 120 °C overnight and calcining at temperature of 500 °C (Saib et al., 2006). Variables which can affect the resultant catalyst are the rate of addition of precursor solution, rate of drying, temperature and duration of heating. Cobalt catalyst has also been prepared using the slurry (wet) impregnation method where the amount of impregnation liquid was in excess of the total pore volume of the support material (Khodakov et al., 2007). Impregnation method can produce small particles but difficult to obtain narrow particle size distribution, as depicted in Figure 1.

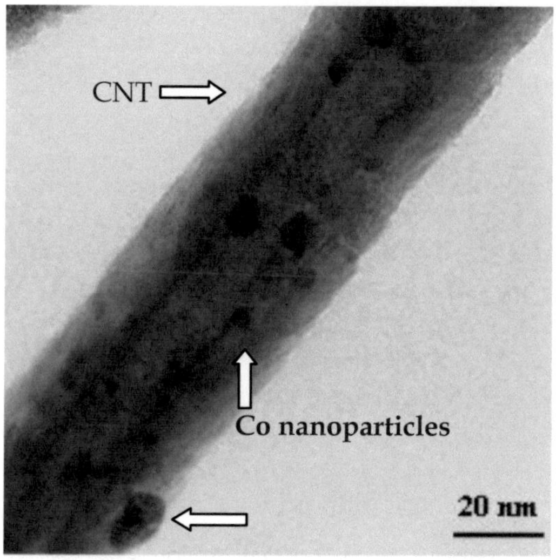

Fig. 1. TEM of 5 wt% Co impregnated on CNTs support. (Adapted from Ali et al., 2011)

2.2 Precipitation

The precipitation method is more common for iron catalyst than cobalt-based FT catalyst. The required amount of the precursor salt i.e. $Co(NO_3)_2.6H_2O$ is dissolved in deionized water and added to the support with constant stirring followed by the addition of precipitating agent i.e. ammonia (Tasfy, 2011) and the resultant slurry is stirred at slightly elevated temperature i.e. 90°C for about 8 hours. The precipitate is dried and then calcined at 500°C.

2.3 Colloidal method

Colloids are synthesized in the presence of surfactants which disperse and stabilize the nanoparticles in an organic solvent. Some of the approaches include polyol method, ethylene glycol method, modified coordination capture method and pseudo-colloidal method. The polyol process involves heating a mixture of catalyst precursor in surfactants, such as oleic acid and oleyl amine in a high-boiling solvent, such as diphenyl ether. The high temperature alcohol reduction of iron(III) acetylacetonate metal precursor resulted in monodispersed iron nanoparticles (Sun & Zeng, 2002). This synthesis process is also called "heating-up" process. The size of the nanoparticles is controlled by changing the concentration of the precursor, the amount and type of surfactant, the aging time and temperature of the reaction. Another synthetic method that produces uniform nanocrystals that is comparable to the "heating up" process is called the "hot injection" method. The "hot injection" method induces high supersaturation and leads to fast homogeneous nucleation reaction followed by diffusion-controlled growth process, which control the particle size distribution. The colloidal method has been shown to generate well-dispersed iron nanoparticles with average size of 5 nm, as depicted in Figure 2.

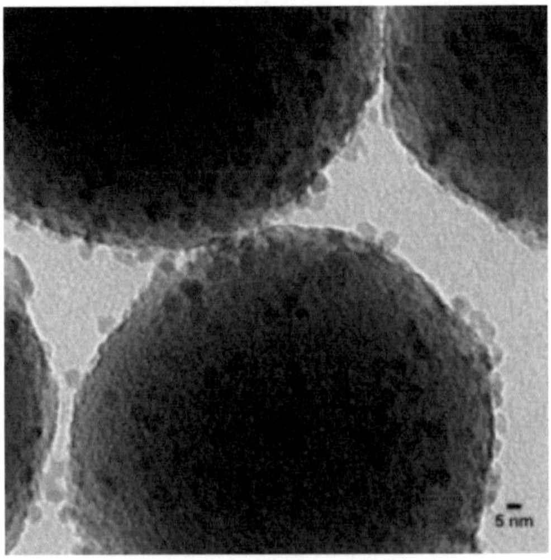

Fig. 2. TEM image of 6 wt% Fe on SiO_2 spheres prepared by the colloidal method (Adapted from Mohd Zabidi, 2010).

2.4 Strong electrostatic adsorption method

Synthesis of uniformly distributed Co nanoparticles remains a great challenge. Strong electrostatic adsorption (SEA) is a catalyst preparation method which is based on basic concept of electrostatic attraction of oppositely charged particles (Jiao & Regalbuto, 2008). Silica and other metal oxides contain hydroxyl groups on its surface. Point of zero charge (PZC) is the pH value of a medium where the hydroxyl groups on the surface of the support remain neutral. In a pH<PZC medium, the hydroxyl groups will be protonated and become positively charged and thus attracting anions. When pH>PZC, the hydroxyl groups will deprotonate and became negatively charged and attracting cations. In other words, pH value plays an important role in the deposition of metal precursor. The SEA method was able to produce nanocatalyst of spherical-shaped Co nanoparticles on SiO_2 support, shown in Figure 3, with Co average size of 3 nm (Ling, 2011).

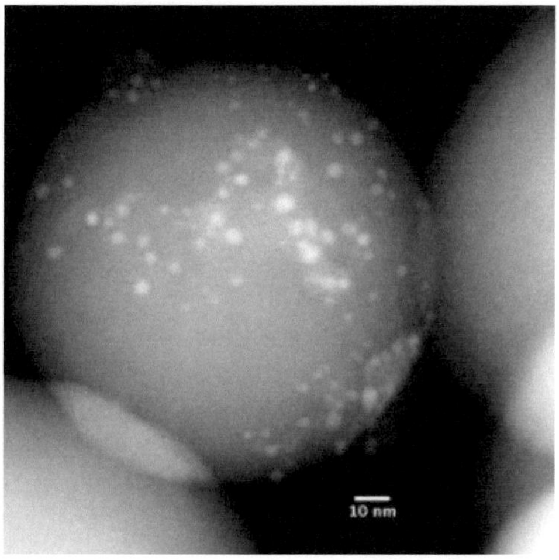

Fig. 3. STEM image of 10 wt% Co on SiO_2 spheres prepared via SEA method (Adapted from Ling, 2011).

3. Basic concepts: Microemulsion

A microemulsion is a liquid mixture of water, a hydrocarbon and a surfactant (Eriksson et al., 2004; Malik et al., 2010). A surfactant is a molecule that possesses both the polar (hydrophilic head) and the non-polar (hydrophobic tail) groups. When the concentration of the surfactant exceeds the critical micelle concentration, molecules aggregate to form micelles. When micelles are formed in an organic medium, the aggregate or tiny droplets is referred as a reversed micelle, in which the polar heads are in the core and the non-polar tails remain outside to maintain interaction with hydrocarbon. It is also referred as water-in-oil microemulsion (Figure 4). The core of the reverse (w/o) microemulsion is of interest as it can be viewed as an elegant nanoreactor which can accommodate chemical reaction. The core interior is hydrophilic thus certain water-soluble can be dissolved inside it. Microemulsion technique offers several advantageous for synthesizing nanoparticles as it enables controls of size, geometry and morphology (Malik et al., 2010).

3.1 Synthesis of nanoparticles via microemulsion technique

The properties of nanoparticles synthesized using the w/o microemulsion are influenced by several factors such as size of water droplets, surfactant concentration and nature of precipitation agent. There are two approaches of synthesizing the nanoparticles via the microemulsion route. The first method involves mixing two sets of microemulsions containing the metal precursor and the precipitating agent or reducing agent. In the second approach, the precipitating agent is added directly to the microemulsion containing the metal precursor.

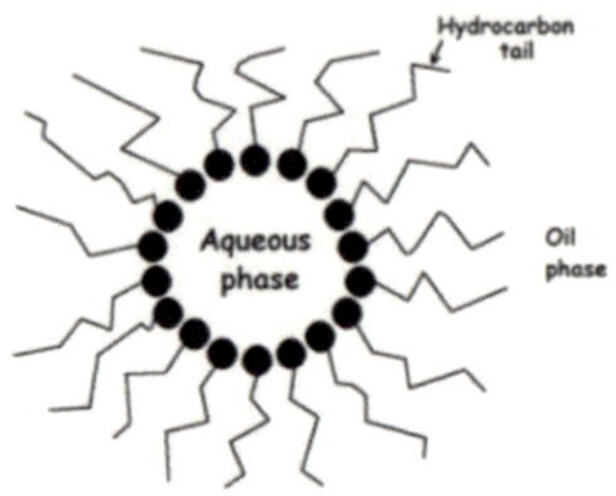

Fig. 4. Reversed micelle (Adapted from Eriksson et al., 2004).

3.1.1 Mixing of two microemulsion systems

The w/o microemulsion synthesis method consists of preparing two sets of microemulsions containing the metal salt and the reducing agent as shown in Figure 5 (Eriksson et al., 2004). The precursor metal salt and reducing agent are dissolved in the aqueous phase whereas the surfactant is prepared in an organic medium. For preparation of supported iron nanocatalyst, the reverse microemulsion method involved preparing two reverse microemulsions. The first reverse microemulsion consisted of $Fe(NO_3)_3.9H_2O$ (aq) and sodium bis(2-ethylhexyl) sulfosuccinate (AOT, ionic surfactant) in hexanol and the second reverse microemulsion was prepared by mixing an aqueous hydrazine solution (reducing

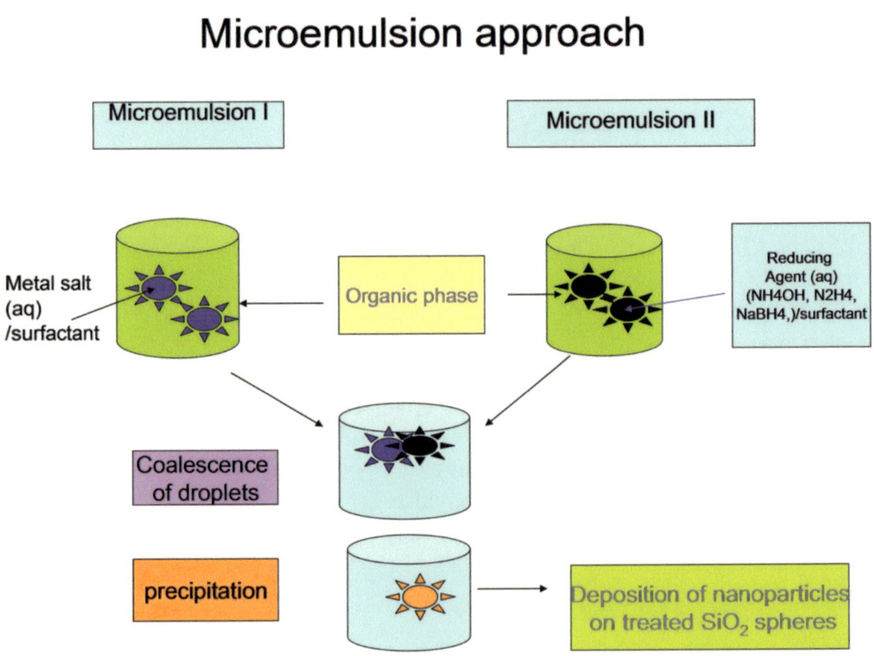

Fig. 5. Reverse microemulsion approach for synthesis of nanoparticles on catalyst support. (Adapted from Eriksson, et al., 2004).

agent) with the AOT solution (Mohd Zabidi, 2010). SiO_2 was chosen as the catalyst support and SiO_2 spheres were then added to the mixture followed by stirring for 3 hours under nitrogen environment. Figure 6 shows the TEM micrograph of a spherical model catalyst prepared using the reverse microemulsion method comprising 6 wt% Fe on SiO_2 support (Mohd Zabidi, 2010). The reverse microemulsion method produced spherical-shaped iron oxide nanoparticles with average diameters of 6.3 ± 1.7 nm, however, the coverage of the SiO_2 surfaces was found to be non-uniform.

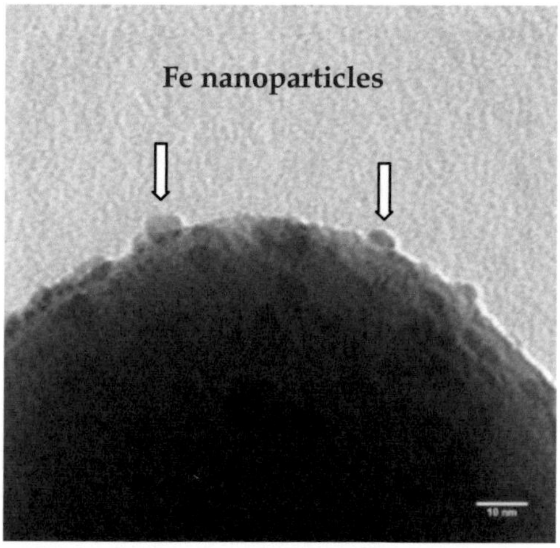

Fig. 6. TEM of 6 wt% Fe on SiO_2 support (Adapted from Mohd Zabidi, 2010)

3.1.2 Direct addition of precipitating reagent to the microemulsion

Another approach of synthesizing nanoparticles from microemulsion is via direct addition of the reducing agent or precipitating agent to the microemulsion containing the metal precursor, as depicted in Figure 7 (Eriksson et al, 2004). It was reported that 5 wt% Fe/SiO_2

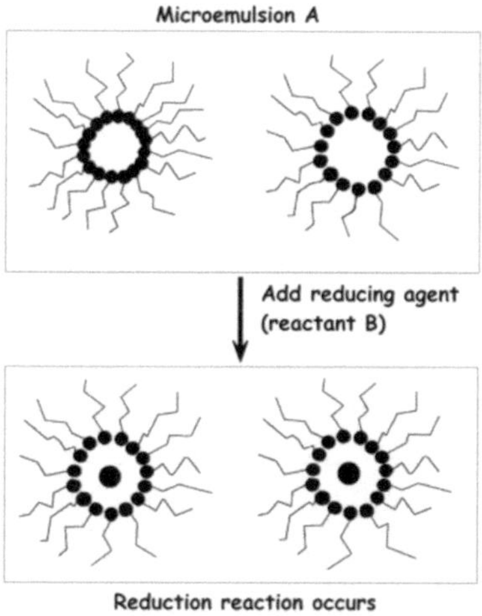

Fig. 7. Direct addition approach for preparation of nanoparticles using the reverse microemulsion method (Adapted from Eriksson et al., 2004).

prepared using the water-in-oil microemulsion resulted in Fe particles with average size of 5.2 nm (Hayashi et al, 2002). At the same Fe loading, the impregnation method produced much larger Fe particles (14.2 nm) which resulted in approximately 30% lower CO conversion in the Fischer-Tropsch reaction.

3.2 Influence of synthesis parameters on the size of nanoparticles

The polydispersity of the system and the size of nanoparticles are influenced by the nature of solvent, surfactant or co-surfactant, presence of electrolyte, concentration of reducing / or precipitating reagents and molar ratio of water to surfactant (Eastoe et al., 2006). The size of water droplets in the microemulsion, the concentration of surfactant and the nature of the reducing/precipitating agent will influence the particle size of metal prepared via microemulsion technique (Eriksson et al., 2004). Increase in water-to-surfactant ratio will increase the size of the water droplets in the microemulsion and hence the size of the resultant metal nanoparticles. Varying the water-to-surfactant ratio (R) from 5 to 50 increased the particle size of cobalt from 5.1 nm to 9.4 nm which resulted in a decrease in turnover frequency (TOF) from 8.6×10^{-3} s^{-1} to 1.2×10^{-3} s^{-1} for FT reaction conducted at 493 K, 2.0 MPa over 10%Co/ITQ catalyst (Prieto et al., 2009).

The number of droplets will increase with increasing presence of surfactant in the system (Eriksson et al., 2004). The surfactant molecules could result in steric hindrance which retard the fast rate of growth of the nuclei. Subsequently, the particles grow at the same rate and led to formation of nanoparticles with narrow particle size distribution.

3.3 Mechanism of formation of nanoparticles

The mechanisms by which nanoparticles are formed have been described by Malik and co-workers using two models (Malik et al., 2010). The first model is based on the Lamer diagram where nucleation occurs when the concentration of the precursor reactant reaches a critical supersaturation value, resulting in a constant number of nuclei thus size of particles increase with concentration (Malik et al, 2010). In the second model, variation in the concentration of precursor does not change the size of particles as the particles are thermodynamically stabilized by the surfactant.

4. Comparison of physicochemical properties between catalysts prepared via different routes

The preparation methods influenced the morphologies of the nanocatalyst particles as depicted by the TEM micrographs in Figure 8. The reverse microemulsion method resulted in relatively larger Co nanoparticles and more agglomeration compared to those synthesized via impregnation and strong electrostatic adsorption (SEA) methods. The size of Co nanoparticles synthesized using reverse microemulsion method ranged from about 5 nm to 15 nm. The coverage of SiO_2 surface was less than that obtained using the other two methods.

Figure 9 shows the H_2-TPR profile of SiO_2-supported cobalt nanocatalyst prepared via the impregnation and reverse microemulsion methods. The first reduction stage (around 350 °C) corresponds to reduction of Co_3O_4 to CoO while the peak around 400 °C was for the reduction of CoO to Co^0 (Fishcer et al., 2011). The Co/SiO_2 sample prepared via the reverse microemulsion method exhibited a second reduction stage at 700 °C, which was much higher compared to that obtained for the sample prepared via the impregnation method indicating poorer reducibility, possibly due to strong metal-support interactions.

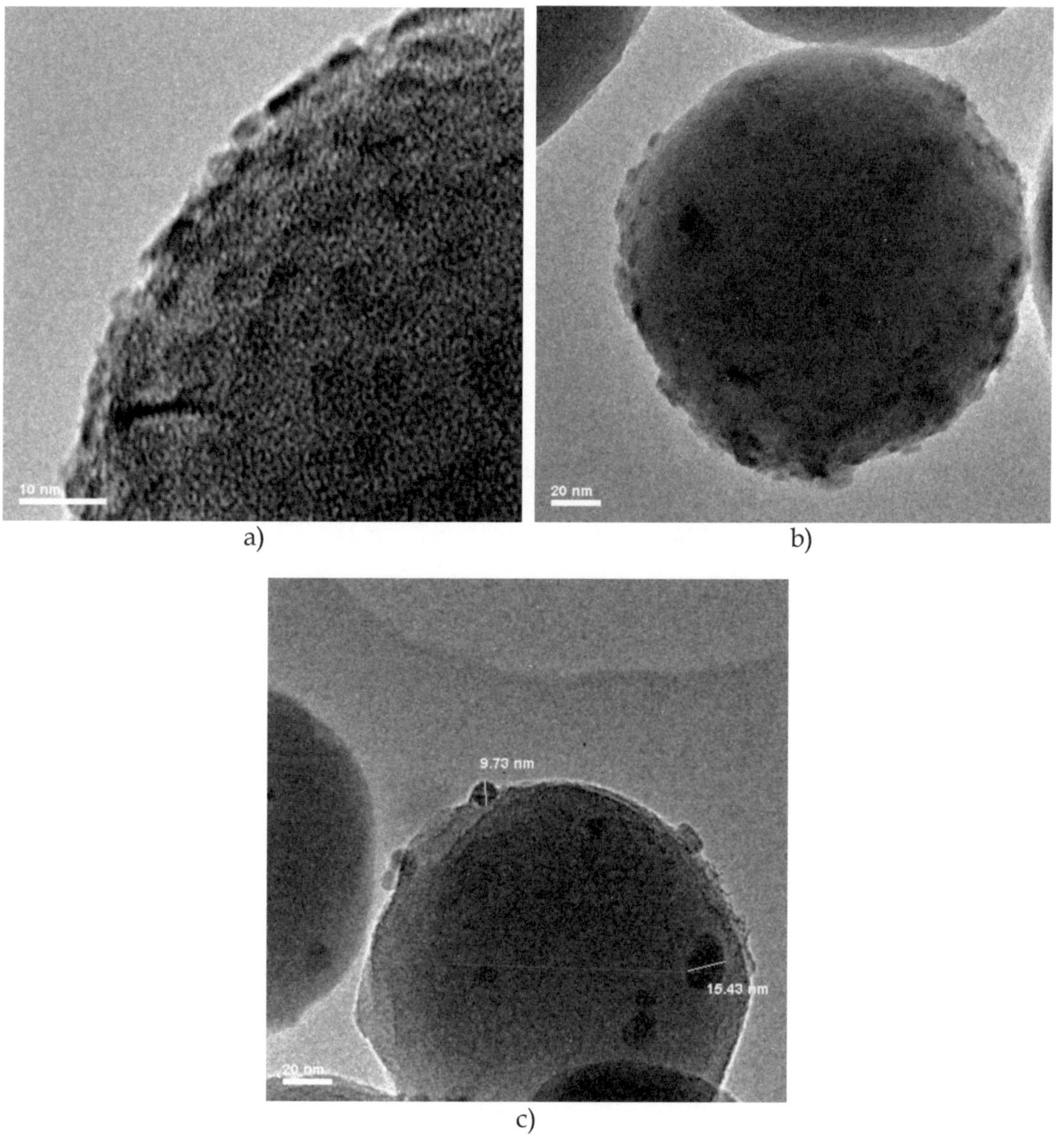

Fig. 8. TEM images of 5 wt% Co on SiO_2 support prepared via (a) impregnation (Adapted from Ali et al., 2011) (b) SEA method (Adapted from Ling et al., 2011) (c) reverse microemulsion (Adapted from AbdulRahman, 2010).

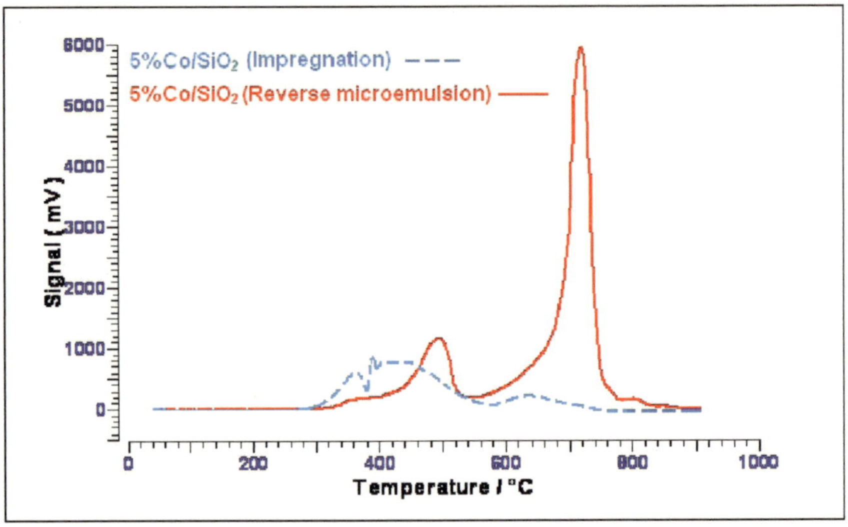

Fig. 9. H_2-TPR profile of 5 wt% Co on SiO_2 spheres prepared via impregnation and reverse microemulsion methods. (Adapted from Abdul Rahman, 2010).

4.1 Catalytic activity

The performance of the nanocatalyst in Fischer-Tropsch synthesis reaction is influenced by the catalyst's preparation route. Table 1 and 2 show the performance of supported cobalt-based and iron-based nanocatalysts, respectively, in the FT reaction. The cobalt

Catalyst	Preparation route	Particle size (nm)	X_{CO} (%)	S_{CH4} (%)	S_{C5+} (%)	Reaction conditions
10%Co/CNTs	Reverse micromulsion	7.8	62	6	92	220°C, 2MPa (Trepanier et al., 2010)
10%Co/CNTs	Impregnation	7.9	54	10	86	220°C, 2MPa (Trepanier et al., 2010)
10%Co/SiO_2	Strong electrostatic adsorption (SEA)	5	2	0.1	80	220°C, 0.1MPa (Ling, 2011)
10%Co/SiO_2	Impregnation	-	10	70	10	220°C, 0.1MPa (Ling, 2011)
5%Co/SiO_2	Impregnation	5	4.7	17.7	1.3	270°C, 0.1MPa (Ali et al., 2011)
5%Co/CNTs	Impregnation	3	15.7	16.4	14	270°C, 0.1MPa (Ali et al., 2011)

Table 1. Comparison of performance of Co-based nanocatalysts in FT reaction

nanoparticles synthesized using the reversed microemulsion technique was found to be more uniform and exhibited narrow size range compared to those synthesized via the impregnation method (Trepanier et al., 2010). The more uniform cobalt clusters also resulted in a decrease in the CH_4 selectivity and increase in C_{5+} selectivity due to effective participation of olefins in the propagation of carbon chain. The CNTs support also served as a better support for Co nanoparticles compared to those of SiO_2 as indicated by the higher value of C_{5+} selectivity likely due to confinement of the Co nanoparticles inside the channels of the CNTs (Trepanier et al, 2010).

Catalyst	Preparation route	Particle size (nm)	X_{CO} (%)	S_{CH4} (%)	S_{C5+} (%)	Reaction conditions
5%Fe/SiO_2	Reverse microemulsion	-	1.5	27.5	10.1	250°C,1.01MPa (Heranz et al., 2006)
5%Fe/SiO_2	Reverse microemulsion	5.2	11.1	18.2	-	260°C,4.0 MPa, (Hayashi et al., 2002)
5%Fe/SiO_2	Impregnation	14.2	3.5	35.5	-	260°C,4.0 MPa, (Hayashi et al., 2002)
6% Fe/SiO_2	Impregnation	8.6	54.0	57.0	20.7	250°C,0.1MPa (Tasfy, 2011)
6% Fe/SiO_2	Precipitation	17	45.7	63.4	9.6	250°C,0.1MPa (Tasfy, 2011)

Table 2. Comparison of performance of Fe-based nanocatalysts in FT reaction

These results also show that the optimum size of Co nanoparticles is around 7.8 - 7.9 nm. Smaller Co nanoparticles which were obtained via other preparation routes were found to result in lower C_{5+} selectivity. SiO_2-supported Co nanocatalysts prepared via the strong electrostatic adsorption method resulted in higher C_{5+} selectivity than those obtained using the impregnation method when FT reaction was conducted at atmospheric pressure (Ling, 2011, Ali et al., 2011).

It has been reported that the iron nanocatalyst prepared via microemuslion exhibited high selectivity towards C_{2+} oxygenates (Hayashi et al, 2002). At the same Fe loading, the size of Fe nanoparticles prepared via impregnation was about three times larger than that of microemulsion which could have led to higher CH_4 production even when the reaction was carried out at 4 MPa pressure. As expected, the selectivity to CH_4 increased by decreasing the reaction pressure. The impregnated iron nanocatalyst exhibited better performance towards C_{5+} than that obtained from the microemulsion nanocatalyst. However, the microemulsion catalyst resulted in much lower CH_4 selectivity compared to that of the impregnated and precipitated Fe nanocatalysts. At atmospheric pressure, the Fe nanocatalyst prepared via impregnation resulted in lower CH_4 selectivity and higher C_{5+} selectivity compared to those obtained over the precipitated Fe nanocatalyst.

5. Conclusion and outlook

The average size and size distribution of the active metal nanoparticles namely, iron and cobalt, are influenced by the catalysts preparation route. Each preparation method has its own advantageous and disadvantageous. The impregnation method seems to be the simplest preparation route, unfortunately, it does not always produce small nanoparticles at narrow particle size distribution. The reverse microemulsion method has several advantageous compared to the conventional impregnation method, as theoretically the size of the metal nanoparticles could be easily controlled through manipulation of the size of the microemulsion droplets by varying the water-to-surfactant ratio. These elegant nanoreactors, however do not always generate nanoparticles at the desired size range with good dispersion due to interaction with the catalysts support.

For the Co-based nanocatalyst, the reverse microemulsion technique was able to produce cobalt nanoparticles at 7.8 nm which showed good performance in terms of achieving high selectivity to C_{5+} and low CH_{4+} selectivity. More improvements are still required before the reverse microemulsion method can substitute the simpler, less elegant impregnation method.

6. Acknowledgment

The financial support provided by Ministry of Science, Technology and Innovation (E-Science Fund No: 03-02-02-SF0036), Ministry of Higher Education Malaysia (FRGS No: FRGS/2/2010/SG/UTP/02/3) and Universiti Teknologi PETRONAS (STIRF: 31/09.10) are acknowledged.

7. References

AbdulRahman, N. (2010). Preparation of Nanocatalyst Using Reverse Microemulsion Method. *Postgraduate Symposium*, Malaysia, 2010

Ali, S., Mohd Zabidi, N. A. & Subbarao, D. (2011). Synthesis and Characterization of Cobalt-Based Nanocatalysts: Effect of Supports, *Proceedings of the 2nd International Conference on Nanotechnology: Fundamentals and Applications,* No.381,pp.1-6, ISBN 978-0-9867183-0-4, Ottawa, Canada, July 27-29, 2011

Bao, B., El-Halwagi, M. & Elbashir, N. O. (2010). Simulation, Integration, and Economic Analysis of Gas-to-Liquid Processes. *Fuel Processing Technology*, Vol.91, No.7, (February 2010), pp. 703-713, ISSN 0378-3820

den Breejen, J.P., Sietsma, J.R.A., Friedrich, H., Bitter, J.H. & de Jong, K.P. (2010). Design of Supported Cobalt Catalysts with Maximum Activity for the Fischer-Tropsch Synthesis. *Journal of Catalysis*, Vol. 270, No.1, (March 2010), pp. 146-152, ISSN 0021-9517

Eastoe, J., Hollamby M. & Hudson, L. (2006), Recent advances in Nanoparticle Synthesis with Reversed Micelles. *Advances in Colloid and Interface Science*, Vol. 128-130,(December 2006), pp. 5-15, ISSN 0001-8686

Eriksson, S., Nylen, U., Rojas, S. & Boutonnet, M. (2004). Preparation of Catalysts from Microemulsions and Their Applications in Heterogeneous Catalysis. *Applied Catalysis A*, Vol. 265, No. 2, (July 2004), pp. 207-219, ISSN 0926-860X

Fishcer, N., van Steen, E., Claeys, M. (2011), Preparation of Supported Nano-sized Cobalt Oxide and fcc Cobalt Crystallites. *Catalysis Today*, Vol. 171, No. 1, (August 2011), pp. 174-179, ISSN 0920-5861

Hayashi, H., Chen, L. Z., Tago, T., Kishida, M. & Wakabayashi, K. (2002). Catalytic Properties of Fe/SiO_2 Catalysts Prepared using Microemulsion for CO Hydrogenation. *Applied Catalysis A.*,Vol. 231, No.1-2, (May 2002), pp. 81-89, ISSN 0926-860X

Heranz, T., Rojas, S. Perez-Alonso, F.J., Ojeda, M., Terreros, P. & Fierro, J. L .G. (2006). Carbon Oxide Hydrogenation over Silica-supported Iron-based Catalysts Influence of Preparation Route. *Applied Catalysis A.,* Vol. 308, No.1, (May 2006), pp. 19-30, ISSN 0926-860X

Hyeon T. (2003). Chemical Synthesis of Magnetic Nanoparticles. *Chemical Communications,* No. 8, (November 2002) pp. 927-934, ISSN 1359-7345

Jiao, L. & Regalbuto, J. R. (2008). The Synthesis of Highly Dispersed Nobel and Base Metals on Silica Via Strong Electrostatic Adsorption: 1 Amorphous Silica. *Journal of Catalysis,* Vol. 260, No.2, (December 2008), pp. 329-341, ISSN 0021-9517

Khodakov, A.Y., Chu, W. & Fongarland, P. (2007). Advances in the Development of Novel Cobalt Fischer-Tropsch Catalysts for Synthesis of Long-Chain Hydrocarbon and Clean Fuels. *Chemical Reviews,* Vol. 107, No.7, (July 2006), pp. 1692-1744, ISSN 1520-6890

Ling, C.K., Synthesis and characterization of Co/SiO_2 prepared via strong electrostatic adsorption (SEA) method, *Postgraduate Symposium,* Malaysia, 2011

Ling, C.K., Mohd Zabidi, N.A. & Mohan, C. (2011). Synthesis of Cobalt Nanoparticles on Silica Support using Strong Electrostatic Adsorption Method. *Journal Defect and Diffusion Forum,* Vol. 312-315, (April 2011), pp. 370-375, ISBN:978-3-03785-117-3

Malik, M. A., Wani, M. Y., & Hashim, M, A. (2010). Microemulsion Method: A Novel Route to Synthesize Organic and Inorganic Nanomaterials. *Arabian Journal Chemistry (in Press: Corrected Proof) (October 2010)*

Mohd Zabidi, N. A. (2010). Supported Nanoparticles for Fuel Synthesis, In: *Carbon and Oxide Nanostructures Synthesis, Characterisation and Applications, Advanced Structured Materials 5,* Yahya, N, pp. 245-262, Springer-Verlag, ISBN 978-3-642-14672-5, Heidelberg, Germany

Moodley, D.J. (2008). On the Deactivation of Cobalt-based Fischer-Tropsch Synthesis Catalysts. *PhD Thesis,* Eindhoven University of Technology, The Netherlands.

Prieto, G., Martinez, A., Concepcion, P. & Moreno-Tost, R. (2009). Cobalt Particle Size Effects in Fischer-Tropsch Synthesis: Structural and *in-situ* Spectroscopic Characterisation on Reverse Micelle-Synthesised Co/ITQ-2 Model Catalysts. *Journal of Catalysis.* Vol. 266, No. 1 (August 2009), 129-144, ISSN 0021-9517

Saib, A.M., Borgna, A., J. van de Loosdrecht,A. van Berge, P.J., Geus, J.W. & Niemantsverdriet, J.W. (2006). Preparation and Characterisation of Spherical Co/SiO_2 Model Catalysts with Well-Defined Nano-sized Cobalt Crystallites and a Comparison of Their Stability Against Oxidation with Water. *Journal of Catalysis,* Vol. 239, No.2 (April 2006), pp. 326-339, ISSN 0021-9517

Sarkar, A., Seth, D., Dozier, A.K., Neathery, J.K., Hamdeh, H.H. & Davis, B.H. (2007). Fischer–Tropsch Synthesis: Morphology, Phase Transformation and Particle Size Growth of Nano-scale Particles, *Catalysis Letter,* Vol. 117, No.1, (July 2007), pp. 1-17, ISSN 1572-879X

Steynberg, A.P., Espinoza, R.L., Jager, B. & Vosloo, A.C. (1999). High Temperature Fischer-Tropsch Synthesis in Commercial Practice. *Applied Catalysis A.,*Vol. 186, No. 1-2, (October 1999),pp. 41-54, ISSN 0926-860X

Sun, S. & Zeng, H. (2002) Size-Controlled Synthesis of Magnetic Nanoparticles. *Journal of American Chemical Society,* Vol. 124, pp 8204-8205, ISSN 0002-7863

Tasfy, S. H. (2011) Performance Characterization of Supported Iron Nanocatalyst in Fischer-Tropsch Reaction, *MSc Thesis,* Universiti Teknologi PETRONAS, Malaysia

Trepanier, M., Dalai, A.K., Abatzoglou, N. (2010). Synthesis of CNT-supported Cobalt nanoparticle Catalysts using a Microemulsion Technique: Role of Nanoparticle size on Reducibility, Activity and Selectivity in Fischer-Tropsch reactions. *Applied Catalysis A.*, Vol. 374, No.1-2, (February 2010), pp. 79-86, ISSN 0926-860X

12

Nanoparticles Preparation Using Microemulsion Systems

Anna Zielińska-Jurek, Joanna Reszczyńska,
Ewelina Grabowska and Adriana Zaleska
Gdansk University of Technology
Poland

1. Introduction

A wide range of techniques have been developed for the preparation of nanomaterials. These techniques include physical methods such as mechanical milling (Arbain et al., 2011) and inert gas condensation (Pérez-Tijerina et al., 2008), along with chemical methods such as chemical reduction (Song et al., 2009), photochemical reduction (Ghosh et al., 2002), electrodeposition (Mohanty, 2011), hydrothermal (Hayashi & Hakuta, 2010), and sol-gel synthesis (Sonawane & Dongare, 2006). Among all chemical methods the microemulsion has been demonstrated as a very versatile and reproducible method that allows to control over the nanoparticle size and yields nanoparticles with a narrow size distribution (Lopez-Quintela, 2003).

Microemulsions are homogeneous in macroscale and microheterogeneous in nanoscale dispersion of two immiscible liquids consisting of nanosized domains of one or both liquids in the other, stabilized by an interfacial film of surface active molecules. The essential distinction between normal emulsion and microemulsion is their particle size and stability. Normal emulsions age by coalescence of droplets and Ostwald ripening. Microemulsions are thermodynamically stable, single optically isotropic and usually form spontaneously. Microemulsions have ultralow interfacial tension, large interfacial area and capacity to solubilize both aqueous and oil-soluble compounds. Depending on the proportion of various components and hydrophilic–lypophilic balance (HLB) value of the used surfactant microemulsions can be classified as water-in-oil (W/O), oil-in-water (O/W) and intermediate bicontinuous structural types that can turn reversibly from one type to the other. The dispersed phase consists of monodispersed droplets in the size range of 5 - 100 nm. The nanodroplet size can be modified by varying concerned parameters, e.g. the type of stabilizer, continuous phase, the precursor content dissolved within the nanodroplets, and the water content, referred to as molar ratio of water to surfactant (*W*). In addition the stability of the microemulsion can be influenced by addition of salt, concentration of reagents, temperature or pressure.

This chapter focuses on nanoparticles preparation using a microemulsion method, which has been employed for the preparation of particles from a diverse variety of materials, including metals (Pt, Pd, Ir, Rh, Rh, Au, Ag, Cu) (Capek, 2004), silica and other oxides (Lee et al., 2005; Fu & Qutubuddin, 2001), polymers (Krauel et al., 2005), semiconductors (Pinna et al., 2001), superconductors (Kumar et al., 1993) and bimetallic nanoparticles (Pt/Pd, Pt/Ru, Pt/Ir, Pt/Rh, Ag/Au, Ag/Cu) with a core-shell or alloy structure (Pal et al., 2007; Castillo et al., 2008). Such dynamic colloidal templates are known to produce particles of smaller size than those obtained using normal precipitation in aqueous systems.

2. Preparation of nanoparticles in microemulsion system

The preparation procedure of metallic nanoparticles in W/O microemulsion commonly consists of mixing of two microemulsions containing metal salt and a reducing agent, respectively as shown in Fig. 1a.

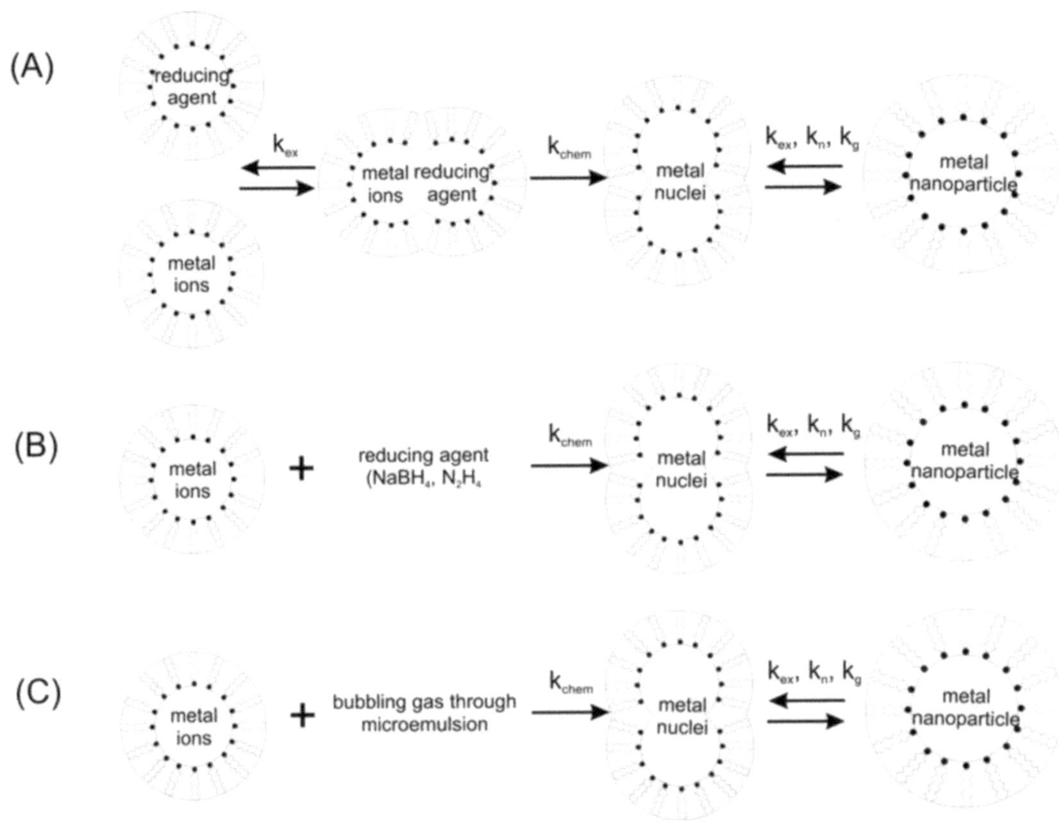

Fig. 1. Schematic illustration of nanoparticles preparation using microemulsion techniques: Particle formation steps. k_{chem} is the rate constant for chemical reaction, k_{ex} is the rate constant for intermicellar exchange dynamics, k_n is the rate constant for nucleation, and k_g is the rate constant for particle growth

After mixing two microemulsions, the exchange of reactants between micelles takes place during the collisions of water droplets result of Brownian motion, the attractive van der Waals forces and repulsive osmotic and elastic forces between reverse micelles. Successful collisions lead to coalescence, fusion, and efficient mixing of the reactants. The reaction between solubilizates results in the formation of metal nuclei. Bönnemann et al. reported that at the initial stage of the nucleation, metal salt is reduced to give zerovalent metal atoms, which can collide with further metal ions, metal atoms, or clusters to form an irreversible seed of stable metal nuclei (Bönnemann & Richards, 2001). Growth then occurs around this nucleation point where successful collision occurs between a reverse micelle carrying a nucleus and another one carrying the product monomers with the arrival of more reactants due to intermicellar exchange. The nucleation reaction and particle growth take place within the micelles and the size and morphology of as-prepared nanoparticles depend on the size and shape of the nanodroplets and the type of the surfactant, whose molecules are attached on the surface of the particles to stabilize and protect them against further growth.

Wongwailikhit et al. prepared iron (III) oxide, Fe_2O_3 using W/O microemulsion by mixing the required amount of H_2O in a stock solution of AOT in *n*-heptane. The solution was left overnight, then the concentrated NH_2OH and $FeCl_3$ were dropped into the W/O microemulsion. Suspension of Fe_2O_3 was filtered and washed with 95% ethanol and dried at 300°C for 3 h. They obtained spherical, monodisperse nanoparticles with diameter of about 50 nm. The particles size was depended on the water content in microemulsion system. The increase of particles size was obtained with increasing the water fraction in W/O microemulsion (Wongwailikhit & Horwongsakul, 2011).

Sarkar et al. prepared pure monodispersed zinc oxide nanoparticles of different shapes. Microemulsion was composed of cyclohexane, Triton X-100 as surfactant, hexanol as co-surfactant and aqueous solution of zinc nitrate or ammonium hydroxide/sodium hydroxide complex. The molar ratio of TX-100 to hexanol was maintained at 1:4. The microemulsion containing ammonium hydroxide/sodium hydroxide was added to microemulsion containing zinc nitrate and stirred. The nanoparticles were then separated by centrifuging at 15,000 rpm for 1 h. The particles were washed with distilled water and alcohol and dried at 50°C for 12 h (Sarkar et al., 2011).

Li et al. prepared nanometer-sized titania particles by chemical reactions between $TiCl_4$ solution and ammonia in microemulsion systems. In order to prepare W/O microemulsions, a cyclohexane was used as the oil phase and a mixture of poly $(oxyethylene)_5$ nonyle phenol ether (NP5, chemical purity) and poly $(oxyethylene)_9$ nonyle phenol ether with weight ratio 1:1 as the nonionic surfactant (NP5-NP9). Two microemulsion systems were prepared, containing a 0.5 M titanium tetrachloride ($TiCl_4$) aqueous solution and a 2.0 M ammonia as the aqueous phase, respectively. The oil phase, surfactant, and aqueous phase were mixed in an appropriate proportion in a beaker at 13°C in a water bath to form the microemulsion. Appropriate amounts of microemulsion I containing 0.5 M $TiCl_4$ aqueous solution and microemulsion II containing 2.0 M ammonia were mixed together, leading to the formation of insoluble titania particles. The precipitates were centrifuged, washed by the use of acetone, followed by vacuum drying for two hours. TiO_2 particles prepared in these systems had average size of about 5 nm and a narrow size distribution. TEM, DTA—TGA, and Raman spectroscopy studies indicate that their phase transition behavior is close to that of the dry gel prepared by the sol-gel method (Li & Wang, 1999).

Another method to prepare nanoparticles is from a single microemulsion as shown in Fig. 1B and 1C. One of the reactants usually a precursor of metal nanoparticles is solubilised inside reverse micelles and the second reactant (often a reducing agent) added directly to the microemulsion system. For the nanoparticles formed in single microemulsions the mechanism is based on intramicellar nucleation and growth and particle aggregation.

This method was applied for the first time by Boutonnet et al. in 1982 for preparation of Pt, Rh, Pd and Ir nanoparticles in W/O microemulsion. Metallic nanoparticles were formed in single microemulsions using hexadecyltrimethylammonium bromide (CTAB) or pentaethyleneglycole dodecyl ether (PEGDE) as a stabilizer. The precursors composed of water-soluble metal salts and hydrogen gas, bubbled through the microemulsion, or hydrazine, were added directly to the microemulsion, as the reducing agent (Boutonnet et al., 1982).

Husein et al. described that intramicellar nucleation and growth dominate when high reactant occupancy numbers are coupled with rigid surfactant layer, while intermicellar nucleation and growth dominate at low occupancy numbers and less rigid surfactant layers. At intermediate values of occupancy the number and surfactant layer rigidity, both intramicellar and intermicellar nucleation and growth contribute to the final particle size and polydispersity (Husein & Nassar, 2008).

Sanchez-Dominguez et al. prepared Pt, Pd and Rh nanoparticles by an oil-in-water microemulsion reaction method. The microemulsion containing metal precursor (Pt-COD, Pd-AAc, Rh-COCl) was prepared by mixing appropriate amounts of surfactant, cosurfactant(s), oil phase and deionized water. The used systems: water/ Tween 80/Span 20/1,2-hexanodiol/ethyl oleate (System A); water/Brij 96V/butyl-S-lactate (System B) and water/Synperonic 10/5/isooctane (System C). Then to the solutions, under vigorous stirring at 25°C, a small amount of an aqueous solution of sodium borohydride was added (Sanchez-Dominguez et al., 2009).

3. Effects of the parameters on the formation of nanoparticles in microemulsion

The formation of nanoparticles in the microemulsion system is a strong function of the intermicellar exchange, which is denoted by the intermicellar exchange rate coefficient (k_{ex}) and affected by many factors such as: the type of continuous phase, the precursor content dissolved within the nanodroplets, and the water content, referred to as molar ratio of water to surfactant (W). The high exchange rate between the micelles yields large numbers of nanoparticles with a relatively small diameter. On the contrary, slow exchange of materials between the micelles leads to formation of a few numbers of nuclei and results in larger final particle size.

Bagwe & Khilar studied the influence of smaller particle size. The micellar exchange rates increase as the chain length of the oil increases from 10^6 $M^{-1}s^{-1}$ for cyclohexane and 10^7 for *n*-heptane to 10^8 $M^{-1}s^{-1}$ for decane. Silver nanoparticles decrease as the chain length of the oil increases from 5.4 nm for cyclohexane and 6 nm for decane to 22 nm for *n*-heptane. Less bulky solvent molecules with lower molecular volumes, such as *n*-heptane or cyclohexane, can penetrate more easily in the surfactant layer, resulting in additional interfacial area and interfacial rigidity. In addition the plasmon absorption peak shifts toward longer wavelengths (red shift) as the particle size increases and the chain length of the oil decreases from decane to *n*-heptane to cyclohexane (Bagwe & Khilar, 2000).

Mori et al. reported that the particle size produced in the cyclohexane is slightly smaller than that in the octane system, although the difference in microemulsion sizes is large at the same molar ratio water to surfactant (*W*). The change in the growth depends on the chain length of the solvent molecules. With increasing of the chain length, the alkane molecules become increasingly coiled and their penetration in the surfactant layer becomes difficult. Thus, the interaction between the surfactant and solvent molecules decreases with an increase in the chain length of the alkane molecules. As a result, the micellar exchange rate increases with the chain length. On the other hand, short chain alkane and cyclohexane molecules can easily penetrate the surfactant layer to generate additional interfacial rigidity. Thus, the micellar exchange rate would decrease, which further affects the formation of silver atoms (Mori et al., 2001).

Petit et al. also reported that larger silver particles were formed in isooctane, bulkier with a larger molecular volume solvent than in cyclohexane (Petit et al., 1993).

The second studied parameter which can influence the nanoparticle size and shape is the type of the surfactants and the addition of the co-surfactants. The surfactants consist of two main entities, a hydrophilic head group and a hydrophobic (or lypophilic) tail group, which form soft aggregates in solvents and are held together by van der Waals and ionic forces. The surfactant acts as a stabilizing agent, effectively dispersing the obtained nanoparticles in the solution, providing sites for the particle nucleation and preventing aggregation of the nanoparticles. In W/O microemulsions surfactants form reverse micelles, nano-sized water pools dispersed within the bulk organic solvent which act as nanoreactors for the chemical reduction of the metallic precursors and metallic nanoparticle preparation.

For the most surfactant-mediated synthesis, the connection between morphology of the surfactant aggregates and the resulting particle structure is more complex (than simply relating the average size and shape of the micelles to the size and shape of the precipitated particles). These molecular-level variables are subject to change with macroscopically manipulated experimental conditions, as shown in Table 1.

Macroscopic parameters	Nano-sized parameters
Identity of included chemical species	Static, size and shape of micelles
Microemulsion composition	Aggregation number
Water-to-surfactant molar ratio	Dynamic interaction, rates and types of merging and dissociation of micelles
pH	Distribution of charged entities around dispersed particles
Ionic strength	
Dissolved species concentrations	Surfactant film curvature and head-group spacing
Method and rate of introduction of species	Effective Coulumb repulsion potential
Temperature and pressure	
Aging time	Van der Waals, hydrogen and hydrophobic interactions
Method and rate of stirring	
Homogenous or heterogeneous nucleation	Screening length

Table 1. Macroscopic and nanoscopic variables in the microemulsion-assisted and particularly reverse-micellar preparation method of nanoparticles (Uskokovic & Drofenik, 2007)

Usually, many surfactants can be used to form microemulsion, including cationic surfactants such as cetyltrimethylammonium bromide (CTAB), anionic surfactants such as bis(2-ethylhexyl)sulfosuccinate (AOT), sodium dodecyl benzene sulfonate (SDBS) and lauryl sodium sulfate (SDS), and nonionic surfactants such as Triton X-100 and sorbitan monooleate Span 80, nonylphenyl ether (NP-5) or polyoxyethylene (9) nonylphenyl ether (NP-9).

The most commonly used surfactant for the formation of reverse micelles is the sodium bis(2-ethylhexyl) sulfosuccinate, also known as Aerosol-OT or AOT, seen in Fig. 2. AOT is a twin tailed, anionic surfactant with a sulfosuccinate head group stabilized as a salt by a sodium cation. The AOT molecule has an inverted conical shaped structure and has proven to be an effective emulsifier, thus finding a wide range of applications as well as numerous intensive studies. The surfactant layer acts as steric stabilizer to inhibit the aggregation of nanoparticles formed. The microemulsion formed by AOT is made up of three kinds of components AOT, water, alkane (without addition of co-surfactants). In this system, micelles consist of a hydrophilic core compartmentalised by the hydrophilic head group of the AOT, forming a "water-pool" characterized by the molar ratio of water to surfactant (W= [H_2O]/[AOT]) and with the hydrophobic alkyl tails extending into the nonpolar continuous phase solvent.

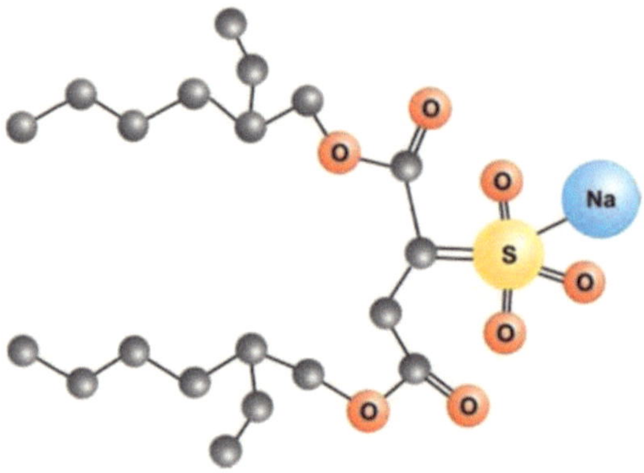

Fig. 2. Model of surfactant molecule bis(2-ethylhexyl)sulfosuccinate (AOT)

AOT microemulsion has extensively been applied for preparation of metallic nanoparticles (Pt, Pd, Cu, Ag, Au, Ni, Zn), metal sulfides (CdS, ZnS) and metal oxides (TiO_2, SiO_2). The resultant particles have high stability, small particle size, and good monodispersity. Due to its higher solubility in organic phase AOT helps to extract metal cations from the aqueous to reverse micellar phase. In addition the particles formed in AOT microemulsion have relatively strong electrostatic interactions with the negatively charged head polar group of AOT molecules, which comes into being a protective effect against aggregation (Zhang et al., 2007).

In the first papers by Lisiecki & Pileni on the preparation of metallic nanoparticles, it was demonstrated that both the bulk solvent and W value have an effect on the preparation of copper and silver nanoparticles in the AOT microemulsion system (Lisiecki & Pileni, 1995).

Table 2 gives a brief summary of various parameters which influence on the size and shape of nanoparticles prepared using the microemulsion method. By going through Table 2, one can observe that AOT has been reported to be the most suitable surfactant for silver nanoparticles preparation, and water-to-surfactant ratio is a crucial parameter in setting up the final size of the obtained nanoparticles.

Zhang et al. (Zhang et al., 2007; Zhang et al., 2006) reported that silver nanoparticles prepared in AOT microemulsion have a smaller average size and narrower size distribution compared to the particles prepared using cationic or nonionic surfactant in the microemulsion system (Zhang et al., 2007). In addition, due to the adsorption of AOT molecules onto the particles, the resultant sol can be preserved for a long time without precipitation and it easily transfers the obtained nanoparticles into nonpolar solvents (Zhang et al., 2006).

TX-100 (Triton X-100) another commonly used surfactant for preparation of nanoparticles in W/O microemulsion has extensively been applied for preparation of monodisperse gold, nickel, copper, and semiconductors nanoparticles such as titanium dioxide and silica but as performed reseraches have shown this surfactant is not sufficient for stabilization of silver nanoparticles in the microemulsion system. This nonionic surfactant reveals strong solvent dependence for the formation of reverse micelles. Because of the presence of long polyoxyethylene group the polar interior of the reverse micellar aggregate would be of different nature; it is also expected that the TX-100 system would show a different phase behavior compared to AOT molecules.

Spirin et al. compared stability of the gold colloids prepared in W/O microemulsion system. They found that gold nanoparticles formed in water/Triton X-100/hexane microemulsions were much smaller than those obtained in AOT-based microemulsion system. It was suggested, that for nonionic surfactant TX-100 gold nanoparticles were formed in the micelle shell rather than in the water pool. In the shell, the gold clusters were stabilized by oxyethylene groups of TX-100 molecules. In addition 0.1 M 1-hexanol used as a co-surfactant in water/TX-100/cyclohexane system, could decrease the particle size (Spirin et al., 2005). For gold nanoparticles obtained in reverse micelles of the water/AOT/cyclohexane system, particles could grow through intermicellar exchange followed by agglomeration. The exchange rate for AOT-based microemulsion is higher than for nonionic TX-100-based microemulsion. Therefore, the collision probability between particles in AOT-stabilized microemulsion can be higher, and as a consequence, gold particles can agglomerate easier than in nonionic TX-100 microemulsion.

Hong et al. investigated the effect of concentration of surfactants and the hydrophilic group chain length of surfactant on the physical properties of nanosized TiO_2 particles. Non-ionic surfactants - Brij 52 (polyoxyethylene glycol hexadecyl ether polyethylene 2-cetyl ether), Brij 56 (polyoxyethylene glycol hexadecyl ether polyethylene 10-cetyl ether), Brij 58 (polyoxyethylene glycol hexadecyl ether polyethylene 20 cetyl ether) were employed in this work. For the Brij series, head group size increases from Brij 52 to Brij 58 (average number of oxyethylene groups increases from 2 to 20), but with a constant tail length (average number of hydrophobic carbons is 16). They have found that the photocatalytic activity and also the size of the particles increased with an increase of hydrophilic group chain length (Hong et al., 2003).

Solanki et al. studied the effects of reaction parameters, including water-to-surfactant ratio (W), type of continuous oil phase in water/AOT/cyclohexane microemulsion system. They found that silver nanoparticles were smaller and narrower in the size distribution at lower water content than that obtained at higher W value. When W values increased from 5 to 8 the particle size increased from 4–9 nm to 50–58 nm (Solanki et al., 2010).

Chen et al. prepared platinum ultrafine particles by the reduction of H_2PtCl_6 with hydrazine in AOT/isooctane microemulsion system. They have found that the hydrodynamic diameters of reverse micelles measured by DLS increased with the increase of W values. When the aqueous phase was the solution of 0.1 M H_2PtCl_6 or 1.0 M hydrazine, it was observed that the reverse micellar sizes were smaller than those obtained using water as the aqueous. It could be attributed to the fact that the dissociation of H_2PtCl_6 or hydrazine in solution caused the increase of ionic strength, which reduced the repulsion between the head groups of ionic surfactant and led to the formation of smaller reverse micelles (Chen et al., 1999). The formation rate was faster at larger W values. This could be attributed to the fact that the number of nuclei formed in aqueous phase increased as the W value increased at constant AOT and H_2PtCl_6 concentrations, leading to the increase in the formation rate.

Inaba et al. prepared titanium dioxide nanoparticles in microemulsion system composed of water, Triton X-100 and isooctane. The TiO_2 nanoparticles showed monodispersity, a large surface area and high degrees of crystallinity and thermostability. The particle size of TiO_2 was controlled by changing the water content of the reverse micellar solution (Inaba et al., 2006).

We have earlier investigated the effect of various surfactants on silver and gold nanoparticle size deposited on TiO_2 surface using the microemulsion system. We have found that the primary gold particles size was dependent on the type of the surfactant (anionic AOT, nonionic TX100 or nonionic Span 80) and reducing agent (N_2H_4, $NaBH_4$), which was used for stabilization and reduction of gold ions in the microemulsion system. For nonionic surfactant Triton X-100 or Span 80 smaller gold nanoparticles with the diameter of about 20 nm were obtained. Our results proved that the size of gold is the key-factor for high level activity under visible-light irradiation. When anionic AOT (sodium bis-2-ethylhexyl-sulfosuccinate) was used, the size of gold nanoparticles averaged 70 nm and it was beneficial related to higher photocatalytic activity observed for larger gold nanoparticles deposited on titania surface. We have also found that the decrease of water to surfactant molar ratio (*W*) from 6 to 3 during preparation in water/AOT/cyclohexane system resulted in the decrease in gold particle size from 68 to 38 nm. The decrease in W value favored the formation of smaller water droplets and led to the decrease in the average gold particles size. Titania size added during precipitation of noble metal nanoparticles can also slightly influence the size of gold in the microemulsion method. Due to the presence of hydroxyl groups on the TiO_2 surface, TiO_2 possesses highly hydrophilic properties which influence micelles water pools properties. The smaller titania particles were used, the smaller metal particles were obtained in W/O microemulsion system. It is expected that smaller particles with higher specific surface area exert more influence on the surface tension of micelle shell than larger TiO_2 nanoparticles. We have also found that gold particles obtained in the air atmosphere were large and polydisperse due to oxidized for $[Au(OH)_4]^-$ formation. Therefore, the obtained Au-TiO_2 photocatalysts prepared in the air atmosphere revealed lower photocatalytic activity in visible region compared to the samples obtained in argon or nitrogen atmosphere (Zielińska-Jurek et al., 2011).

Ganguli et al. discussed the effect of various surfactants on the morphology of nanomaterials. They reported that cationic surfactants lead to anisotropy and preparation of nanorods of several divalent metal carboxylates. They explained that an isotropic growth occurs when using the non-ionic surfactant TX-100 leading to spherical nanoparticles of (~5 nm). On changing the non-ionic surfactant from TX-100 to Tergitol, larger cubes of size of about 50 nm were formed. They explained that the positively charged surfactants assemble on the surface of the negatively charged nickel oxalate and thus favor the anisotropic growth (rods). In the absence of such positively charged surfactants, an isotropic growth leads to spheres and nanocubes. Thus the choice of the surfactant becomes critical to the size, shape and stability of the nanoparticles obtained in microemulsion (Ganguli et al., 2010). In a mixed cationic-anionic surfactant solution, such as the synergism, the formation of a worm-like micelle is favored, which directs the growth of the one-dimensional nanostructure (Petit et al., 1993).

Another important factor which determines size and shape of nanoparticles is the addition of electrolytes. Generally addition of salt influences the degree of dissociation of emulsifier, the solubility of emulsifier in the aqueous phase and the micelle aggregation number. In the interesting review on preparation of silver nanoparticles Zhang et al. summarized that the presence of electrolyte is favorable for the formation of silver nanowires (Zhang et al., 2007).

Chiang et al. reported that addition of 2.5 M of NaCl in the water pool of reverse micelles in water-AOT-isooctane microemulsion induces a marked change in the particle shape with appearance of cylinders, trigons and cubics (Chiang et al., 2004). They have also found that the control of the final Au nanoparticle size and shape was related to the molar ratio of the reduction agent to the precursor and the sequence of the addition of metal salt into the mixed reverse micelles. A decrease in the molar ratio of the reduction agent to the precursor and direct injection of precursor to mixed reverse micelles containing the reduction agent resulted in the formation of anisotropic gold nanoparticles, such as cylinders and trigons (Chiang et al., 2004).

Jang et al. reported that the diameter of polypyrrole nanotubes were affected by different microemulsion parameters such as, the weight ratio of aqueous $FeCl_3$ solution/AOT, type of nonpolar solvent, and reaction temperature. Polypyrrole nanotubes were formed through chemical oxidation polymerization in microemulsion consist of 74.0 wt % of hexane, 22.4 wt % of AOT and 3.6 wt % of aqueous $FeCl_3$ solution at 15°C, as shown in Fig. 3. It was found that an aqueous $FeCl_3$ solution determines the formation of rod-shaped AOT micelles by decreasing the second critical micelle concentration responsible for structural transitions of spherical micelles and increasing the ionic strength of the solvent. The average diameter of polypyrrole nanotubes was approximately 94 nm and their length was more than 2 μm (Jang & Yoon, 2005).

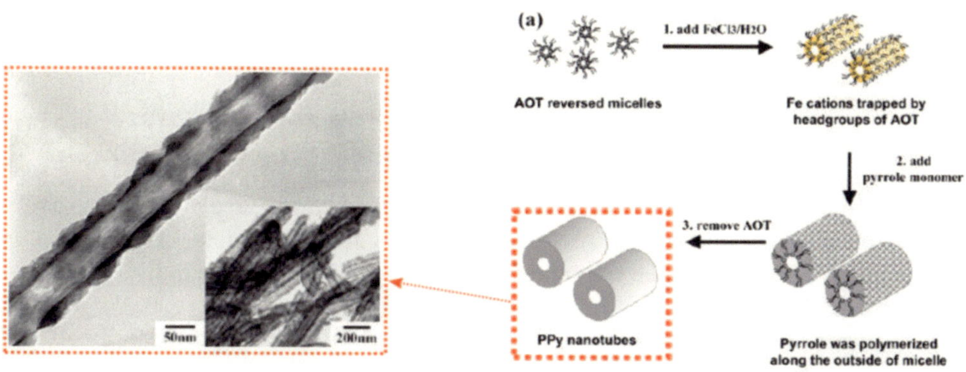

Fig. 3. TEM image and schematic route for the preparation of polypyrrole nanotubes in W/O microemulsion (Jang & Yoon, 2003)

Metal/ Metal oxides	Microemulsion system (surfactant /co-surfactant/oil)	Precursor	Reducing Agent	Particle diameter and shape	Parameters determining the size and shape of nanoparticles	References
Ag	Brij 30/AOT/ *n*-heptane	$AgNO_3$	Brij 30	3-6 nm and 20-30 nm spherical nanoparti-cles	Weight ratios [Brij30]/([AOT]+[Brij30])=1, 0.75, 0.5, 0.25,0 Aqueous solution contained 1, 2, or 3 wt% $AgNO_3$	(Andersson et al., 2005)
Ag	AOT/ cyclohexane AOT/ isooctane	$AgNO_3$	$NaBH_4$	5-55 nm	Water to surfactant molar ratio *W*=5 or 8 Molar ratio of reducing agent to reagent (R=0.66-1.5)	(Solanki et al. 2010)
Ag	SDS/isoamyl alcohol/ cyclohexane	$AgNO_3$	NH_2OH	6.5 nm spherical nanoparti-cles	Molar ratio of N_2H_4/Ag^+ =3/1 Molar ratio of W =H_2O/SDS = 5	(Zhang et al. 2008)
Pt, Pd, Rh	Tween 80, Span 20/ 1,2-hexanodiol/ ethyl olate Brij 96V/ butyl-S-Lactate Synperonic-10/isooctane	Pt-COD, Pd-AAc, Rh-COCl	$NaBH_4$	3.1-6.3 nm undefined shape	molar ratio of $NaBH_4$/metal =2/1 Oil phase content (5-20 %wt.) Surfactant content (9-25 %wt.) Water content (60-83,5 %wt.) Metal content (0.08-2.58 [g/kg of microemulsion])	(Sanchez-Dominguez et al., 2009)

Table 2. Survey of recent publications on metal/metal oxide particles prepared using microemulsion method.

4. Bimetallic, core-shell nanoparticles and alloys

Bimetallic composite nanoparticles, composed of two different metal elements, are of greater interest than monometallic nanoparticles from both scientific and technological point of view. The structure of bimetallic nanoparticles is defined by the distribution modes of the two elements and can be oriented in random alloy, alloy with an intermetallic compound, cluster-in-cluster and core-shell structures. The bimetallic nanoparticles have unique catalytic, electronic, and optical properties distinct from those of the corresponding metallic particles. The structure of bimetallic combinations depends on the preparation conditions, miscibility, and kinetics of the reduction of metal ions.

Silver and gold have almost identical lattice constants (0.408 for Au and 0.409 for Ag) and are completely miscible over the entire composition range, which leads to a strong tendency toward alloy formation. Hence single-phase alloys can be achieved with any desired composition and the absorption spectra of alloy nanoparticles exhibit only one surface plasmon band, whose absorption maximum depends on the alloy composition. The optical properties of metal nanoparticles depend on surface plasmon resonance. The origin of surface plasmon resonance in noble metal nanoparticles is the free conduction electrons in the metal surface (d electrons in silver and gold). The mean free path of the electrons in the gold and silver is 50 nm. Therefore, for the particles smaller than 50 nm, no scattering is expected from the bulk. Hence all the spectral properties are the function of the surface and not of bulk. According to Mie theory the total extinction coefficient of small metallic particles is the summation of all electronic and magnetic multipole oscillations, contributing to the absorption and scattering of the interacting electromagnetic field. Now for the particles much smaller than the wavelength of the absorbing light only the dipole term is assumed to contribute to the absorption. The electric field of the incoming electromagnetic radiation induces the formation of a dipole in the nanoparticles. A restoring force in the nanoparticles tries to compensate for this, resulting in a unique resonance wavelength. Alloy nanoparticles have received a special attention due to the possibility of tuning their optical and electronic properties over a broad range by simply varying the alloy composition. The characteristics of the bimetallic nanoparticles are listed in Table 3.

Xia et al. obtained well-dispersed Ag/Ni core-shell nanoparticles using reduction of silver nitrate and nickel nitrate with sodium borohydride in water-in-oil (W/O) microemulsions of water/polyoxyethylene (4) nonylphenol (OP-4) and polyoxyethylene (7) nonylphenol (OP-7)/*n*-heptane. To prepare Ag/Ni nanoparticles the required amount of sodium hydroxide aqueous solution was added to the mixture of nickel nitrate and silver nitrate (in a molar ratio of 2:1) aqueous solutions to form fine precipitates of Ag_2O and $Ni(OH)_2$. Ammonia solution was added drop by drop to this mixture until the precipitates completely dissolved to form $[Ag(NH_3)_2]^+$ and $[Ni(NH_3)_6]^{2+}$, then pure water was added. Next, under continuous mechanical stirring, OP-4, *n*-heptane, OP-7 was added, into the above $[Ag(NH_3)_2]^+$ and $[Ni(NH_3)_6]^{2+}$solution, and remained for 30 min in water-bath at 25°C. The second microemulsion was prepared in a similar way, except that metal precursor was replaced by sodium borohydride aqueous solution. The first microemulsion was added dropwise into the second one, and left for 1 h in water-bath at 25°C under continuous mechanical stirring. The products were washed several times with water and anhydrous ethanol, and then dried under vacuum at room temperature for 4 h (Xia et al., 2010).

Ahmed et al. obtained bimetallic Cu/Ni nanoparticles using CTAB (cetyltrimethyl ammoniumbromide) as the surfactant, 1-butanol as the co-surfactant and isooctane as the oil phase and hydrazine/NaOH as the precipitating agents followed by the reduction in hydrogen atmosphere. They prepared four microemulsions with different aqueous phases containing $Cu(NO_3)_2 \cdot 3H_2O$, $Ni(NO_3)_2 \cdot 6H_2O$, $N_2H_4 \cdot H_2O$ and NaOH. Microemulsions A and B were created by mixing two solutions, A- containing metal precursors, B – reducing agents. The obtained green precipitate was washed with 1:1 chloroform/methanol mixture and dried in the air, then heated to 500°C for 5 h. (CuNi). They had also prepared $CuNi_3$ and Cu_3Ni nanoparticles using the same method (Ahmed et al., 2008).

Chemical composition of nanoparticles	Preparation method	Precursor	Reducing agent	Particle diameter and shape	Parameters determining the size and shape of nanoparticles	References
Ag/Ni nanoparticles	To prepare Ag/Ni nanoparticles the required amount of sodium hydroxide aqueous solution was added to the mixture of nickel nitrate and silver nitrate (in a molar ratio of 2:1) aqueous solutions to form fine precipitates of Ag_2O and $Ni(OH)_2$. Ammonia solution was added drop by drop to this mixture until the precipitates completely dissolved to form $[Ag(NH_3)_2]^+$ and $[Ni(NH_3)_6]^{2+}$, then pure water was added. Ag/Ni bimetallic nanoparticles were prepared by dropwise addition microemulsion containing metal precursors in water cores into the microemulsion containing the reducing agent ($NaBH_4$)	$AgNO_3$ $Ni(NO_3)_2$	$NaBH_4$	50-100 nm core-shell structure and spherical shape The gold shells formed on these Ni cores were driven by the spontaneous reaction between the Ni atoms on the surface of the Ni nanoparticles and subsequently added Au^{3+} ions	Molar ratio H_2O/surfactant = 3-11 The molar ratio of Ni to Au = 0.25 - 0.75. For higher dosage of Ni^{2+} and Ag^+ good core–shell structures were obtained Adjusting the dosage of Ni^{2+} and Ag^+ led to the percentage of large size particles increased	(Xia et al., 2010)
Ag/Au nanoparticles	Ag/Au nanoparticles were prepared by the microemulsion method using Triton X-100 as a surfactant, cyclohexane as oil phase and 1-hexanol as a co-surfactant. Reverse micelle systems containing a reducing agent and a metal compound solution were mixed under stirring	$AgNO_3$ $HAuCl_4$	$NaBH_4$	20-30 nm undefined shape	Molar ratio H_2O/surfactant = 1-7 Metal ion concentration = 0.1-0.05 M Ag and Au ion concentration	(Pal et al., 2007)
Ni/Au nanoparticles	$NiCl_2$ aqueous solution to Brij30 and n-octane solution. The second microemulsion containing $NaBH_4$ with the proportions identical to those in the first one was then added dropwise. The reaction was pertained under argon atmosphere. The gold shells formed on these Ni cores were driven by the spontaneous reaction between the Ni atoms on the surface of the Ni nanoparticles and subsequently added Au^{3+} ions. The reverse microemulsion solution containing $HAuCl_4$ was added drop by drop to the earlier prepared microemulsion containing Ni^{2+}	$NiCl_2$ $AuCl_4$	$NaBH_4$	8 nm core-shell structure	Molar ratio of Ni to Au =0.25 - 0.75 Molar ratio W =H_2O/surfactant = 3 - 11	(Chen et al., 2009)

Chemical composition of nanoparticles	Preparation method	Precursor	Reducing agent	Particle diameter and shape	Parameters determining the size and shape of nanoparticles	References
Ag/Au core-shell bimetallic clusters	0.8 mL of $NaBH_4$ and 5 mL of citric acid was added to $AgNO_3$ solution. A solution of 0.4 mM Ag^+ and 0.1–0.4 mM $AuCl_4^-$ was prepared by dissolving the corresponding salt in water. The reaction mixture was stirred for 15 min at 100°C, then gold metal salt solution was added to the solution of Ag nanoparticles. Eventually, heating was stopped and stirring continued for 10 min	$AgNO_3$ $HAuCl_4$	$NaBH_4$	3-4 nm core-shell bimetallic clusters the absorption spectra of bimetallic nanoparticles suggested the formation of core–shell structure	gold metal salt concentration (0.1–0.4 mM $AuCl_4$)	(Chen et al., 2006)
Ag/Au alloy nanoparticles	Ag–Au alloy nanoparticles were obtained by reduction $HAuCl_4$ and $AgNO_3$ with $NaBH_4$ in the presence of sodium citrate at room temperature. Five 125 mL flasks were filled with 100 ml of deionized water and 50 ml of 0.01 M sodium citrate. Varying mole fractions of 26 mM $HAuCl_4$ and 58 mM $AgNO_3$ were added to each solution for a total metal salt concentration of 0.005 mM.	$AgNO_3$ $HAuCl_4$	$NaBH_4$	10 nm Ag–Au alloy nanoparticles optical absorption spectra reveal that the nanoparticles have been prepared for alloy structure nanoparticles	initial Au/Ag molar ratios (0:1, 0.25:0.75, 0.5:0.5, 0.75:0.25, and 1:0)	(Chen et al., 2006)
Ag/Au nanoparticles	W/O microemulsion solutions hydrazine/ $HAuCl_4$/$AgNO_3$ Solutions containing $HAuCl_4$ and $AgNO_3$ were obtained by mixing the W/O microemulsion solution containing $HAuCl_4$ and that containing $AgNO_3$. The preparation of bimetallic nanoparticles was achieved by mixing equal volumes of two W/O microemulsion solutions at the same molar ratio of water to Aerosol OT (v_0) and concentration of Aerosol OT, one containing an aqueous solution of metal salts and the other containing an aqueous solution of hydrazine	$HAuCl_4$ $AgNO_3$	hydrazine	particle size analysis indicated that the resultant bimetallic nanoparticles were monodisperse and had a mean diameter of 4 - 22 nm, increasing with an increase in the $W=([H_2O]/[AOT])$ value and Ag content	$W=([H_2O]/[AOT])$ value and Ag content	(Chen & Chen, 2002)

Chemical composition of nanoparticles	Preparation method	Precursor	Reducing agent	Particle diameter and shape	Parameters determining the size and shape of nanoparticles	References
Ag/Au alloy nanoparticles	For the preparation of Ag–Au alloy nanoparticles $AgNO_3$ solution (0.01 M), $HAuCl_4$ solution (0.01 M) and 10% (v/v) polyacrylamide were used. To the polyacrylamide solution equal volumes of $AgNO_3$, $HAuCl_4$, hydrazine hydrate and tri-sodium citrate (3% of polyacrylamide volume) were added and the reaction was carried out under microwave for 1 min..	$HAuCl_4$ $AgNO_3$	hydrazine	histograms show that the particles obtained through these methods are in the range of 5–50 nm change in the colour of the solution indicates the formation of alloy nanoparticles with different composition	different ratios of Ag and Au ion concentrations and hydrazine hydrate as a reducing agent	(Pal et al., 2007)
Ag(Au) bimetallic core–shell nanoparticles	I - different amount of $HAuCl_4$ was added to water, 5 µl of 0.1M CTAB was added to each solution. After that, different amount of 0.1 M ascorbic acid were added to solutions. At last, 500 µl of Ag colloids was added to the three solutions. II - 10 µl of 6 mM $HAuCl_4$ and 5 µl of 0.1M CTAB were added to water, and the three solutions called D, E and F. 3 µl of 0.1 M ascorbic acid was added to each solution. Ag colloids were added to the solutions D, E and F.	$HAuCl_4$ $AgNO_3$	ascorbic acid	20-50 nm Ag(Au) bimetallic core–shell nanoparticles the amount of $AuCl_4^-$ had an effect on the shape of nanoparticles	amount of 6 mM $HAuCl_4$ (10, 5 and 20 µl) amount of 0.1 M ascorbic acid (3, 1.5 and 6 µl) amount of Ag colloids (800, 500 and 200 µl)	(Qian, & Yang, 2005)

Table 3. Survey of recent publications on bimetallic nanoparticles prepared using microemulsion method

Based on the use of W/O microemulsions we have proposed a novel method for preparation of the photocatalysts obtained by gold nanoparticles deposition or gold and silver (bimetallic nanocomposites) deposition on the TiO_2 surface.

Ag/Au-modified titanium dioxide nanoparticles were prepared by adding silver precursor into water/AOT/cyclohexane microemulsion containing gold precursor in water cores. The water content was controlled by fixing the molar ratio of water to the surfactant (*W*) at 4. Stirring was carried out for 1 h under argon and then gold and silver ions were reduced by dropwise addition of microemulsion containing a reducing agent. The addition of $NaBH_4$ or N_2H_4 to $HAuCl_4$ solution, changed the color from yellow to pinkish red for smaller nanoparticles with the diameter of about 10 nm and grey for larger nanoparticles with the diameter of about 100 nm, respectively. Finally, Ag/Au-TiO_2 nanoparticles were prepared by adding TiO_2 precursor titanium tetraisopropoxide (TIP) or powdered into the microemulsion containing gold nanoparticles in water cores. Precipitated Ag/Au/TiO_2 nanoparticles were centrifuged (2000 rpm for 5 min), washed with ethanol, acetone and water to remove the remaining surfactant, dried at 80°C for 48 h and calcinated at 450°C for 2 h.

We have found that bimetallic nanoparticles deposited on titania surface enhanced their photocatalytic activity in visible region. The photodegradation rate of model organic pollutant (0.21 mM phenol aqueous solution) under visible light equaled to 0.46 and 1.30 μmol dm^{-3} min^{-1} for TiO_2 nanoparticles modified by 1.5 mol% of Au or 4.5 mol% of Ag introduced into microemulsion system, respectively. The introduction of the same amount of both silver and gold precursor resulted in the increase of phenol degradation rate up to 3.57 μmol dm^{-3} min^{-1}. Higher silver amount was more beneficial to the photocatalytic activity of the obtained Ag/Au-TiO_2 nanoparticles than higher gold amount (Zielińska-Jurek, 2011).

In order to obtain Ag/Cu-TiO_2 nanoparticles two microemulsions were prepared by mixing the aqueous solution of metal ions (Ag^+, Cu^{2+}) and $N_2H_4 \cdot H_2O$ into the 0.2 M AOT/cyclohexane solution (Zielińska-Jurek, 2010). Cu or Ag/Cu bimetallic nanoparticles were prepared by dropwise addition microemulsion containing the reducing agent (hydrazine) into the microemulsion containing metal precursor in water cores as was shown in Fig. 4. The molar ratio of hydrazine and silver nitrate or copper nitrate was held constant at the value of 3. Water content was controlled by fixing the molar ratio of water to surfactant (*W*) at 2. Then TiO_2 precursor titanium tetraisopropoxide (0.2 M TIP) was added into the microemulsion containing metal nanoparticles. The concentration of metal precursors, which varied from 0.1 to 6.5 mol%, was related to the concentration of TIP in the microemulsion system. During the precipitation nitrogen was bubbled continuously through the solution.

The Ag/Cu-TiO_2 precipitated particles were separated, washed with ethanol and deionized water several times to remove the organic contaminants and the surfactant. The particles were dried at 80°C for 48 h and were then calcinated at 450°C for 2 h. It was found that the obtained nanocomposites contained highly and uniformly dispersed Ag/Cu nanoparticles on the TiO_2 surface. The maximum in the photocatalytic activity under visible light was observed for the sample containing 0.5 mol % of Cu and 1.5 mol% of Ag.

The rate of phenol decomposition average was 2.41 μmol dm^{-3} min^{-1} and was higher than for Ag/Cu-TiO_2 containing 1.5 mol% of Cu and 0.5 mol% of Ag. It indicates that the presence of silver was more beneficial for visible light activation of TiO_2 doped photocatalysts than a higher copper amount. The Au/Cu-TiO_2 nanoparticles revealed lower photocatalytic activity compared to Ag/Cu-TiO_2 photocatalysts containing the same metal amount loading on TiO_2 surface The photocatalysts modified with silver revealed higher antimicrobial activity than pure TiO_2 obtained in the microemulsion system or the samples containing only copper nanoparticles deposited on TiO_2 surface. It indicated that silver possesses higher antimicrobial activity than copper nanoparticles prepared in the microemulsion system. The best antimicrobial activity revealed Ag/Cu- TiO_2 with the highest silver amount average 6.5 mol% and the sample Ag/Cu-TiO_2 containing 1.5 mol% of Ag and 0.5 mol% of Cu, which exhibited also the highest efficiency of phenol degradation under visible light.

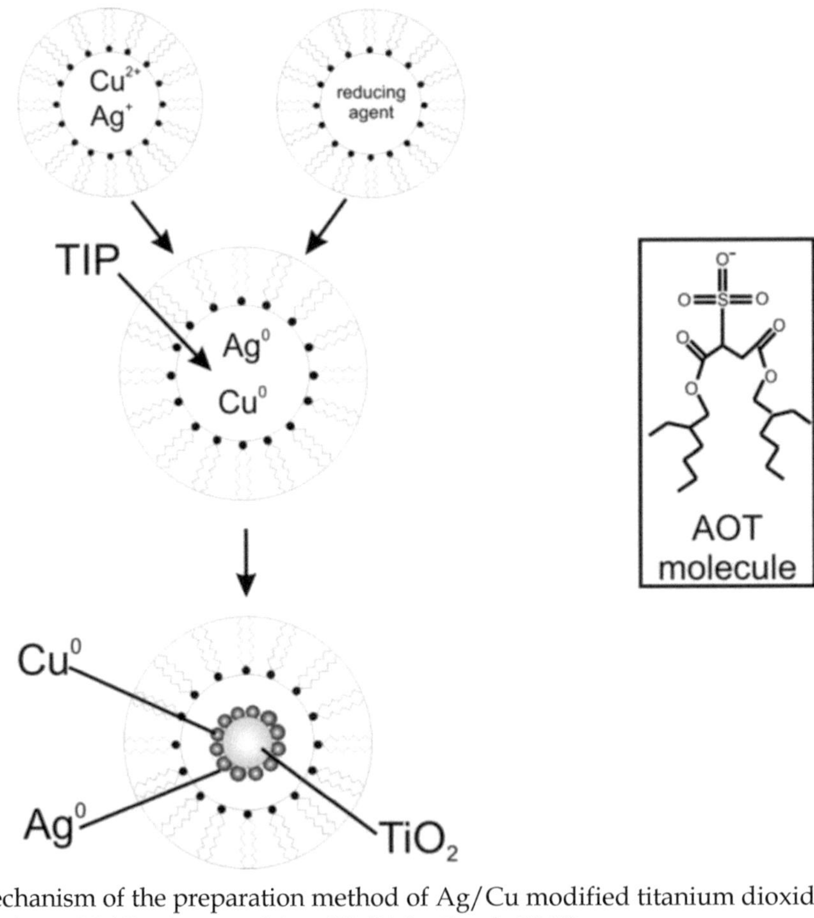

Fig. 4. Mechanism of the preparation method of Ag/Cu modified titanium dioxide nanoparticles in W/O microemulsion (Zielińska-Jurek, 2010)

5. Challenges in nanoparticles preparation using microemulsion as a template

It has been shown that microemulsion as a liquid structure having a high surface area can be used as chemical nanoreactors to obtain a wide range of nanoparticles of different chemical nature, size and shape. However, recovery of the nanoparticles and their separation from the microemulsion system, as well as the recovery and recycling of the organic solvents remains a challenge. Conventional separation techniques such as ultracentrifugation, solvent evaporation, addition of a suitable solvent to cause phase separation or precipitation (ethanol, acetone, water) have been applied to recover nanoparticles from the W/O microemulsion system.

An interesting approach is the addition of CO_2 as an antisolvent, since it provides an inexpensive, benign, nontoxic means to efficiently control the nanoparticles stability. Supercritical carbon dioxide exhibits a dielectric constant even lower than most organic solvents and can be used as non-toxic continuous phase in W/O microemulsions (Hutton, 1998). Jonhston et al. have found that an ammonium carboxylate perfluoropolyether (PFPE) surfactant can be used for stabilization of microemulsions in supercritical CO_2 (Jonhston et al., 1996). The PFPE-based microemulsion using supercritical CO_2 as continous phase were subsequently used as nanoreactors for preparation of nanoparticles. Holmes et al. prepared CdS nanoparticles in PFPE-scCO_2 microemulsion at 45°C and 345 bar. The particle size was about 1.8 nm depending on the water to surfactant molar ratio (*W*) (Holmes et al., 1999).

Dong et al. prepared CuS nanoparticles by mixing two separate water in carbon dioxide microemulsions stabilized using sodium salt of bis(2,2,3,3,4,4,5,5-octafluoro-1-pentyl)-2-sulfosuccinate (AOT). The size of copper sulfide nanoparticles ranged from 4 to 6 nm (Dong et al., 2002). Wu et al. obtained TiO_2 nanoparticles from its precursor titanium (IV) isopropoxide (TTIP) by a combination of supercritical fluid microemulsion and supercritical-drying techniques, in which TTIP hydrolized in reverse micelles (H_2O being surrounded by CO_2) formed by surfactant Zonyl FSJ using a medium of supercritical carbon dioxide. TiO_2 particles prepared in these systems have average size of about 2-7 nm (Wu et al., 2008). Ohde et al. reported that nanometer-sized silver and copper metal particles can be prepared by chemical reduction of Ag^+ and Cu^{2+} ions dissolved in the water core of water in supercritical fluid carbon dioxide microemulsion. Sodium cyanoborohydride and *N,N,N',N'*-tetramethyl-*p*-phenylenediamine were used as reducing agents for preparation of metal nanoparticles in the microemulsion. They found that diffusion and distribution of the oxidized form of the reducing agent between the micellar core and supercritical CO_2 appeared to be the rate-determining step for the formation of silver nanoparticles in this system.

Meziani et al. prepared silver nanoparticles in the Rapid Expansion of a Supercritical Solution into a Liquid SOLVent (RESOLV). Perfluorinated surfactant-stabilized water-in-CO_2 microemulsion was used to dissolve silver salt for the rapid expansion. An aqueous $AgNO_3$ solution was added to the syringe pump followed by the addition of PFPE-NH_4 to result in a W value of 10. After loading of CO_2 to the pressure of 2500 psia, the mixture in the syringe pump was equilibrated with stirring for 2 h. The rapid expansion was carried out at 4000 psia and 35°C through a 50-micron fused silica capillary nozzle into a room-temperature aqueous solution of $NaBH_4$. The room-temperature receiving solution also contained PVP polymer (5 mg/mL) as a protection agent for the formed nanoparticles. Sodium hydroxide (NaOH) was used to adjust the basicity of the room-temperature receiving solution. TEM images yielded average sizes of 3.1 nm (size distribution standard deviation 0.8 nm) for the Ag nanoparticles in the supernatant and 10 nm (size distribution standard deviation 2 nm) for the Ag nanoparticles in the precipitate (Meziani et al., 2005). Drawbacks of this technique are: the cost of using compressed CO_2, a significant limitation of using $scCO_2$ as continuous phase in microemulsion, low solubility of most solutes in neat supercritical CO_2. Therefore, different co-surfactants (1-pentanol, 2-propanol) are often used to improve the solubility for hydrophilic components.

Another interesting approach is the use of photo-destructible surfactants as an attractive alternative to commonly used surfactants (AOT, TX-100, CTAB, SDS, NP-9, etc.) for stabilization of nanoparticles in W/O microemulsion. A recent development in this field is the work of Eastoe, which demonstrated that photodestructible surfactants can be used to induce destabilization and phase separation in microemulsions. They studied the surface properties and UV-driven destruction of surfactant sodium 4-hexylphenylazosulfonate (Eastoe, 2006).

Vesperinal et al. obtained gold nanoparticles in water-in-oil –microemulsion with the addition of a photodestructible surfactant hexylphenylazosulfonate (C6PAS). They used UV light to induce destabilization of the microemulsion and flocculation of nanoparticles. The changes in dispersion stability occur owing to UV-induced breakdown of the hydrophilic hexylphenylazosulfonate into insoluble hydrophobic photoproducts (hexylbenzene and 4-hexylphenol). They observed that in the photo-induced flocculate gold particle shapes were irregular and the particle sizes were in the range 15–120 nm (Vesperinal et al., 2007). By formulation of an appropriate mixture containing the photolyzable surfactant (sodium hexylphenylazosulfonate, C6PAS) and inert surfactants Salabat et al. photoflocculated/separated silica nanoparticles obtained in the organic solvent. During photodestruction of C6PAS non-surface-active oil-soluble hexylbenzene, and weakly surface-active hexylphenol (as well as sodium sulfur salts) were obtained. In this regard, it is an important goal for advanced projects to prepare size-controlled chemically clean nanoparticles with narrow size distribution (Salabat et al., 2007).

6. Conclusions

A large number of different nanomaterials have been prepared in water-in-oil microemulsions. The generated particle sizes can be controlled by the nanodroplet size of the inner phase of the microemulsion. The size of the particles can be controlled by:

- surfactant/co-surfactant type,
- solvent type,
- concentration of the reagents,
- ionic additives,
- water/surfactant molar ratio.

Among various preparation of nanoparticles methods, the use of microemulsion is an effective route for yielding a wide range of monodisperse nanoparticles of different size and shape. We have shown that the microemulsion method is preferable since it allows studying noble metal nanoparticles of different sizes deposited on the same titania material. Our results proved that the size of gold is the key-factor for high level activity under visible-light irradiation. However, titania material also plays a crucial role in obtaining material with high activity and the degree of influence of gold size on photoactivity (Zielińska-Jurek et al., 2010). The recovery of nanoparticles and their separation from the reaction medium is still a key step when using microemulsion system for preparation of nanocomposites.

7. References

Ahmed, J.; Ramanujachary, K.V.; Lofland, S.E.; Furiato, A.; Gupta, G.; Shivaprasad, S.M. & Ganguli, A.K. (2008). Bimetallic Cu–Ni nanoparticles of varying composition ($CuNi_3$, CuNi, Cu_3Ni). *Colloids and Surfaces A: Physicochem. Eng. Aspects*, Vol.331, pp. 206–212

Andersson, M.; Pedersen, J.,S. & Palmqvist, A.E.C (2005). Silver Nanoparticle Formation in Microemulsions Acting Both as Template and Reducing Agent. *Langmuir*, Vol.21, pp. 11387- 11396

Arbain, R.; Othman, M. & Palaniandy, S. (2011). Preparation of iron oxide nanoparticles by mechanical milling. *Minerals Engineering*, Vol.24, pp. 1-9

Bagwe, R.P. & Khilar, K.C. (2000). Effects of Intermicellar Exchange Rate on the Formation of Silver Nanoparticles in Reverse Microemulsions of AOT, *Langmuir*, Vo.6, pp. 905-910

Bönnemann, H. & Richards, R.M. (2001). Nanoscopic Metal Particles 2 Synthetic Methods and Potential Applications, *Eur. J. Inorg. Chem*. 245522480

Boutonnet, M.; Kizling, J.; Stenius, P. & Maire, G. (1982). ThePreparation of monodisperse colloidal metal particles from microemulsions. *Colloids Surf.*, Vol.5, pp. 209–225

Capek, I. (2004). Preparation of metal nanoparticles in water-in-oil (wyo) microemulsions. *Advances in Colloid and Interface Science*, Vol.110, pp. 49–74

Castillo, N.; Díaz Barriga, L.; Pérez, R.; Martínez-Ortiz, M.J. & Gallardo, A.C. (2008). Structural and chemical characterization of pTX-pD1-X bimetallic nanoparticles supported on silica. *Rev. Adv. Mater. Sci.* Vol.18, pp. 72

Chen, D.; Liu, S.; Li, J.; Zhao, N.; Shi, C.; Du, X. & Sheng, J. (2009). Nanometre Ni and core/shell Ni/Au nanoparticles with controllable dimensions synthesized in reverse microemulsion, *Journal of Alloys and Compounds*, Vol.475, pp. 494–500

Chen, D.H. & Chen, C.J. (2002). Formation and characterization of Au–Ag bimetallic nanoparticles in water-in-oil microemulsions. *J. Mater. Chem.*, Vol.12, pp. 1557-1562

Chen, D.H., Yeh, J.J. & Huang, T.C. (1999). Synthesis of Platinum Ultrafine Particles in AOT Reverse Micelles. *Journal of Colloid and Interface Science*, Vol.215, pp. 159–166

Chen, H.M.; Liu, R.S.; Jang, L.Y.; Lee, J.F. & Hu, S.F. (2006). Characterization of core–shell type and alloy Ag/Au bimetallic clusters by using extended X-ray absorption fine structure spectroscopy, *Chemical Physics Letter*, Vol.421, pp. 118-123

Chiang, C.L.; Hsu, M.B. & Lai, L.B. (2004). Control of nucleation and growth of gold nanoparticles in AOT/Span80/isooctane mixed reverse micelles. *Journal of Solid State Chemistry*, Vol.177, pp. 3891–3895

Dong, X.; Potter, D. & Erkey, C. (2002). Synthesis of CuS nanoparticles in water-in-carbon dioxide microemulsions, *Ind. Eng. Chem. Res.*, Vol.41, pp. 4489-4493

Eastoe, J. (2006). Photo-destructible Surfactants in Microemulsions, *Progr. Colloid. Polym. Sci.*, Vol.133, pp. 106–110

Fu, X. & Qutubuddin, S. (2001). Synthesis of titania-coated silica nanoparticles using a nonionic water-in-oil microemulsion, *Colloids and Surfaces A: Physicochemical and Engineering Aspects*, Vol.179, pp. 65–70

Ganguli, A.K.; Ganguly, A. & Vaidya S. (2010). Microemulsion-based synthesis of nanocrystalline materials, *Chem. Soc. Rev.*, Vol.39,pp. 474–485

Ghosh, S.K.; Kundu, S. & Pal, T. (2002). Evolution, dissolution and reversible generation of gold and silver nanoclusters in micelle by UV-activation. *Bull. Mater. Sci.* Vol.25, pp. 581-582

Hayashi, H. & Hakuta, Y. (2010). Hydrothermal Synthesis of Metal Oxide Nanoparticles in Supercritical Water. *Materials*, Vol.3, pp. 3794-3817

Holmes, J. D.; Bhargava, P. A.; Korgel, B. A. & Johnston, K. P. (1999). Synthesis of cadmium sulfide Q particles in water-in-CO_2 microemulsions. *Langmuir*,Vol.15, pp. 6613-6615

Hong, S.S.; Sig Lee M. & Lee, G.D. (2003). Photocatalytic decomposition of *p*-nitrophenol over titanium dioxide prepared by reverse microemulsion method using nonionic surfactants with different hydrophilic groups *React. Kinet. Catal. Lett.*, Vol.80, pp 145-151

Husein, M.M. & Nassar, N.N. (2008). Nanoparticle Preparation Using the Single Microemulsions Scheme. *Current Nanoscience*, Vol.4, pp 370-380

Inaba, R.; Fukahori, T.; Hamamoto, M. & Ohno, T. (2006). Synthesis of nanosized TiO_2 particles in reverse micelle systems and their photocatalytic activity for degradation of toluene in gas phase *J. Mol. Catal. A: Chem.* Vol.260, pp. 247–254.

Jang, J., Yoon, H., (2003) Facile fabrication of polypyrrole nanotubes using reverse microemulsion polymerization *Chem. Commun.* Vol. *6*, pp. 720-721

Jang, J., Yoon, H., (2005) Formation mechanism of conducting polypyrrole nanotubes in reverse micelle systems, *Langmuir* Vol. *21*, pp. 11484-11489

Johnston, K. P.; Harrison, K. L.; Clarke, M. J.; Howdle, S. M.; Heitz, M. P.; Bright, F. V.; Carlier, C. & Randolph, T. W. (1999). Water in- carbon dioxide microemulsions: An environment for hydrophiles including proteins. *Science*, Vol.271, pp. 624

Krauel, K.; Davies, N.M.; Hook, S. & Rades, T. (2005). Using different structure types of microemulsions for the preparation of poly(alkylcyanoacrylate) nanoparticles by interfacial polymerization. *J. Controlled Release*, Vol.106, pp. 76-87

Kumar, P.; Pillai, V. & Shah, D.O. (1993). Preparation of Bi-Pb-Sr-Ca-Cu-O oxide superconductors by coprecipitation of nanosize oxalate precursor powders in the aqueous core of water-in-oil microemulsions *Appl. Phys. Lett.*, Vol.62, pp. 765-768

Lee, M., S.; Park, S.S., Lee, G.D., Ju, C.S. & Hong, S.S. (2005). Synthesis of TiO_2 particles by reverse microemulsion method using nonionic surfactants with different hydrophilic and hydrophobic group and their photocatalytic activity. *Catal Today*, Vol.101, pp. 283-290

Li, G.L. & Wang, G.H. (1999). Synthesis of nanometer-sized TiO_2 particles by a microemulsion method. *Nanostructured Materials*, Vol.11, pp. 663-668

Lisiecki, I. & Pileni, M.P. (1995). Copper Metallic Particles Synthesized "in Situ" in Reverse Micelles: Influence of Various Parameters on the Size of the Particles, *J. Phys. Chem.*, Vol.99, pp. 5077-5082

Lopez-Quintela, M.A. (2003). Synthesis of nanomaterials in microemulsions: formation mechanism and growth control. *Curr. Opin. Coll. Int. Sci.* Vol.8, pp. 137-144

Meziania, M.J.; Pathaka, P.; Beachama, F.; Allardb, L.F. & Sun, Y-P. (2005). Nanoparticle formation in rapid expansion of water-in-supercritical carbon dioxide microemulsion into liquid solution, *J. of Supercritical Fluids*, Vol. 34, pp. 91–97

Mohanty, U.S. (2011). Electrodeposition: a versatile and inexpensive tool for the synthesis of nanoparticles, nanorods, nanowires, and nanoclusters of metals, *J. Appl. Electrochem.*, Vol. 41, pp. 257-270

Mori, Y.; Okastu Y. & Tsujimoto, Y. (2001). Titanium Dioxide Nanoparticles Produced in Water-in-oil Emulsion. *J. Nanopart. Res.* Vol. 3, pp. 219–225

Pal, A., Shah, S. & Devi, S. (2007a). Preparation of silver, gold and silver–gold bimetallic nanoparticles in W/O microemulsion containing TritonX-100, *Colloids and Surfaces A: Physicochem. Eng. Aspects*, Vol.302, pp. 483–487

Pal, A., Shah, S. & Devi, S. (2007b). Synthesis of Au, Ag and Au-Ag alloy nanoparticles in aqueous polymer solution, *Colloid Surface A*, Vol.302, pp. 51-57

Pérez-Tijerina, E.; Gracia Pinilla, M.; Mejía-Rosales, S.; Ortiz-Méndez,U., Torres, A. & José-Yacamán, M. (2008). Highly size-controlled synthesis of Au/Pdnanoparticles by inert-gas condensation, *Faraday Discuss.*, Vol.138, pp. 353-362

Petitt, C.; Lixonf, P. & Pileni M.P. (1993). In Situ Synthesis of Silver Nanocluster in AOT Reverse Micelles. *J. Phys. Chem.*, Vol.97, pp. 12974-12983

Pinna, N.; Weiss, K.; Sack-Kongehl, H.; Vogel, W.; Urban, J. & Pileni, M.P. (2001). Triangular CdS Nanocrystals: Synthesis, Characterization, and Stability. *Langmuir*, Vol.17, pp 7982-7987

Qian, L. & Yang, X. (2005). Preparation and characterization of Ag(Au) bimetallic core-shell nanoparticles with new seed growth method, *Colloid Surface A*, Vol.260, pp. 79-85

Salabat, A.; Eastoe, J.; Vesperinas, A.; Tabor, R.F. & Mutch, K.J. (2008). Photorecovery of Nanoparticles from an Organic Solvent. *Langmuir*, Vol.24, pp. 1829-1832

Sanchez-Dominguez, M.; Boutonnet, M. & Solans, C. (2009). A novel approach to metal and metal oxide nanoparticle synthesis: the oil-in-water microemulsion reaction method, *J Nanopart Res*, Vol.11, pp. 1823–1829

Sarkar, D.; Tikku, S.; Thapar, V.; Srinivasa, R.S. & Khilar, K.C. (2011). Formation of zinc oxide nanoparticles of different shapes in water-in-oil microemulsion, *Colloids and Surfaces A: Physicochemical and Engineering Aspects*, Vol.381, pp. 123–129

Solanki, J.N. & Murthy, Z.V.P. (2010). Highly monodisperse and sub-nano silver particles synthesis via microemulsion technique, *Colloids and Surfaces A: Physicochem. Eng. Aspects*, Vol.359, pp. 31–38

Solanki, J.N.; Sengupta, R. & Murthy, Z.V.P. (2010). Synthesis of copper sulphide and copper nanoparticles with microemulsion method, *Solid State Sciences*, Vol.12, pp. 1560-1566

Sonawane, R.,S. & Dongare, M.,K. (2006). Sol–gel synthesis of Au/TiO_2 thin films for photocatalytic degradation of phenol in sunlight, *J. Mol. Cat. A*,Vol.243, pp. 68–76

Song, K.C.; Lee; S.M., Park; T.S. & Lee, B.S. (2009). Preparation of colloidal silver nanoparticles by chemical reduction method. *Korean J. Chem. Eng.* Vol.26, pp. 153-155

Spirin, M.G.; Brichkin, S.B. & Razumov, V.F. (2005). Synthesis and Stabilization of Gold Nanoparticles in Reverse Micelles of Aerosol OT and Triton X-100. *Colloid J.* Vol.67, pp. 485–490

Uskokovic, V. & Drofenik, M. (2007). Reverse micelles: inert nano-reactors or physic-chemically active guides of the capped reactions. *Adv. Coll. Interf. Sci.*, Vol.133, pp. 23-34

Vesperinas, A.; Eastoe, J.; Jackson, S. & Wyatt, P. (2007). Light-induced flocculation of gold nanoparticles. *Chem. Commun.*, Vol. , pp. 3912–3914

Wongwailikhit, K. & Horwongsakul, S. (2011). The preparation of iron (III) oxide nanoparticles using W/O microemulsion. *Materials Letters*, Vol.65, pp. 2820–2822

Wu, C.I.; Huang, J.W.; Wen, Y.L.; Wen, S.B.; Shen, Y.H. & Yeh, M.Y. (2005). Preparation of TiO_2 nanoparticles by supercritical carbon dioxide. *Materials Letters*, Vol.62, pp.1923–1926

Xia, L.; Hu, X.; Kang, X.; Zhao, H.; Sun, M. & Cihen, X. (2010). A one-step facile synthesis of Ag–Ni core–shell nanoparticles in water-in-oil microemulsions. *Colloids and Surfaces A: Physicochem. Eng. Aspects*, Vol.367, pp. 96–101

Zhang, W.; Qiao, X.; Chen, J. & Wang, H. (2006). Preparations of silver nanoparticles in water-in-oil AOY reverse micells. *J. Colloid Interf Sci*, Vol.302, pp. 370-373

Zhang, W.; Qiao, X. & Chen, J. (2007). Synthesis of silver nanoparticles—Effects of concerned parameters in water/oil microemulsion. *Mater. Sci. Eng. B*, Vol.142, pp. 1–15

Zhang, W.; Qiao, X.; Chen, J. & Chen Q. (2008). Self-assembly and controlled synthesis of silver nanoparticles in SDS quaternary microemulsion. *Materials Letters*, Vol.62, pp. 1689–1692

Zielińska-Jurek, A.; Walicka, M.; Tadajewska, A.; Łącka, I.; Gazda, M. & Zaleska, A. (2010). Preparation of Ag/Cu-doped titanium (IV) oxide nanoparticles in W/O microemulsion. *Physicochem. Probl. Miner. Process.* Vol.45, pp. 113-126

Zielińska-Jurek, A.; Kowalska, E.; Sobczak, J.W.; Lisowski, W.; Ohtani, B. & Zaleska, A. (2011). Preparation and characterization of monometallic (Au) and bimetallic (Ag/Au) modified-titania photocatalysts activated by visible light. *Appl. Catal. B: Environ.*, Vol.101, pp.504–514